Methods of
Soil and Plant Analysis

Methods of Soil and Plant Analysis

Susanta Kumar Pal
M.Sc (Ag), Ph.D
Professor
Department of Agricultural Chemistry and Soil Science
Bidhan Chandra Krishi Viswavidyalaya
Mohanpur-741252, Nadia
West Bengal, India

NEW INDIA PUBLISHING AGENCY
101, Vikas Surya Plaza, CU Block, LSC Market
Pitam Pura, New Delhi 110 034, India
Phone: + 91 (11) 27 34 17 17 Fax: + 91 (11) 27 34 16 16
Email: info@nipabooks.com
Web: www.nipabooks.com

Feedback at feedbacks@nipabooks.com

ISBN: 978-93-95763-85-1

Composed, Designed by NIPA

Dedicated to
my departed parents
and elder brothers

DILIP KUMAR DAS
Professor and Former Head
Department of Agricultural Chemistry and Soil Science
Bidhan Chandra Krishi Viswavidyalaya
Mohanpur, Nadia, West Bengal, India

Foreword

Soil Science is an important and basic science in agriculture which deals with different domains of soil research namely, soil formation, genesis and classification, soil physics, soil chemistry, soil fertility and plant nutrition, soil biology, etc. Characterization as well as our understanding of soils requires that they are precisely analyzed and described. While the physical properties of soils determine their adaptability to cultivation, chemical properties tell about their chemical environment and nutrient status to the crop production - the most important use of soils on this densely populated planet. Determination of different soil physical and chemical properties in the field or in the laboratory following suitable analytical methods is first step towards appropriate soil managements and scientific recommendations for increasing crop production.

The present book entitled "Methods of Soil and Plant Analysis" is certainly a good attempt taken by Dr. Susanta Kumar Pal, Professor of Agricultural Chemistry and Soil Science and is really appreciable. I have gone through the entire manuscript and found that the author has taken keen interest in writing this book consisting of 19 chapters. This analytical book on soil science has been written involving basic chemistry and explaining the text with incorporation of some pictures and figures which are unique feature of the book. The author has paid equal emphasis on writing each chapter, of which the chapter 18 and 19 are an important inclusion to the existing analytical books related to soil science.

The author has also put some questions at the end of each chapter relating analytical determination of soil parameters which will definitely be useful to the students, researchers and teachers of agriculture in general and soil science in particular.

I am confident that this well written and compiled analytical book will be useful for post graduate students and researchers, and also important addition to university library.

Dilip Kumar Das

Preface

This book can be considered as an enlarged and revised edition of my earlier book "Soil Sampling and Methods of Analysis". The earlier edition was widely accepted and found to be useful to the students, teachers and researchers working in this field. By this time, I received lot of suggestions and requests from the readers for its improvement and inclusion of few more chapters related to this field. Keeping these views in mind, both publisher and author have agreed upon to publish this book as a new one wherein 8 new chapters have been included.

This book is comprised of 27 chapters besides references and 16 appendices at the end. Each chapter is ended with some relevant questions and answers on that topic for the benefit of the students. In addition to 'Soil Physical and Chemical Analysis' this book covers two more major aspects namely 'Soil Mineralogical Analysis' and 'Plant Analysis'. The typographical errors committed inadvertently in the earlier book have been taken care of here.

Hope, like earlier one, this book will definitely fulfil the need of the students, teachers and researchers engaged in the field of agriculture, geology, earth science, etc. Like past, any sort of suggestions from the readers for its improvement in future will highly be appreciated.

The author gratefully acknowledges the suggestions received from the readers of all corners of this country and abroad and each source from where the technical matters are drawn.

Kalyani
Dated: 28th August, 2018

S.K. Pal

Contents

Section II: Soil Mineralogical Analysis

Section IV: Plant Analysis

1

Collection and Processing of Soil Samples

Soil is a mass of large variability and thus its sampling is the most challenging problem for any soil analyst. Practically, soil sampling is a two step process which includes (i) collection of soil from homogeneous unit and (ii) processing which in turn includes drying, grinding, sieving, mixing, partitioning and storing.

1.1 Sources of Error in Soil Chemical Analysis

Different sources of error involved starting from soil sampling to chemical analysis of a sample can be divided into three main groups:

(i) **Sampling error:** This error results from variability among the different samples drawn from same soil unit with respect to the objective of sampling.

(ii) **Sub-sampling error:** This error is due to variability introduced among sub samples of the same soil sample.

(iii) **Analytical error:** This error introduced due to variability from one determination to another on same sub sample.

The error encountered in any analysis by a person can be grouped into two categories: (A) Systematic error, and (B) random error.

(A) **Systematic** or **Determinate Errors** are the errors which can be avoided and magnitude of these errors can be measured. These errors include: (i) **personal error-** error due to variation among individuals, (ii) **instrumental and reagent error-** due to manufacturing defects of instruments, uncalibrated weights or measuring devices and impurity in reagents, (iii) **errors of methods-** due to incompleteness of reactions or side reaction during main determination.

(B) On the other hand, **Random** or **Indetrminate Errors** are the variation in successive measurements under an almost identical condition that an analyst encountered even with utmost care. The analyst has little control on these errors.

1.2 Accuracy and Precision

The **accuracy** in any estimation may be defined as the concordance between estimated value and its actual value. On the other hand, **precision** in any measurement may be defined as the concordance of a series of measurements of the same quantity. Therefore, accuracy expresses the correctness of a measurement with or without much precision, while precision indicates the reproducibility (under varying laboratory condition) or repeatability (under same laboratory condition, i.e. rapid successions) of a measurement with or without much accuracy. The difference between accuracy and precision of any measurement can easily be understood from the following example.

Say, a substance contains 25.54 percent of a constituent. Two analysts analyse the same sample and their replicated results are as follows:

Analyst A- (i) 25.50%, (ii) 25.47%, (iii) 25.68%, (iv) 25.63%

The results range from 25.47 to 25.68% and the arithmetic mean is 25.57%.

Analyst B- (i) 25.35%, (ii) 25.42%, (iii) 25.37%, (iv) 25.38%

The results range from 25.37 to 25.42% and the arithmetic mean is 25.38%.

The results of Analyst A are accurate (slight deviation from actual value), but the precision is less (range is wide). On the other hand, the results of Analyst B are precise (range is narrow), but are not accurate (large deviation from actual value).

1.3 Composite Soil Sample

A composite soil sample is the representative of an individual soil volume. In fact composite soil sample is taken to reduce the time and cost incurred in the soil analysis of different soil samples drawn from same volume. Thus, analysis of a composite sample is the average of different soil samples drawn from the same soil volume.

During taking of a composite sample following points must be considered:

(i) Equal amount of soil should be taken from each digging/core.

(ii) Each sample must be taken from equal cross section (depth).

(iii) With respect to the objective of sampling one composite sample will be taken from a homogeneous soil unit, e.g. in case of sampling from the experimental plots each plot receiving different treatment is an individual soil unit.

(iv) Enough core/sample should be taken to make one composite soil sample. As the number of core/sample increases the variability decreases. Usually, 20-30 cores/sample irrespective of size of the field are sufficient.

(v) The core/sample should be taken at random in a zigzag direction.

(vi) There should be no chemical interaction of soil material composited.

1.4 Soil Sampling Tools

Depending upon the purpose and soil moisture status different soil sampling equipments are used. These are (i) Tube auger, (ii) Screw type auger, (iii) Post-hole auger, (iv) Dutch auger, (v) Spade and (vi) Khurpi (Fig. 1.1).

(i) **Tube auger:** Suitable for soft and moist soil free from gravel and stones.

(ii) **Screw type auger:** It is used in hard or relatively dry soil as well as stony and gravelly soil, but unsuitable for very dry and sandy soil.

(iii) **Post-hole auger:** It is useful for sampling in excessively wet area like rice field.

(iv) **Dutch auger:** This auger is suitable for moist and wet soil having moderately fine to fine textured soil, but not for coarse textured soil as well as dry soil.

(v) **Spade and khurpi:** These are used for sampling in soft and moist soil of the surface layer.

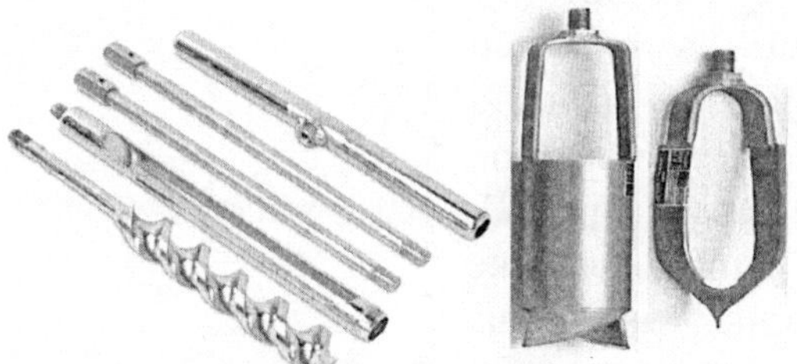

Screw type- & Tube auger Post –hole auger Dutch auger

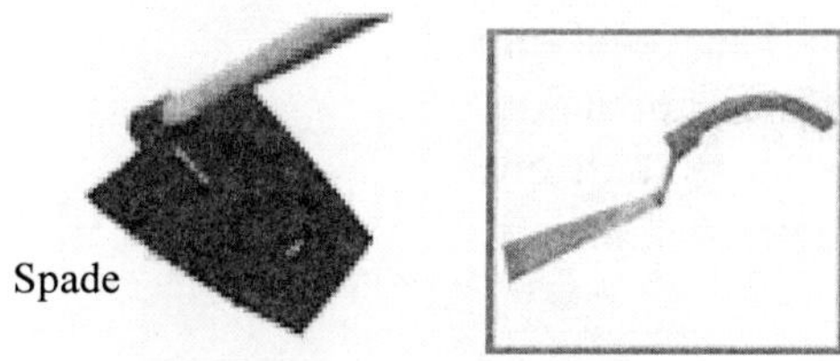

Spade

Khurpi

Fig.1.1: Soil sampling tools

1.5 Collection of Soil Sample

1.5.1 Collection of Disturbed Soil Sample

1.5.1.1 Collection of Soil Sample for Fertility Evaluation from Farmers' fields

Equipments and Materials

Spade and Khurpi (for moist friable soil), Screw type auger (for hard, dry, stony soil), polythene sheet or clean cloth, polythene packet of 1 kg capacity, cloth bag, wax pencil or marker, paper tag.

Procedure

- Before starting of sampling traverse the field and if you have no soil map then prepare a soil map referring any bench mark point such as road, permanent irrigation/drainage channel, etc.

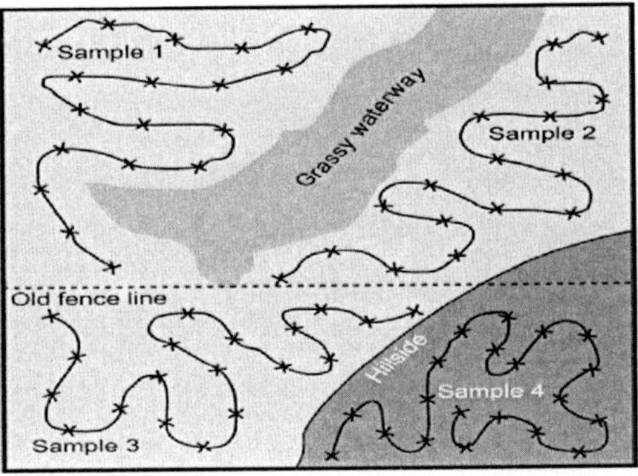

Fig.1.2: Field divided according to the variability

- Keeping variability in soil texture (feeling by finger), topography (visual observation), soil organic matter content (usually indicated by soil colour), fertility status (as indicated by crop growth) in mind determine area under one composite sample (Fig.1.2). Variability demands separate sampling. Otherwise, in this subcontinent the area of a field which is usually restricted by 1 ha receiving equal soil management does not require more than one composite sample.

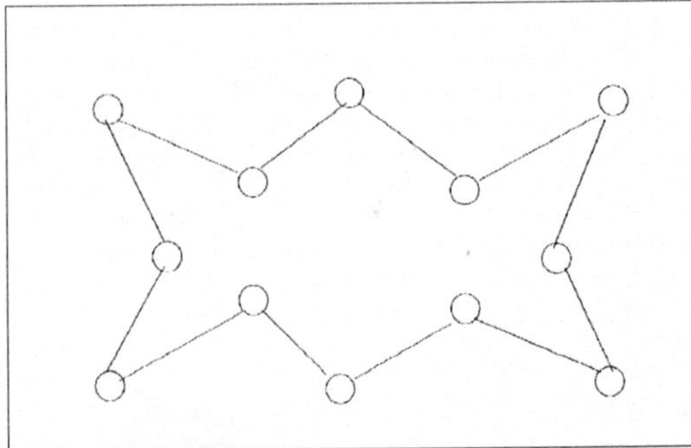
Fig. 1.3: Sampling spots

- Identify abnormal spots (if any) developed due to high $CaCO_3$ content, high salinity or alkalinity or acidity or presence of hard pan below the surface. Separate sampling from abnormal spots is needed for its management.
- Now move in a zigzag path from any corner of the field to the direction of cultural operation which is usually done along the length of the field or the slope and select 10 to 30 spots at random leaving about 2-3 m from the borders.
- Take sample from individual spots after scrapping off surface litter, but without removing soil.
- Make a 'V' knotch by spade and take sample from both sides of the knotch at uniform thickness (2 cm) with the help of khurpi (Fig.1.3).

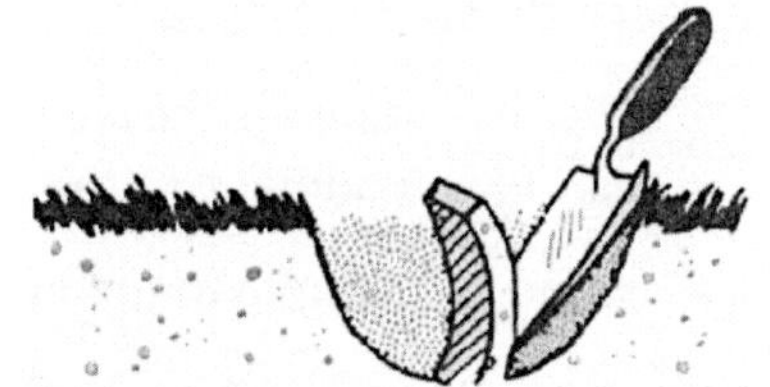

Fig.1.4: Sampling from 'V' knotch

- Depth of sampling depends on rooting pattern of the crop grown. Usually, for most of the field crops sampling depth remains restricted to 6″ or 15 cm from the surface.
- Thoroughly mix soils collected from various spots by hand on a clean piece of cloth or polythene sheet.
- Put composite sample in a polythene bag mentioning following information on the paper tag:

  (i) Farmer's name and address (ii) date of sampling (iii) plot No. and mouza (iv) land situation (high, medium, low) (v) previous crop history (crop, variety, amendment, manure and fertilizer applied) (vi) crop to be cultivated (vii) any special information.
- Pour polythene bag in a cloth bag labeled with wax pencil or marker and send it to laboratory for further processing and analysis.

1.5.1.2 Collection of Soil Sample from Experimental Plots

- Treat each experimental plot receiving different treatment as individual soil volume or unit and take separate composite sample for each unit.
- For experimental plot take soil samples by soil auger or spade and khurpi at every 2 steps leaving about 50 cm strip from the borders.
- Collect soil sample following the same steps as described for that of farmer' field.

1.5.1.3 Collection of Sample for Plantation Crop and/Problem Soil

- Dig a pit of 180 cm depth (90 cm depth is sufficient for problem soil) making one side of the pit vertical.
- Collect 500g soil sample scrapping uniformly through out the depth of 15 cm from the surface. Take a separate sample of the salt crust (in case of problem soil), if visible.
- Collect samples in the same manner from 15-30, 30-60, 60-90, 90-120, 120-150 and 150-180 cm depths.
- If any hard pan is there in the pit take sample separately mentioning the depth and thickness.
- Put the samples in the cloth bags separately indicating the depth of sampling, name and address of the farmer, field No., etc.
- Send the sample to the laboratory for further processing and analysis.

1.5.1.4 Processing of Sample

Processing of sample includes drying, sieving, mixing, partitioning and storing.

Equipments and Materials

Drying cabinet, agate-mortar and rubber pestle, sieve, cloth or paper, spatula or knife, glass jar with lid for storing sample, log book

Drying

Dry soil in shade (not in direct sunlight) where there is good air circulation facility. Some chemical analyses require immediate analysis without drying (NO_3-N, exchangeable Fe^{2+}, etc). However, result of chemical analysis is expressed on oven dry weight basis.

Grinding

- Remove large stones, gravels, stubbles, straw and grasses before grinding.
- Grind soil aggregates lightly without breaking the primary coarse particles (sand and silt) with the help of rubber pestle in an agate-mortar.

Sieving

- For routine soil testing and other chemical analysis pass the soil through 2 mm sieve (Appendix I).
- If the soil contains stones and gravels in significant amount then for research purpose the proportion of this fraction to the total soil weight must be taken into account in expressing result of chemical analysis.

- Avoid the practice of discarding large volume of soil aggregates and taking a small proportion of sieved soil.

Mixing

- Spread the soil over a square cloth or paper and roll the soil from one corner to the opposite corner holding two diagonal corners of cloth or paper.
- Follow the same process holding other two diagonal corners.
- Repeat the total process at least for 5 times.

Partitioning

- After thoroughly mixing keep the soil at the middle of the cloth or paper in the form of dome.
- Slightly flattening the top, divide the dome into two equal halves with the help of sharp knife or spatula (Fig.1.5).

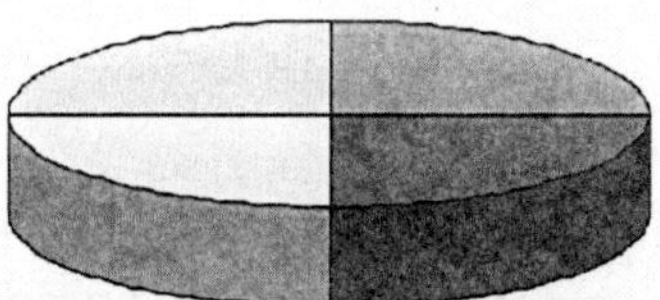

Fig.1.5: Quartering technique (Discard any diagonal two parts)

- Again, divide each half equally into two parts and ultimately get four equal quarters.
- Discard any two opposite quarters.
- Mix other two quarters left on the paper, again form the dome and follow the same process until you get the desirable quantity (500 g) of soil left on the paper.

Storing

In most of the cases, soils are analyzed immediately after processing and then discarded. But in some cases where there is a need of long time storage for further laboratory analysis or as reference sample, soil are stored in a completely air tight jar properly labeled on and inside it. The entry of the full details of the sample must be done in a log book.

1.5.1.5 Precautions

- Each composite sample should represent one homogeneous unit/volume.
- No sample should be taken from recently fertilized field.
- Take separate samples from abnormal spots like, area near tree, old manures pile, depressed area, etc.

- Crop grown in rows take sample from the area between rows.
- Avoid contamination of any chemicals with the soil sample at any stage of sampling.
- Dry sample in shade only (except the cases which require immediate analysis).
- For grinding never use iron mortar-pestle.

1.5.2 Collection of Undisturbed Soil Sample

Undisturbed soil samples are required for analysis of some soil physical properties *viz*, bulk density, pore space, retension of water, transmission of water at suction, etc.

Equipments and Materials

Core sampler consists of two cylinder fitted one inside the other (The inside cylinder is the sample holder. The outer core extends at both the ends-upper end to accept the hammer and lower end to form the cutting edge), spade, knife.

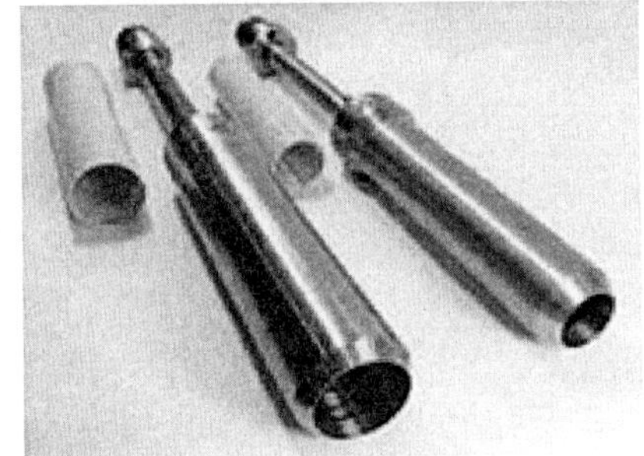

Fig. 1.6: Core sampler

Procedure

- With the help of hammer drive the sampler into a vertical soil surface slightly enough to fill the sampler but avoiding the compression of the soil in the confined space of sampler.
- Using spade dig out the sampler without disturbing the natural position of soil within the sampler.
- Separate the two cylinders retaining the soil mass undisturbed in the sampler (inner core).
- Trim the soil extending beyond each end of sampler with the help of sharp knife.
- Randomly collect 8-10 such cores of desired depth from an individual soil unit of 1 hectare.

Questions and Hints

Q.1 What is composite soil sample?

Hint: Composite soil sample is the representative of the soil sampling volume. Composite soil sample gives a mean analytical value from where the samples were drawn. A composite sample is drawn actually to minimize the cost and time involved in the analysis of large number of sample from an individual soil unit / volume taking variability into consideration.

Q.2 How many cores are required for a composite sample?

Hint: To have a representative soil sample from an individual soil unit / volume more the number of cores less will be the variability. However, 20–30 cores are considered to be sufficient for a composite sample from maximum 1 hectare of land.

Q.3 For fertility evaluation what is the best time of sampling?

Hint: Usually autumn sampling i.e. after kharif crop is the most suitable time for collection of sample. However, depending upon the requirement, sampling may be done at anytime.

Q.4 What is the interval of sampling from farmers' field?

Hint: Alternate year sampling.

Q.5 How can you minimize errors during sampling?

Hint: See section 1.2

Q.6 In which condition drying of sample should be avoided?

Hint: For determination of some soil chemical parameters *viz.* NO_3-N, exchangeable Fe^{2+}, etc, analysis should be done immediately without drying of sample to avoid oxidation which causes increase or decrease in their contents.

Q.7 Where undisturbed soil sample is required?

Hint: For estimation of some soil physical properties like bulk density, pore space, retention of water, etc, undisturbed soil sample is required to have the idea about the above properties of soil in situ, because the values of disturbed sample bear less meaning at least for agricultural purposes.

Q.8 Where from samples should not be taken?

Hint: (a) ploughed field, (b) recently fertilized field, (c) unusual spots (depressed area, manure heap, contaminated area) of the field, (d) near border.

Q.9 How will you collect sample from a field of standing crop?

Hint: Collect samples from the area between the rows of the standing crop.

Q.10 What is difference between accuracy and precision?

Hint: See section 1.2

Section -I

SOIL PHYSICAL ANALYSIS

2

Soil Texture and Textural Class

2.1 Particle Size Distribution Analysis

Soil is composed of inorganic particles and organic molecules of varying size and shape. The expression of the proportion of the different size inorganic particles is termed as '**soil texture**'. It is one of the most important basic characters being little changed by cultivation or other practices. Particles are arbitrarily distinguished into following groups on the basis of their diameters.

Table 2.1: Size limits of soil separates

U.S.D.A*		I.S.S.S**	
Name of separates	Diameter (mm)	Name of separates	Diameter (mm)
Very coarse sand	2.0-1.0		
Coarse sand	1.0-0.5	Coarse sand	2.0-0.2
Medium sand	0.5-0.25		
Fine sand	0.25-0.10	Fine sand	0.2-0.02
Very fine sand	0.10-0.05		
Silt	0.05-0.002	Silt	0.02-0.002
Clay	<0.002	Clay	<0.002

*U.S.D.A-United States Department of Agriculture

** I.S.S.S-International Society of Soil Science

The analytical procedure by which the particles are separated is called '**mechanical analysis**'. There are two methods of mechanical analysis commonly used in the laboratory: one is International Pipette Method (Robinson, 1922) and other is Hydrometer Method (Bouyoucos, 1927) based upon the principle of changes in density of the suspension with time at a given depth (Day, 1965).

2.1.1 International Pipette Method

Principle

Pretreatment and Dispersion

Pretreatment and dispersion of soil sample are mainly done to get maximum dispersion and to maintain it during the process of particle size distribution analysis. Pretreatment involves removal of cementing agents and flocculating ions which helped soil particles to form aggregate or to floccule. Cementing agents are organic matter, oxides of iron and aluminium, carbonates and the flocculating agents are the divalent exchangeable cations. The objective behind dispersion is to create an atmosphere of similar charge (monovalent) with large water hull around each soil particle (particularly finer particles) so that particles start to repel each other.

Sieving and Sedimentation

Sieving and sedimentation are the most common method used for isolating the soil particles. Sieving is used for separating coarse fraction while sedimentation technique is used for silt and clay fractions. Principle behind the sedimentation technique is that the rate of settling of particles in a viscous medium depends on the size, shape and density of the particle. In a given medium, say water, larger particles fall more rapidly than smaller one of same density and consequently settle out of suspension more quickly.

The relation between the particle radius and settling velocity was first established by Stokes (1851).

$$v = \frac{2}{9}\left(\frac{d_p - d_l}{\eta}\right).g.r^2 \qquad 2.1$$

Where, v is the falling velocity of particle in cm/sec, d_p is the density of the particle in g /cc, d_l is the density of the liquid in g/cc, g is the acceleration due to gravity in cm/sec^2, r is the radius of particle in cm and ç is the viscosity of the liquid in g sec/cm.

Say, a particle drops down h cm in t second; v of Eq. (2.1) can be substituted by h / t and can be written as:

$$v = \frac{h}{t} = \frac{2}{9}\left(\frac{d_p - d_l}{\eta}\right).g.r^2$$

$$\text{or}, t = \frac{9\eta h}{2(d_p - d_l).g.r^2}$$

Thus, after t second if a sample is taken up to h cm depth from the suspension surface, no particle other than that of radius less than r will remain. This is the basic principle behind separating out different size particles.

Table 2.2 Sedimentation time for different size particles settling through a depth of 10 cm in water

Temperature (°C)	Settling time		
	2μ Hr. Min.	5μ Hr. Min.	20μ Min. Sec.
20	8 0	1 17	4 48
21	7 49	1 15	4 41
22	7 38	1 13	4 35
23	7 27	1 11	4 28
24	7 17	1 10	4 22
25	7 7	1 8	4 16
26	6 57	1 7	4 10
27	6 48	1 5	4 4
28	6 39	1 4	4 0
29	6 31	1 3	3 55
30	6 22	1 1	3 49
31	6 14	1 0	3 44

Equipments and Materials

Special pipette (20 ml) with stand for mechanical analysis, sieve shaker, horizontal reciprocating type shaker, set of 6 sieves having opening diameter 1.0, 0.5, 0.25(60 mesh), 0.177 (80 mesh), 0.105 (140 mesh) and 0.047(300 mesh) mm and 4 - 5″diameter frame with cover and pan, large size evaporating dish, 1000 ml graduated cylinders, 500 ml beaker with watch glass, water bath, Buchner funnel, desiccator, oven, stop watch, balance, Whatman No. 40 filter paper and thermometer.

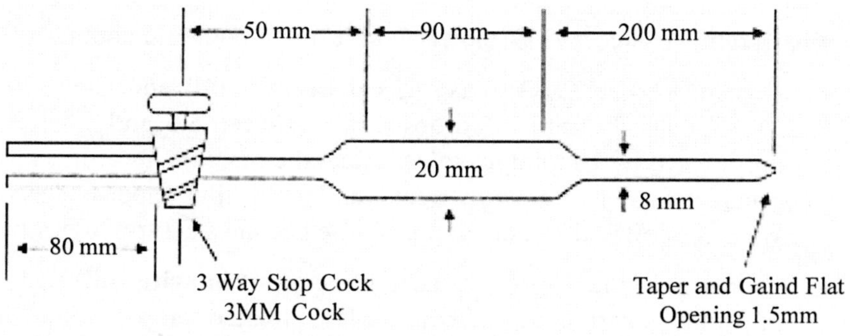

Fig. 2.1: Special pipette

Reagents

- 30% Hydrogen peroxide (H_2O_2), commercially available
- 5 % Calgon solution- Dissolve 50 g of calgon in water and make up volume to litre (Calgon is a commercial preparation of sodium hexametaphosphate containing sufficient sodium carbonate) to give a pH of 8.3 in a 5% aqueous solution.
- Hydrochloric acid (2N): Dilute 83 ml of concentrated hydrochloric acid to 500 ml with water.
- Citrate-bicarbonate buffer: Dissolve 88.4 g of sodium citrate in water and make up volume to 1 litre to get 0.3 **M** sodium citrate solution (solution A). Prepare 1**M** solution of sodium bicarbonate containing 84 g of sodium bicarbonate in 1 litre of solution (solution B). Mix solution A and solution B in 8:1 ratio (volume basis).
- Sodium dithionite ($Na_2S_2O_4$)

Procedure

Pretreatment

- Take approx. 20 g of 2 mm air dry soil (50 g for sandy soil and 10 g for clayey soil) in a 500 ml beaker and add 50 ml of water to wet it.
- **To remove organic matter** treat the soil with 20 ml of 30% H_2O_2, cover the beaker with watch glass for overnight.
- Next day put the beaker on hot water bath at about 70-80°C adding 5 ml H_2O_2 at one hour interval with occasional stirring.
- Continue the process till large bubbles cease to evolve(oxidation of organic matter completed)
- To remove excess H_2O_2 slowly boil it for an hour and cool it.
- **To remove $CaCO_3$ (In case of calcareous soil)** add about 25 ml of 2N HCl slowly until pH drops down between 3.5 and 4.0 (tested by litmus). Allow 10 minutes to complete the reaction and filter through a Buchner funnel fitted with Whatman No.40 filter paper under suction. Wash the soil with water till the filtrate becomes neutral to litmus.
- **To remove iron oxides (in case of red and laterite soil)** add 180 ml of citrate –bicarbonate buffer to the H_2O_2 treated sample. Stir and slowly add 3.5 g of Na-dithionite ($Na_2S_2O_4$). Put the beaker on a hot water bath at about 80°C for 20 minutes with occasional stirring. Filter through a Buchner funnel fitted with Whatman No.40 filter paper. Wash the soil at least 5 times with 20 ml of water at a time.

- Transfer the soil on filter paper to a previously weighed 250 ml beaker by using a rubber tipped policeman.
- Dry the sample overnight in an oven at 105°C, cool it in a desiccator and weigh. It is the base weight of the sample to be required in calculating various size fractions.

Dispersion

- Add 10 ml of calgon solution to the dry soil in the beaker and leave it for overnight.
- Transfer the suspension to the cup of a high speed stirrer and make the volume to about 300 ml.
- Stir the suspension for 10 minutes, wash the stirrer blades.

Fractionation

a) Separation of coarser fraction

Wet Sieving

- Allow to stand the thoroughly dispersed and stirred soil suspension for 1 to 2 minutes.
- Decant the suspended portion into the sieve of 0.02 mm (**for ISSS system**) or 0.05 mm (**USDA system**) fitted in a special apparatus of wet sieving (Fig. 2.2).
- Add water to the residue in the beaker, stir the mixture and allow settling for 1 to 2 minutes and again decanting the suspended portion into the sieve as before.
- Transfer total soil particles from beaker into the sieve using jet of water from wash bottle.
- Agitate the residue on the screen with water jet but don't rub the screen.
- Transfer the washings to a one litre sedimentation cylinder and make up the volume to the mark with distilled water.

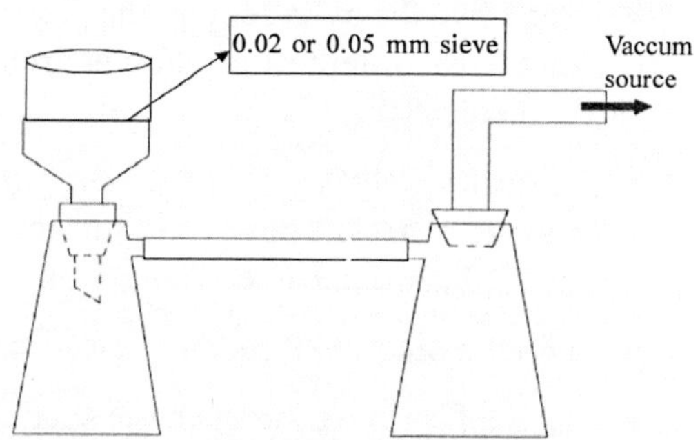

Fig. 2.2: Special apparatus for wet sieving

Dry Sieving

- Put the sieve with its content in a previously weighed evaporating dish, dry for 2 hours at 105°C in oven, cool and weigh. This will give the weight of total sand fraction. To separate out fine sand from coarse sand fraction (**ISSS**) sieve the total sand fraction through 0.2 mm sieve. Material on the sieve is the coarse sand fraction and the difference of it from total sand is the fine sand fraction.
- To separate out different size sand fractions (**USDA**) arrange the sieves on the pan in the sequence of 1.0, 0.5, 0.25, 0.1 and 0.05 mm opening size from top to bottom.
- Transfer the total sand fraction from evaporating dish to the top sieve (1 mm), put the cover in place and fasten the nest of sieves firmly in the shaker.
- Shake the sieves for 3 minutes and transfer the separates one at a time into previously weighed evaporating dish, separating from the top and proceeding downward.
- Weigh the evaporating dish and calculate the cumulative weight of sand.
- Weigh the content (residual silt) in the pan and include it as a part of the coarse silt.

b) Separation of Silt and Clay

- Record the temperature of the suspension in the sedimentation cylinder.
- Mark on the cylinder at a 10 cm depth from the surface of the suspension.
- Insert the rubber cork tightly in the mouth of cylinder; stir the suspension end over end several times till to have a homogeneous suspension and no sedimentation left at the bottom of the cylinder.
- Set down the cylinder into the position of pipette stand on the flat surface of the sedimentation cabinet, remove the cork and record the time immediately.
- Adjust the sliding clamp of sampling apparatus required to immerse the pipette at 10 cm below the surface at the appropriate time for sampling of $\leq 20\mu$ soil particle (Table 2.2).
- Dip the pipette 30 second before sampling time.
- At appropriate time open the suction stopcock and be ready to close it at the moment when the liquid fills the pipette (10 ml).

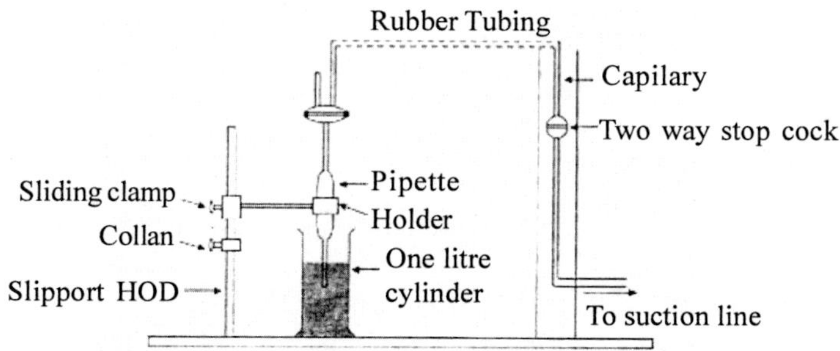

Fig. 2.3: Sampling apparatus using special pipette

- Raise the pipette and by reversing the stopcock drain the suspension into a previously weighed beaker of 100 ml. Rinse the pipette with distilled water in the same manner and add the washing to the beaker.
- Dry the beaker in the oven at 105 °C for 12 hours, cool in the desiccator and weigh it. This weight is the weight of international silt plus clay.
- Take sample of ≤ 2μ soil particle in the same manner restirring the suspension. Count time from the latest stirring and measure sampling depth from existing surface of the suspension not from the level used for previous sampling.
- Collect the suspension in another previously weighed beaker, dry it in oven, cool and take the weight. This is the weight of clay fraction only.
- To subtract the weight of dispersing agent present in the suspension aliquot, dilute 10 ml of dispersing agent to 1 litre with distilled water. Take aliquot using same pipette, drain it to previously weighed beaker, dry and weigh.

Calculation

Say,

Weight of oven dry soil after removal of cementing agents (base weight) = A g

Weight of International silt (20 - 2μ) + clay = B g

Weight of clay fraction (< 2μ) = C g

Weight of dispersing agent in aliquot = D g

Volume of aliquot = V ml

$$\textbf{International silt + clay (\%)} = (\mathbf{B} - \mathbf{D}) \times \left(\frac{1000}{V}\right) \times \left(\frac{100}{A}\right) = \mathbf{X}$$

$$\textbf{Clay (\%)} = \textbf{(C - D)} \times \left(\frac{1000}{V}\right) \times \left(\frac{100}{A}\right) = \textbf{Y}$$

International silt = X - Y

% USDA silt (50 - 2μ) = [100 – (% sand + % clay)] or (X – Y) + residual silt

2.1.2 Hydrometer Method (Bouyoucos, 1927)

Principle

The principle of hydrometer method is similar to that of pipette method. Fundamentally, both the methods depend on Stokes' law. In case of pipette method, suspension density is measured directly from the weight of particles in a definite volume of suspension, while in hydrometer method the suspension density is measured as a function of buoyant force acting on the hydrometer bulb immersed in the suspension.

If Ψ is the suspension density, P_l and P_s are the liquid and particle density in g/cc, respectively and C is the concentration of the suspended solid, in g/litre, then

$$\psi = P_l + \frac{C}{1000} \times \left(1 - \frac{P_l}{P_s}\right)$$

For the particular values of P_l and P_s, hydrometer scales are calibrated in terms of C.

Equipments and Materials

Standard hydrometer- ASTM (American Society of Testing Materials) No. 152H with Bouyoucos scale in g/litre, electrical stirrer, inside graduated specially made hydrometer cylinder of 1 litre, water bath, 500 ml beaker with watch glass, thermometer.

Reagents

- 6% Hydrogen peroxide (H_2O_2): Dilute 100 ml of 30% H_2O_2 to 500 ml with distilled water.
- 5% Calgon solution- Dissolve 50g of calgon (sodium hexametaphosphate: sodium carbonate - 4:1) in water and make up volume of 1 litre.
- Amyl alcohol

Procedure

- Take 50 g air dry soil (passed through 2 mm sieve) in a 500 ml beaker. Weigh equal amount of soil and oven dry at 105°C for determination of oven dry weight of soil.
- Moisten the soil with water and add 50 ml of 6% H_2O_2, cover with watch glass and leave it for sometime.
- Place it on hot water bath at 65 - 70°C; stir it occasionally until organic matter is completely oxidized (as indicated by absence of effervescence).
- Remove the beaker, cool and transfer the content in a dispersing cup.
- Add 100 ml of 5 % Calgon and about 400 ml of distilled water.
- Allow the sample to soak for overnight.
- Stir the contents for 5 minutes for sandy soil and 15 minutes for fine textured soil.
- Transfer the suspension from the dispersing cup to the hydrometer cylinder and make up the volume to the mark with distilled water.
- Mix the content thoroughly putting palm of one hand on the mouth of the cylinder and turning upside down and back for several times.
- Record the temperature of the suspension.
- Place the cylinder on a table and note the time immediately.
- Put 2 drops of amyl alcohol if the surface of the suspension if frothy.
- Dip the hydrometer into the suspension 30 seconds before the appropriate time and take readings. For USDA scheme of classification take readings at 40 seconds (silt plus clay) and at 2 hours (clay) and for **ISSS** scheme

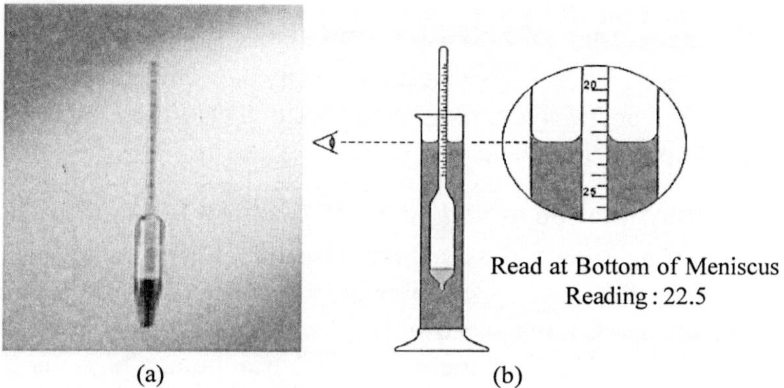

Fig. 2.4: (a) Hydrometer; (b) hydrometer reading

of classification take readings at 4 minutes (silt plus clay) and at 2 hours (clay).

- To calibrate the hydrometer follow same steps but without soil and note the reading (R_1).

  [*Note:* The hydrometer is calibrated at 67°F. If the suspension temperature is above or below 67°F, then for each degree above or below 67°F, the correction will be added or subtracted. The correction is equal to the difference of temperature between suspension and 67°F multiplied by 0.2].

- To separate out sand fractions follow the same steps as in International pipette method.

Calculation

Say, hydrometer readings at 4 minutes and 2 hours are X_1 and X_2, respectively. Temperature of the suspension is T°F, then

(i) Corrected hydrometer reading at 4 minute = $(X_1 - R_1) + (T - 67) \times 0.2$

(ii) Corrected hydrometer reading at 2 hours = $(X_2 - R_1) + (T - 67) \times 0.2$

(iii) If weight of soil taken = W g

$$\text{Percent silt plus clay in the soil} = [(X_1 - R_1) + (T - 67) \times 0.2 \times \frac{100}{W} = A$$

$$\text{Percent clay in the soil} = [(X_2 - R_1) + (T - 67) \times 0.2] \times \frac{100}{W} = B$$

So, percent silt in the sample = A - B

Percent sand in the sample = 100 – A

2.2 Determination of Soil Textural Class

Textural class of a soil can be determined by two ways, (i) putting particle size analysis data on triangular diagram for textural classification of USDA or ISSS and (ii) Rapid feel method.

2.2.1 Determination of Soil Textural Class by Triangular Diagram

Because of variation in the size limit of separates in two classification systems, there are two different triangular diagrams for textural classification. The equilateral triangles are divided into 12 compartments which represent 12 different textural classes. In both the equilateral triangular diagrams the base side represents sand percent; right side represents silt percent and left side represents clay percent (Fig.2.5 & 2.6).

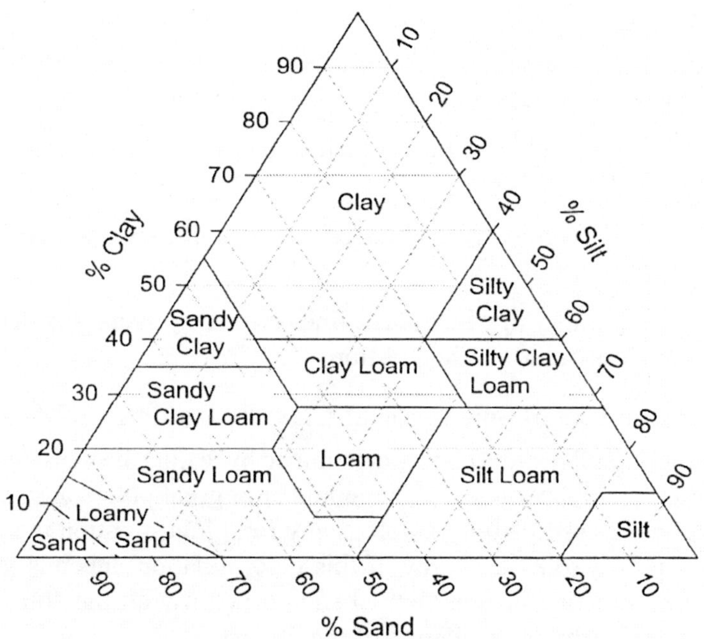

Fig. 2.5: The triangular diagram for textural classification by USDA

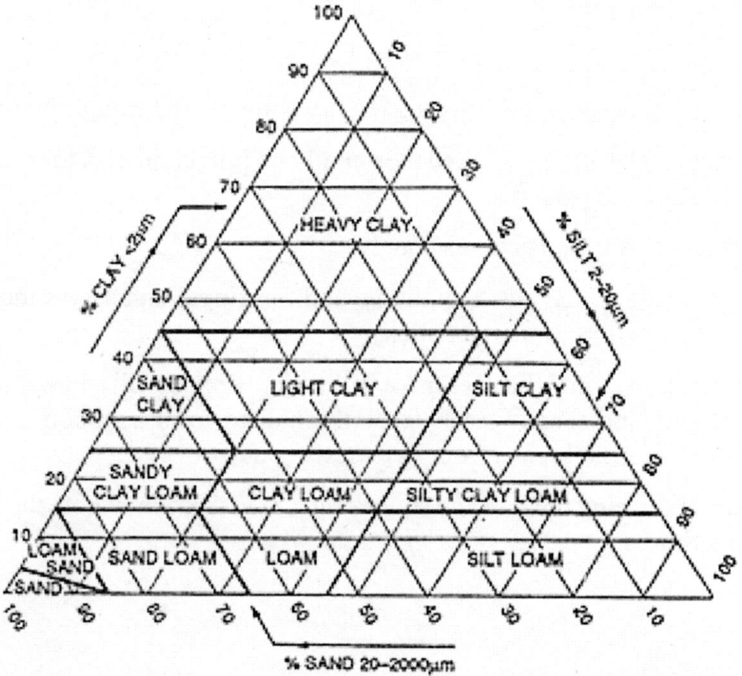

Fig. 2.6: The triangular diagram for textural classification by ISSS

Procedure

- Locate clay percent on clay coordinate and draw a line inward the triangle parallel to the sand coordinate.
- Locate silt percent on silt coordinate and draw a line parallel to the clay coordinate.
- Locate sand percent on sand coordinate and draw a line parallel to the silt coordinate.
- The compartment in which three lines intersect together is the name of textural class of the soil in question.

2.2.2 Determination of Soil Textural Class by Rapid Feel Method

In field condition, soil textural class is determined by feeling the moist soil rubbing between thumb and forefinger. This method is of great practical value in soil survey and land classification as well as any field investigation in which soil texture may play a role. Accuracy of this determination depends largely on experience, so practice of this method should begin with the soil of known textural class to calibrate the feeling of the fingers.

Equipments and Materials

2 mm sieve, porcelain dish

Procedure

- Take a small quantity (15–20 g) of dry soil passed through 2 mm sieve.
- Slowly add water and mixed thoroughly on porcelain dish to moisten the soil, but not to glisten.
- Work with it until becomes dough.
- Rub the moist soil between thumb and forefinger and assess the feeling (whether gritty, slippery or sticky).
- Using palm of both hands form a ball with the moist soil and with the help of thumb and forefinger squeeze the ball to form a ribbon as long as possible until it breaks from its own weight.
- During forming the ball and squeezing it to make ribbon note the feelings each time.

Interpretation of Observations

Observation	Textural class
1. Soil does not cohere into a ball, fall apart	Sand
2. Soil forms a ball, but does not form a ribbon	Loamy sand
3. Soil ribbon is dull and breaks off less than 2.5 cm long, and	
(a) grittiness is prominent	Sandy loam
(b) smooth slippery feel prominent	Silty loam
(c) slightly gritty and slightly smooth feeling	Loam
(d) less gritty than silty loam	Silt
4. Soil exhibits moderate stickiness, forms ribbon of 2.5-5.0 cm long, and	
(a) grittiness is prominent	Sandy clay loam
(b) smooth slippery feel prominent	Silty clay loam
(c) slightly gritty and slightly smooth feeling	Clay loam
5. Soil exhibits sufficient stickiness, forms ribbon longer than 5.0 cm, and	
(a) grittiness is prominent	Sandy clay
(b) smooth slippery feel prominent	Silty clay
(c) slightly gritty and slightly smooth feeling	Clay

Questions and Hints

Q.1 What dispersing agents are used in mechanical analysis other than calgon?

Hint: Other dispersing agents are sodium hydroxide, ammonium carbonate with sodium hydroxide, sodium oxalate, and sodium carbonate.

Q.2 How does calgon act?

Hint: Calgon which is actually a mixture of sodium hexametaphosphate and sodium carbonate, acts in the following way to disperse the finer fractions:

(a) Adsorption of phosphate ions on the positively charged sites of clay particles eliminate the chances of edge - to - edge and edge - to - face attraction.

(b) Adsorption of highly hydrated sodium ions on the negatively charged clay surface replacing strongly adsorbed H, Ca and Mg helps to maintain the potential on the surface above the critical level in order to prevent flocculation.

Q.3 What is the basic principle behind dispersion?

Hint: The basic principle behind dispersion is to create an atmosphere of similar charge (monovalent) with large water hull around each soil particle (particularly finer particles) so that particles start to repel each other.

Q.4 Why organic matter needs to be removed before dispersion?

Hint: Organic matter is one of the important cementing agents present in soil in variable quantities.Its removal is essential for mechanical analysis;

otherwise dispersion of aggregates or floccules into primary particles like, sand, silt and clay will not be possible.

Q.5 Between two methods of mechanical analysis which one is more accurate?

Hint: International pipette method is more accurate than the hydrometer method, but it is more tedious and time consuming than the other.

Q.6 Whether hydrometer method is applicable for saline, organic and calcareous soil?

Hint: No; since removal of salt, organic matter and $CaCO_3$ is not done here.Usually no pretreatment is recommended for this method.

Q.7 What are the flocculating agents present in soil?

Hint: The most common flocculating agents in soil are exchangeable Ca and Mg.

Q.8 What is primary and secondary soil particle?

Hint: Sand, silt and clay are termed as primary soil particle because each grain is consisted of one type material having uniform properties, while ped, clod are called as secondary soil particle because they are formed by aggregation or clustering of number of primary particles due to the action of cementing and flocculating agents.

Q.9 What is the shortcoming of the hydrometer method?

Hint: As removal of organic matter is not usually mandatory for this method, it does not exclude the weight of organic matter from the calculation of soil separates.

Q.10 Whether two hydrometer readings can be taken from single dipping? If not, why?

Hint: No; presence of hydrometer in the suspension will affect the settling velocity of the particles due to the interaction (adherening force) between glass and particles.

3

Density and Porosity of Soil

3.1 Determination of Bulk Density of Soil

3.1.1 Determination of Bulk Density of an undisturbed Soil

Bulk density of soil is expressed as the ratio of the mass (weight) of soil particle to their total volume including the pore space between the soil particles. Alternately, it is the weight of unit volume of dry soil. It is usually expressed in the unit of gram per cubic centimeter (g/cc) or mega gram per cubic meter (Mg/m^3). In fine textured soil bulk density varies from 1.00 to 1.60 g/cc, whereas in coarse textured soil from 1.20 to 1.80 g/cc.

$$\text{Bulk density} = \frac{\text{Weight of soil}}{\text{Volume of soil}} \tag{3.1}$$

Here, the volume of soil means as it exists in situ i.e. in field condition.

Equipments and Materials

Core sampler (Fig.1.5) consists of two cylinder fitted one inside the other (The inside cylinder is the sample holder; the outer core extends at both the ends; upper end to accept the hammer and lower end to form the cutting edge), knife, slide caliper, aluminium moisture box, analytical balance, oven, desiccator.

Procedure

- Drive the sampler into a vertical soil surface slightly enough to fill the sampler but avoiding the compression of the soil in the confined space of sampler.
- Using spade dig out the sampler without disturbing the natural position of soil within the sampler.
- Separate the two cylinders retaining the soil mass undisturbed in the sample holder (inner core).

- Trim the soil extending beyond each end of sampler with the help of sharp knife. Now the volume of sample holder is equal to the volume of soil.
- Transfer the total soil from sample holder to a previously weighed moisture box and weigh it.
- Dry it in the oven at 105°C until a constant weight is reached (10 to 15 hours).
- Cool the moisture box and its content at room temperature in the desiccator and weigh it again.
- Determine length and inner diameter of the sampler with the help of slide caliper.

Calculation

Say,

Weight of aluminium box = W_1 g

Weight of aluminium box + field moist soil = W_2 g

Weight of aluminium box + oven dry soil = W_3 g

Weight of oven dry soil = $(W_3 - W_1)$ g = Y g

Weight of water in soil = $(W_2 - W_3)$ g = Z g

$$\text{Thus, water content of soil (\%)} = \frac{Z}{Y} \times 100 = P$$

$$\text{Bulk density of soil}(g/cc) = \frac{Y}{\frac{\pi.d^2.h}{4}} = \frac{4Y}{\pi.d^2.h}$$

Where, d = inner diameter of the sampler, in cm

h = height of the sampler, in cm

Alternative Method

Determination of Bulk density by Paraffin Clod Method

Bulk density of undisturbed soil sample can also be determined taking soil clod (soil aggregate). Soil clod collected from the field is weighed before and after paraffin wax coating. According to Archimedes' principle loss in weight of paraffin coated soil clod immersed in water equals to the volume of wax coated soil clod.Then, the bulk densities of undisturbed soil sample (clod) are determined using following formulae:

$$\text{Dry bulk density}\left(\rho_b\right) = \frac{\text{Weight of oven dry soil clod}}{\text{Volume of moist soil clod}}$$

$$\text{Wet bulk density}\left(\rho / \text{b}\right) = \frac{\text{Weight of moist soil clod}}{\text{Volume of moist soil clod}}$$

Equipments and Materials

Balance, heater, hot-air oven, aluminium moisture box, wooden bridge, spade, paraffin wax, 25 cm long thread, soil clods, 500 ml beaker.

Procedure

- Collect few firm soil clods of about 4 cm in diameter from the field using spade and brush off loose soil particles.
- Take about 25 cm long fine thread and weigh it.
- Tie a soil clod firmly with one end of the thread and weigh soil clod along with the thread.
- Dip soil clod in molten paraffin wax for few seconds and take it off to harden in air.
- Repeat the above process for few times to completely coat the clod with a thin layer of paraffin wax (ascertain until bubble formation ceases).
- Weigh the paraffin wax coated soil clod.
- Place the wooden bridge on the floor of the balance across one pan and put the beaker ¾th filled with distilled water.
- Dip paraffin wax coated soil clod suspended from the hook of the balance in water without touching wall and bottom of the beaker and weigh it.
- To determine moisture content in soil weigh another soil clod both at fresh and oven dried (105°C for 24 hours) condition.

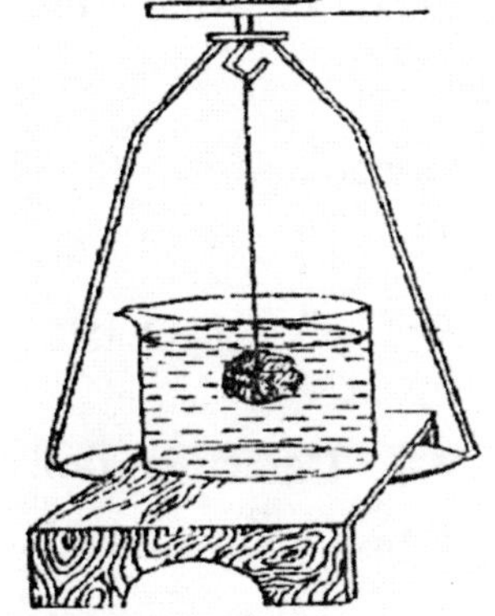

Fig. 3.1: Immersion of soil clod in water

Calculation

Say,

Weight of the thread = W_A g

Weight of the thread + fresh soil clod = W_S g

Weight of the thread + fresh soil clod + paraffin wax in air = W_{SW} g

Weight of the thread + fresh soil clod + paraffin wax in water = W_W g

Weight of another fresh soil clod = W_F g

Weight of oven-dry soil clod = W_O g

Weight of fresh soil clod (g) = (W_S- W_A) = A

Moisture content in soil (%) = $(\frac{W_F - W_O}{W_O} \times 100)$ = B

Weight of oven-dry soil clod (g) = $\frac{[100 \times A]}{[100 + B]}$ = C

Weight of paraffin wax on soil clod (g) = (W_{SW}-W_S)

Volume of paraffin wax on soil clod (cc) = $\left(\frac{W_{SW} - W_S}{0.9}\right)$ [density of paraffin wax is 0.90 g/cc]

Volume of moist soil clod + paraffin wax (cc) = (W_{SW}-W_W) [considering density of water 1.0 g/cc]

Therefore, volume of moist soil clod (cc) = $[(W_{SW} - W_W) - \left(\frac{W_{SW} - W_S}{0.9}\right)] = K$

Dry bulk density (g/cc) = $\frac{C}{K}$

Wet bulk density (g/cc) = $\frac{A}{K}$

3.1.2 Determination of Bulk Density of a disturbed Soil

Equipments and Materials

Apparent specific gravity bottle, balance, burette

Procedure

- Weigh the empty apparent specific gravity bottle without stopper.
- Fill the bottle with air dry processed soil sample upto the brim tapping at least 20 times on the table after each small addition of soil and weigh.
- Remove the soil and fill the bottle with distilled water with the help of burette and record the volume of water needed to fill the bottle.

Calculation

Let,

Weight of empty bottle = W_1 g

Weight of bottle + soil = W_2 g

Volume of water needed to fill the bottle = V ml

$$\text{Bulk density}(\text{g}/\text{cc}) = \frac{W_2 - W_1}{V}$$

3.1.3 Determination of Bulk Density of Gravelly and Rocky Soils

Presence of rocks and gravels prevents soil sampling for bulk density determination by core method. This method involves separation of coarse materials >2.00 mm in size (diameter) from excavated soil.

Equipments and Materials

Plastic wrap, sealable plastic bag, large syringe (140 cc), garden trowel, 2-mm sieve, balance, aluminium moisture box, hot-air oven, water.

Procedure

- Choose a spot as level as possible for filling it evenly with water. Dig a bowl shaped hole of about 3″ deep and 5″ in diameter using trowel. Keep soil and gravel in a plastic bag.
- Separate gravel from soil using 2-mm sieve. Collect soil in a sealable plastic bag and keep aside gravel for next step.
- Line the hole with plastic wrap in such a way that some excess plastic is left around the edge of the hole.
- Place the sieved rocks and gravel in the centre of the hole on the plastic wrap in such a way that the pile of rocks do not protrude above the level of the soil surface (Fig. 3.2).
- Using syringe pour water in the plastic-lined hole. The level of water should be just even with the soil surface. Record the volume of water needed, since it represents the volume of soil (including pore space) removed.
- Weigh the soil sample along with the plastic bag and minus the weight of an empty plastic bag.
- After thoroughly mixing soil sample with fingers take some amount of subsample loose soil in preweighed moisture box.
- Dry the sample in hot-air oven at 105°C temperature for about 24 hours for moisture content determination.

Fig. 3.2

Calculation

Say,

Weight of moist soil = W_M g

Weight of moist subsample soil = W_N g

Weight of oven-dry subsample soil = W_O g

$$\text{Gravimetric water content in soil (g/g)} = \left(\frac{W_N - W_O}{W_O}\right) = A$$

Volume of water required to fill the hole in 5th step = Volume of soil removed from the hole (cc) = K

$$\text{Weight of oven-dry soil (g)} = \frac{W_M}{[1+A]} = B$$

$$\text{Bulk density (g/cc)} = \frac{B}{K} = C$$

Volumetric water content (cc/cc) = A × C

$$\text{Soil porosity (cc/cc)} = \left(1 - \frac{C}{2.65}\right) = Z \text{ [taking particle density as 2.65 g/cc]}$$

Soil water-filled pore space (%) $\dfrac{\text{Volumetric water content}}{\text{Soil porosity}} \times 100 =$

$$\frac{A \times C}{Z} \times 100$$

3.2 Determination of Particle Density of Soil

Particle density of soil is the ratio of mass (weight) of solid particles to their total volume, excluding the pore space between particles. Like bulk density, it is also usually expressed in the unit of grams per cubic centimeter (g/cc) or mega gram per cubic meter (Mg/m^3). For mineral soils the value usually varies from 2.60 - 2.75 g/cc. For general calculation average particle density of arable soil is considered to be 2.65 g/cc. Particle density of organic matter (1.2 - 1.7 g/cc) is much less than that of inorganic particles of soil.

$$\text{Particle density} = \frac{\text{Mass of solid particles}}{\text{Volume of solid particles}} \tag{3.2}$$

Volume of solid particles is calculated indirectly by volume of fluid (usually water) displaced by the solid particles. Again, volume of water is calculated from the mass and density of water.

Fig. 3.3: Pycnometer

Equipments and Materials

Pycnometer fitted with a glass stopper having a vertical capillary opening (Fig. 3.3), analytical balance, hot plate.

Procedure

- Weigh a clean, dry pycnometer and pour oven dry soil to nearly one quarter of its volume.
- Weigh the pycnometer with soil after properly cleaning the outside of the pycnometer with dry cloth duster.
- Fill about half of the volume of pycnometer with distilled water.
- To expel entrapped air within the suspension gently boil the water for few minutes (5 minutes).
- Cool the pycnometer and its content to room temperature.
- Fill the total volume of pycnometer with preboiled cooled distilled water.
- Insert the stopper carefully and clean outside of pycnometer with dry cloth.
- Weigh the pycnometer with its content.
- Remove soil from pycnometer and fill it with preboiled cooled distilled water and weigh it.
- Record the room temperature.

Calculation

Say,

Weight of empty pycnometer = W_A g

Weight of pycnometer + oven dry soil = W_S g

Weight of pycnometer + soil + water = W_{SW} g

Weight of water filled pycnometer = W_W g

Density of water at room temperature (Appendix II) = D_W g/cc

$$\text{Particle density of soil}\,(g/cc) = \frac{W_S - W_A}{\left\{\dfrac{(W_W - W_A) - (W_{SW} - W_S)}{D_W}\right\}}$$

$$= \frac{D_W \text{x} (W_S - W_A)}{(W_W - W_A) - (W_{SW} - W_S)}$$

3.3 Determination of Porosity of Soil by Indirect Method

Porosity or total pore space of soil is the fraction of soil volume not occupied by soil particles. Mathematically, it is the ratio of volume of pore space to total volume of soil. It is governed by the orientation of the soil solids. Porosity gives idea about the storage capacity of fluid (air and water).

$$\% \text{ Pore Space (f)} = 100 - (100 \times \frac{D_b}{D_p})$$

$$= (1 - \frac{D_b}{D_p}) \times 100 \qquad (3.3)$$

Where, D_b and D_p are the bulk density and particle density of soil, respectively.

[*Note:* Air filled porosity of any field soil sample can be calculated by subtracting the volumetric water content of the soil from total porosity. Air filled porosity expresses the volume of air present per unit volume of soil, i.e. the relative air content of a soil. Volumetric water content of any soil is the product of gravimetric water content and bulk density considering density of water is 1 g/cc].

Procedure

- Determine bulk density and particle density of soil as discussed in the section 3.1 and 3.2.
- Calculate the porosity of soil using the equation No. 3.3.

Interpretation of Result

% pore Space	Rating
<30	Low
30 - 50	Medium
>50	High

[*Note:* Capillary porosity or the percent of micro pores (<0.08 mm) and non capillary porosity or the extent of macro pores (>0.08 mm) can be estimated by determining the percent volume occupied by capillary water in a soil, after draining excess water under gravity from the saturated soil. It is actually the field capacity (f_c)]

% Non capillary porosity (f_{nc}) = Total porosity (f) – Capillary porosity (f_c)

Questions and Hints

Q.1 Why does top soil show lower particle density value than the sub surface soil?

Hint: Top soil usually contains more organic matter than the subsurface soil. Again, mass per unit volume of organic matter is always less than that of mineral particles. That's why organic matter content affects particle density of the soil. Higher the organic matter content lower is the value of particle density of a soil and reverse is true.

Q.2 Whether texture and structure have any account on the particle density value of a soil?

Hint: If soils developed from same parent material, then texture and structure have no influence on particle density value of the soils.

Q.3 If a soil compacted with a roller what would its effect on particle density and bulk density values?

Hint: No effect on particle density value of the soil, but bulk density value will increase; as the particle density value is not at all affected by pore space, but bulk density value increases due to reduction of pore volume.

Q.4 Why sandy soil has higher bulk density than clayey soil?

Hint: Coarse textured soils have more number of macropores (> 0.08 mm) found in between granules than the fine textured soils. On the other hand, fine textured soils contain more number of micropores (< 0.08 mm) found within the granules. However, sandy soil or coarse textured have less total pore space than the clayey soil or fine textured soils.

Q.5 How does organic matter content in the soil influence its bulk density?

Hint: The mass per unit volume of organic matter is less than that of mineral particles. Organic matter has also some indirect effect on the bulk density value of the soil. Organic matter helps to form soil aggregate and thereby increases total pore space. Thus, the value of bulk density decreases with increase in the organic matter content of soil and reverse is true.

Q.6 What possible variation in bulk density value do you expect between a long time cultivated soil and a nearby uncropped soil?

Hint: Tillage may temporarily loosen the surface and thereby reduces bulk density value, but long term intense tillage causes reduction (due to higher oxidation) of organic matter content and weakening of soil structure; thus increases bulk density value than the nearby uncropped soil. Again, cropped field usually experiences running of heavy mechinaries for tilling of soil and harvesting of crops, thus forms compact layer (higher bulk density) just below the plough layer.

Q7. What are the precautions need to be taken for determination of bulk density by clod method?

Hint: 1. The temperature of molten paraffin wax should be just few degrees above the the melting point of it (60°-70°C) for its quick solidification in air and less likely penetration in the pores of the clod. 2. Coating should be as thin as possible.

4

Surface Area

4.1 Determination of Specific Surface Area of Soil

Specific surface area of a soil is defined as surface area per unit weight of soil and is expressed as $m^2 g^{-1}$. The specific surface of soil greatly influences the physical and chemical properties such as retention of water at high suction, swelling, plasticity, soil strength, cation exchange capacity and availability of nutrients. The basic principle of determination of specific surface is based on the amount of gas or polar liquid required to form a monomolecular layer on the soild. Knowledge of molecular size (diameter) and the mass of the adsorbate enables one to calculate specific surface of the adsorbent.

Principle

Dyal and Hendricks (1950) introduced a method for estimating specific area based upon the adsorption of ethylene glycol (EG) to form a monomolecular layer over the entire soil surface. The method involves adding EG to pretreated soil or clay sample and evaporating the excess EG in an evacuated system. The quantity of EG retained at the moment the evaporating rate decreases is proportional to the surface area. Later on Carter *et al.* (1965) suggested the use of ethylene glycol monoethyl ether (EGME) having high vapour pressure at room temperature than EG for determining surface area of the layer silicate minerals and soils.

Equipments and Materials

Vacuum desiccators with vacuum pump, aluminium can of a diameter of 6-7 cm and a height of 2 cm with lid, culture chamber consisting of a glass dish with cover having 20 cm diameter and a height of 7.5 cm, balance.

Reagents

- Ethylene glycol monoethyl ether (EGME) (2-ethoxyethanol)
- Anhydrous calcium chloride ($CaCl_2$) or phosphorus pentaoxide (P_2O_5)

- Calcium chloride-EGME solvate: Take approximately 120 g of 40 mesh $CaCl_2$ in 1liter beaker and dry in an oven at 210°C for 1 hour or more. Weigh 20 g of EGME in 400 ml beaker and add 100 g of dry cool $CaCl_2$ into it. Mix the content thoroughly with a spatula. Spread the cool solvate uniformly over the bottom of the culture chamber and store in a sealed desiccators.

Procedure

A. Sample Pretreatment

- Remove organic matter from the soil sample as described in section 2.1.1.
- Saturate the sample with Ca by leaching or repeated shaking and centrifugation with an excess of 1.0 M $CaCl_2$. Remove excess $CaCl_2$ with three successive washing with distilled water. Air-dry the sample and pass through 60 mesh sieve. For determination of only external surface area heat the sample at 600°C temperature for 2 hours to collapse the interlayer space.

B. Sorption of Pretreated Sample

- Weigh approximately 1.1 g of soil or 0.3 g clay separately into two tared aluminium cans and spread uniformly over the bottom of the cans.
- Place the cans with lids beneath in a vacuum desiccator containing approximately 250 g of P_2O_5 or $CaCl_2$.
- Evacuate the desiccator using vacuum pump for an hour, close the stopcock and dry to constant weight (it takes about 6-7 hours).
- Weigh the dried sample accurately minimizing the chances of adsorption of atmospheric water as far as practicable.
- Wet one sample by dropwise addition of approximately 3 ml of reagent grade EGME to form soil- or clay- adsorbate slurry.
- Place both the cans containing wetted and non-wetted sample with their respective lids beneath in the culture chamber on a hardware cloth support (0.5 to 1.0 cm opening) over $CaCl_2$-EGME solvate. Place the cover on the culture chamber.
- Place the entire culture chamber in a vacuum desiccators containing $CaCl_2$ and allow the non-wetted and wetted sample-solvate slurry for 30 minutes to equlibrate.
- Evacuate the desiccators with vacuum pump for 45 minutes and allow the desiccators to stand for 4 to 6 hours at room temperature.

- Release the vacuum, open the desiccators and culture chamber and put the lid on the aluminium can to prevent the sample from adsorbing atmospheric water.
- Record the weight of can plus lid plus sample.
- Return the cans with lids beneath, to the culture chamber and the entire culture chamber to the desiccators.
- Evacuate the desiccators for 45 minutes and weigh the sample at 2 to 4 hours intervals.
- Repeat the evacuation until a constant weight of sample is attained.
- Use the mean of two successive weights that agree within a few tenth of a milligram to calculate the amount of EGME retained to form monomolecular layer by the sample.

Calculation

Weight of aluminium can + lid = X g

Weight of aluminium can + lid + soil = Y g

Weight of soil = (Y-X) g

Weight of aluminium can + lid + EGME wetted soil = Z_1 g

Weight of aluminium can + lid + EGME non- wetted soil = Z_2 g

Actual amount of EGME retained by the sample to form monomolecular layer

= $(Z_1 - Z_2)$ g

$$\text{Specific surface of soil} = \frac{(Z_1 - Z_2)}{(Y - X) \times 0.000286} \ \mathrm{m^2 g^{-1}}$$

[0.000286 g of EGME is required to form a monomolecular layer on one square meter of solid surface]

Questions and Hints

Q.1 What are advantages of using EGME over EG for estimation of specific surface?

Hint: Vapour pressure of the adsorbate has an important role in gravimetric estimation of surface area. EGME has higher vapour pressure at room temperature than does EG. Hence, the former evaporates more rapidly from the adsorbate treated sample, requiring lesser time to evaporate the free liquid and to attain equilibrium with a monomolecular layer. In addition to this, the EGME method has greater precision than EG method due to fewer times of handlings of samples and consequently lesser opportunity of atmospheric water adsorption.

Q.2 How one can measure both the external and interlayer surface area of soil?

Hint: Air-dry soil sample gives the total surface area, while heating the sample at 600°C before sorption gives only external surface area as heating causes collapse of interlayer swelling.

Q.3 What can you do to minimize adsorption of atmospheric water during weighing operations?

Hint: (i) During release of vacuum allow air to flow back into desiccators through a tube filled with 8-mesh anhydrous $CaCl_2$.

(i) After releasing vacuum put the lid on the aluminium can promptly.

(ii) Weigh rapidly.

Q.4 Why $CaCl_2$-EGME is used along with anhydrous $CaCl_2$ in the vacuum desiccators during evacuation?

Hint: Mixture of an anhydrous and solvated form or mixture of two solvated forms of salts has a definite vapour pressure at a given temperature. The EGME-$CaCl_2$ solvate is stable at 70°C. Use of $CaCl_2$-EGME solvate with $CaCl_2$ assures an EGME vapour pressure near that of an adsorbed monomolecular layer.

5
Soil Aggregate Analysis

In natural condition primary soil particles cohere to each other to form a cluster or an aggregate through the cementing action of various agents like organic matter, carbonates, oxides of iron and aluminium, etc. Aggregate analysis tells one about the structural status, pore geometry which in turn influences the movement of air and water in soil, amount of water stable secondary particles (aggregate) as well as the extent of finer particles aggregated to the coarser particles and the erodibility status of surface soil against wind and water.

Methods of Analysis

Among the several methods of aggregate analysis the commonly used methods practiced in agriculture and related fields are wet sieving and dry sieving method of aggregate analysis.

5.1 Wet Sieving Method of Soil Aggregate Analysis

Principle

The method developed by Yoder (1936) based on the principle of determination of proportion of aggregates which are stable against disruptive forces of water. Wet (short period of wetting) soil sample is sieved through a nest of sieves in Yoder's apparatus, which raises and lowers the nest of sieves in water. The amount of aggregates present in each sieve was then measured after drying which includes coarse fraction (sand and gravel) also. Subtracting the amount of coarse fraction from the amount of aggragate in the respective sieve, the actual estimate of aggregate is obtained. To express the agggation staus of a soil, the following indices are used:

Mean Weight Diameter (van Bavel, 1949)

Mean weight diameter (MWD) of the soil aggregates is the sum of the products of mean diameter of a given size fraction of aggregates ($\bar{X}i$) multiplied by the proportion by weight of that fraction to the total soil sample (w_i).

$$\text{MWD} = \sum_{i=1}^{n} \bar{X}_1 W_1 \quad (5.1)$$

where, $w_i = Wi/\sum W_i$, and W_i = the weight of the soil aggregate on i^{th} sieve

Geometric Mean Diameter (Mazurak, 1950)

The weight of the aggregates in a given size fraction is multiplied by the logarithm of the mean diameter of that fraction. The sum geometric mean diameter (GMD) is the sum of these products for all size fractions is divided by the total weight of the sample.

$$\text{GMD} = \frac{\sum_{i=1}^{n} W_i . log\, \bar{X}_i}{\sum_{i=1}^{n} W_i} \quad (5.2)$$

Structural Coefficient or Stability Coefficient (Russel, 1938)

$$\text{S.C.} = \frac{D-S}{S} \quad (5.3)$$

where, D is the percentage of particle < 0.25 mm as determined by the method of mechanical analysis. S is the percentage of aggregates < 0.25 mm in diameter as determined by wet sieving of untreated soil. Higher the value better is the structure.

Equipments and Materials

Yoder apparatus, two sets of standard sieve of 5″ diameter and 2″ height with opening diameter of 5.0, 2.0, 1.0, 0.5, 0.25 and 0.1 mm (4, 10, 18, 35, 60 and 140 mesh, respectively) and one sieve of 8 mm (2.5 mesh) opening diameter, physical balance, oven, desiccator, watch glass, aluminium moisture box.

Procedure

- Collect an air dry soil clod from the field with help of spade.
- Break the clod into smaller aggregates by hands in such a way that the aggregates pass through 8 mm sieve but retain on 5 mm sieve.
- Remove large gravels and roots as far as practicable.
- Weigh 50 g of aggregates (5–8 mm) in three watch glasses separately.
- Keep one in oven for overnight at 105°C for moisture determination and use other two for aggregate analysis in duplicates.

- Arrange two sets of sieve in series in such a way that the topmost sieve has the largest sieve opening (5 mm) and the bottommost sieve has the smallest sieve opening (0.1 mm).
- Now uniformly spread the aggregates on the topmost sieve.
- Add 10 ml of salt free water. After about 5 minutes again spray 5-100 ml of water to make it saturated and wait for 3-5 minutes.
- Transfer the nest of sieve to the wet sieving drum already filled with salt free water. Adjust the level of water in drum in such a way that aggregates just dip into water when the sieves are in highest position.
- Lower down the sieves to the lowest position and allow wetting aggregates for 10 minutes.
- Switch on the sieve shaker and oscillate it for 30 minutes with a frequency of 30-35 cycles/minute and a stroke length of 3.5 cm.
- Remove the nest of sieves and keep it sometime in an inclined position to drain water and hardening of aggregates in air.
- Dry the materials on the sieves at 70-75°C in oven for about 30-40 minutes.
- Transfer the materials on each sieve separately in aluminium boxes and dry them at 105°C in oven for overnight.
- Cool the boxes in desiccator and take weight of the materials (M_i) separately.
- To estimate the amount of coarse particles of >0.25 mm (sand and gravels) on each sieve transfer the material to 250 ml beaker separately and disperse with 0.5 **N** Sodium hexametaphosphate in soil: solution ratio of 1:3 in mechanical shaker for 10 minutes.
- Transfer the dispersed aggregates again on the same sieves in which they were previously retained.
- Oscillate the shaker again for 30 minutes.
- Coarse particle of >0.25 mm will retain on the sieve. Oven dry and record the weight of coarse particles on each sieve (N_i).

Fig. 5.1 Wet sieve shaker (Yoder type)

Calculate the amount of aggregate on each sieve by subtracting the weight of coarse fraction (N_i) from the oven dry aggregated materials on the respective sieve (M_i).

Calculation

Say,

Weight of air dry sample = A g

Weight of oven dry sample = B g

Moisture content (%) = [(A - B) / B] x 100

Example of Sieve Analysis

Sieve No.	Aggregate size range (mm)	Mean diameter of aggregate (mm) $\bar{X}_i$	Wt. of retained aggregate (g)		Wt. of aggregated particles (g) $W_i = M_i - N_i$	$W_i \bar{X}_i$	$W_i \log \bar{X}_i$
			Aggregated clod (M_i)	S and (N_i)			
1	8.0-5.0	6.50	M_1	N_1	W_1	$W_1\overline{X_1}$	$W_1 . log\overline{X_1}$
2	5.0-2.0	3.50	M_2	N_2	W_2	$W_2\overline{X_2}$	$W_2 . log\overline{X_2}$
3	2.0-1.0	1.50	M_3	N_3	W_3	$W_3\overline{X_3}$	$W_3 . log\overline{X_3}$
4	1.0-0.5	0.75	M_4	N_4	W_4	$W_4\overline{X_4}$	$W_4 . log\overline{X_4}$
5	0.5-0.25	0.375	M_5	N_5	W_5	$W_5\overline{X_5}$	$W_5 . log\overline{X_5}$
6	0.25-0.1	0.175	M_6	N_6	W_6	$W_6\overline{X_6}$	$W_6 . log\overline{X_6}$
			$\sum_1^6 M_i$		$\sum_{i=1}^6 W_i$	$\sum_{i=1}^{6} W_i . \bar{X}_i$	$\sum_{i=1}^6 W_i .log\bar{X}_i$

a) Mean Weight Diameter (MWD) = $\sum_{i=1}^{6} W_i . \bar{X}_i / \sum_{i=1}^{6} W_i$

b) Geometric Mean Diameter (GMD) = $\dfrac{\sum_{i=1}^{6} W_i . \log \bar{X}_i}{\sum_{i=1}^{6} W_i}$

c) Change in mean weight diameter (CMWD) = MWD of dry sieving – MWD in wet sieving

d) Structural Coefficient or Stability Coefficient = $\dfrac{D-S}{S}$

Where, D is the percentage of particle < 0.25 mm obtained from mechanical analysis. S is the percentage of aggregates < 0.25 mm in diameter as determined by wet sieving aggregate analysis ($W_6 \times 100 / \sum_{i=1}^{6} W_i$)

5.2 Dry Sieving Method of Soil Aggregate Analysis

Principle

Aggregate analysis by dry sieving method measures size and stability of dry aggregates in which primary particles weakly held together as in arid region soil. The principle behind this method lies on the determination of proportion of aggregates which are stable against vibrating action of sieves simulating the scouring effect of wind.

Equipments and Materials

Rotary sieve shaker, sieves of 8″ diameter and 2″ height with opening diameter of 5.0, 2.0, 1.0, 0.5, 0.25 and 0.1 mm (4, 10, 18, 35, 60 and 140 mesh, respectively) and one sieve of 8 mm (2.5 mesh) opening diameter, physical balance, oven, desiccator, brush, aluminium moisture box.

Procedure

Same as wet sieving method of analysis except following few points-

- Uniformly distribute 50 g air dry soil aggregate (5-8 mm) on the topmost sieve, but do not moisten the aggregates like wet sieving method.
- Fit the bottommost sieve in a pan and cover the topmost sieve with a lid.
- Oscillate the rotary sieve shaker for 10 minutes.
- Collect dry aggregates on each sieve in separate aluminium boxes.
- Dry the aggregates in oven and after cooling in desiccator weigh them separately.
- Follow same steps for estimating the amount of coarse fraction in the dry aggregates on each sieve as wet sieving method.

Fig. 5.2: Rotary type shaker

Calculation

Same as section 5.1

Questions and Hints

Q.1 What is 'soil aggregate'?

Hint: Primary soil particles cohere to each other to form a cluster or secondary particle through the cementing action of various agents like organic matter, aluminium, oxides of iron and carbonates, etc. is termed as soil aggregates.

Q.2 Why dry sieving method is considered to be better than wet sieving in arid region soil?

Hint: Soil aggregates of arid region are so weakly held together in moist condition that the mechanical action of sieving is sufficient to destroy them. Dry sieving gives an important index for characterizing the susceptibility of soils to wind erosion.

Q.3 What is CMWD and what is its significance?

Hint: CMWD is the difference between MWD in dry sieving and that of wet sieving.

Higher value of CMWD indicates the instability of aggregates and susceptibility to crusting and erosion. The soils having lower CMWD has a better structure than the others.

Q.4 Between podsol and laterite surface soil of humid region which one is better aggregated?

Hint: Laterite surface soil is better aggregated than podsol. Surface soil of podsol contains less organic matter, clay and oxides of iron and aluminium due to illuviation; on the other hand high amount of oxides of iron and aluminum in laterite soil favours better aggregation.

Q.5 Between arid and humid region soils which one is less aggregated?

Hint: Arid region soils are less aggregated than humid region soil because arid region soil contains lesser amount of finer particles (slow chemical weathering at low soil miosture), low organic matter (due to rapid oxidation at high temperature), low microbial activity and root growth (due to low soil moisture).

6

Soil Moisture

6.1 Determination of Soil Moisture

6.1.1 Determination of Soil Moisture Content by Gravimetric Method (Direct method)

Traditionally, water content in soil is expressed as the ratio of the weight of water present in the soil to the weight of dry soil. When this ratio is multiplied by 100, it becomes the percentage of water in the soil sample on dry weight basis.

Equipments and Materials

Soil auger, aluminium moisture box with air tight lid, balance, oven and desiccator

Procedure

- With the help of soil auger dig out 10-100 g soil sample from the desired depth and immediately keep it into a previously weighed aluminium box with tight fitting lid to avoid any evaporation loss.
- Weigh the box with wet soil sample and record it.
- Place the aluminium box in the even with lid off and dry it at 105°C for 24 hours.
- Do not put any other wet sample in the oven during this period particularly during last few hours.
- Next day remove the aluminium box from the oven along with its lid and keep in desiccator to cool it at room temperature.
- Weigh the aluminium box and its lid plus dry soil.

Calculation

Let,

Weight of empty aluminum box with its lid = X g

Weight of aluminum box with its lid plus wet soil sample = Y g

Weight of aluminum box with its lid plus dry soil sample = Z g

$$\text{Percent moisture in soil} = \frac{Y-Z}{Z-X} \times 100$$

6.1.2 Determination of Soil Moisture Content by Neutron Moisture Meter Method (Indirect Method)

Soil moisture content by this method can be determined *in situ* without disturbing the soil system. The basic principle behind the method is to estimate the hydrogen nuclei (predominantly of the water molecule except in organic soil) present in a particular volume of soil. The high energy, 2 to 4 million electron volts (MeV) fast (1600 km/sec) neutrons emitted from a source lose its kinetic energy (0.03 eV) and get thermalized or slow down (2.7 km/sec) after several collision with hydrogen nuclei. Usually the source of fast moving neutron may be a mixture of Radium-Beryllium or Americium-Beryllium. Though the former is more hazardous it is preferred than the later because of its longivity (higher half life).

$^{9}Be_4 + {}^{4}He_2 \rightarrow {}^{12}C_6 + {}^{1}n_0 + 5.74$ Mev

(α particle) (Fast neutron)

The amount of thermalized neutrons is then detected by a detector usually filled with BF_3 gas. Boron nucleus absorbing thermalized neutrons emits αparticle which creates electrical pulse on a scaler.

$^{10}B_5 + {}^{1}n_0 \rightarrow {}^{4}He_2 + {}^{7}Li_3 + 2.8$ Mev

Equipments and Materials

Neutron moisture meter assembly consisting of probe, detector, scaler, access tube made up of 20 gauze steel or aluminium and having same outside diameter of the soil auger, moisture box, oven, desiccator.

Procedure

Calibration

- Drill a hole with the help of soil auger and carefully insert the access tube keeping about 10 cm space above the soil surface.

- Close the opening of the tube with a rubber cork to prevent entry of water or debris. In case of high water table situation plug the lower end of the access tube with the rubber stopper.
- Place the probe on the top of the access tube and measure the counts for one minute (or as specified by the manufacturer). This is called standard counts (SC).
- Lower down the probe in the access tube to a desired depth at which soil moisture content is to be estimated and record the counts. This is known as observed counts (OC).
- Take readings at different depths at an interval of more than 15 cm and at least 25 cm from the surface.
- Immediately collect soil samples from the corresponding soil depths and measure soil water content gravimetrically.
- Calculate volumetric water content (θv) by multiplying gravimetric water content with its corresponding bulk density of the soil depth.
- Calculate count ratio by dividing observed count by standard count.

$$\text{Count Ratio}(\text{CR}) = \frac{\text{Observed counts}(\text{OC})}{\text{Standard counts}(\text{SC})} \tag{6.1}$$

Draw the calibration curve by plotting count ratio against volumetric water content.

Moisture Determination

- Record standard and observed count by placing probe accordingly.
- Calculate count ratio.
- From the calibration curve determine the soil water content for any count ratio.

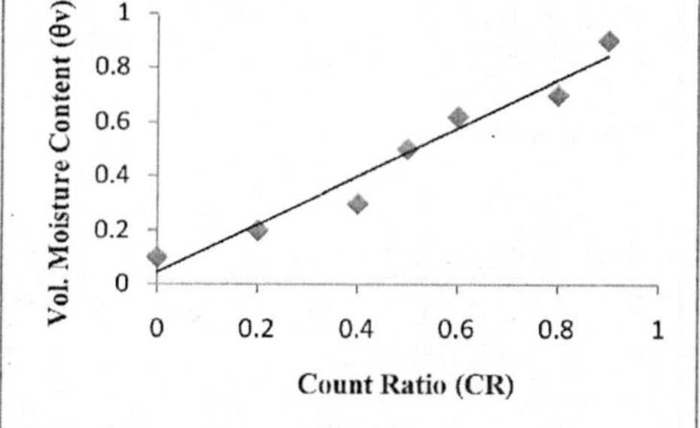

Fig. 6.1 Calibration curve of neutron moisture meter

6.1.3 Determination of Soil Moisture using Time Domain Reflectometry (TDR)

Principle

Time Domain Reflectometry is another indirect nondestructive method of measuring soil moisture. It involves measurement of propagation of electromagnetic (EM) waves or signals. Propagation constants for EM waves in soil, such as velocity and attenuation, depend on soil properties especially water content and electrical conductivity. A waveguide or probe of known length (L) is embedded in soil and travel time for a TDR generated EM ramp to traverse the probe length is determined. From the travel time analysis the soil's apparent or bulk dielectric constant (or dielectric permittivity) which is independent of soil texture, temperature and salinity is computed. The apparent dielectric constant (K_a) of soil surrounding the probe is a function of the propagation velocity (v).

$$v = \frac{c}{\sqrt{K_a}} \text{ or } (K_a) = \left(\frac{c}{v}\right)^2 = \left(\frac{ct}{2L}\right)^2 \quad ---(6.2)[as\ v = \frac{2L}{t}]$$

where, c is the speed of light in vacuum (3×10^8 m s^{-1}), t is the travel time for the EM pulse to traverse twice the distance of waveguide or probe (forward and reflected run, *i.e.* 2L). The travel time is evaluated based on the EM length of probe, which is characterized on TDR output screen by diagnostic changes in the waveform: t_A marks the entry of the signal to the probe and t_B marks the reflection at the end of the probe (Fig.6.2). The EM length of the probe ($\Delta t = t_B - t_A$) increases with increase in water content (and dielectric constant), a consequence of reduced propagation velocity.

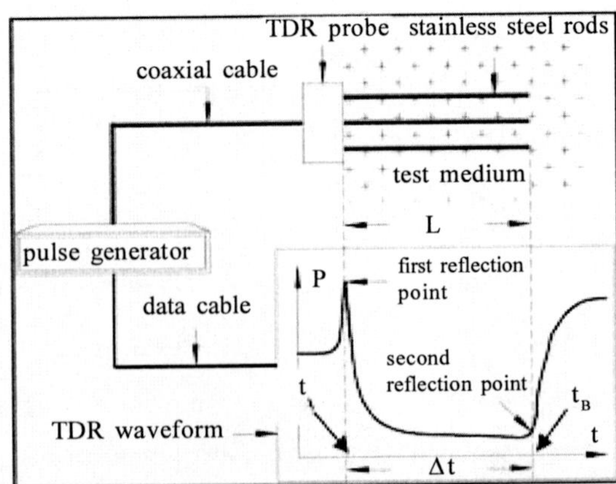

Fig. 6.2: Principle of TDR measurement

Because of unique polar structure of water molecule, its dielectric constant is many times greater than that of air and solid phase of soil (Table 6.1). Therefore, the apparent dielectric constant of soil and other porous materials depends on the volumetric water content (θ) of the tested object. Topp *et al.* (1930) proposed an empirical polynomial equation to correlate K_a with θ for mineral soils.

$$\theta = 4.3 \times 10^{-3} K_a^{3} - 5.5 \times 10^{-4} K_a^{2} + 2.92 \times 10^{-2} K_a - 5.3 \times 10^{-7} \qquad (6.3)$$

Table 6.1: Dielectric Constants of Common Materials

Material	Dielectric constant
Water	81
Air	1
Dry soil	3.0–5.0
Plastic	2.1–2.6
Rubber	2.3–4.0
Dry paper	3.0
Dry wood	2.0–4.0

Equipment and Materials

- The TDR sensor (usually a probe) contains a waveguide consisting of two or three parallel wires which is connected via a coaxial cable (with a constant impedance (50Ω) so that no reflection is generated that can weaken the signal pulse) to a voltage pulse generator which sends precisely defined voltage pulses into the sensor. As the pulse travels along the waveguide its progress varies depending on the moisture content of the soil. When the pulse reaches the end of the waveguide it is reflected. This reflection is visualized in a TDR waveform using an oscilloscope connected to the sensor.
- Installation kit (stainless steel rods with holder, hammer).

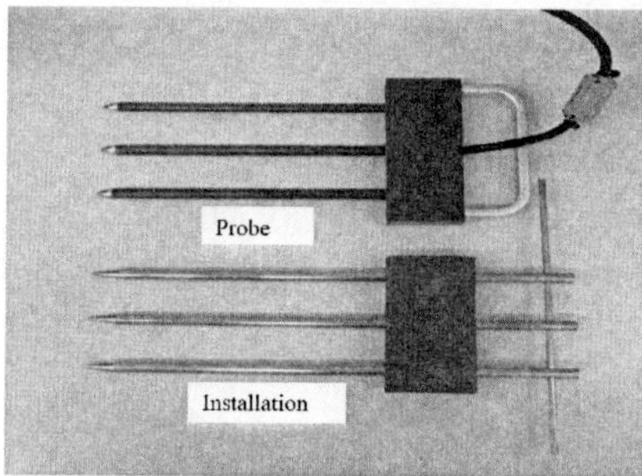

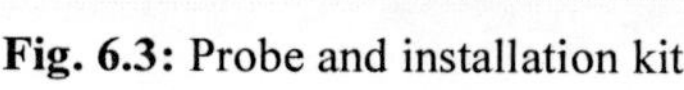

Fig. 6.3: Probe and installation kit

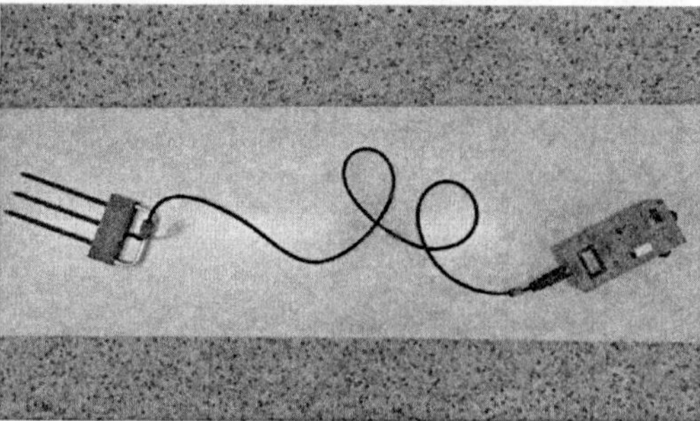

Fig.6.4: TDR sensor system

Procedure

- Use stainless steel rods and hammer to prepare the holes for waveguide or probe.
- Insert the probe into the hole and ensure a good contact with the adjacent soil.
- Connect the probe to the TDR instrument.
- Initiate EM signal transmission and measure travel time.
- Repeat the measurement for several times.

Calculation

Calculate apparent dielectric constant, K_a using equation 6.2

Now, calculate volumetric water content substituting K_a in the equation 6.3

[*Note:* Soil moisture content can also be measured using Frequency Domain (FD) Capacitance and Frequency Domain Reflectometry (FDR)].

6.1.4 Determination of Soil moisture using Frequency Domain (FD) Capacitance and Frequency Domain Reflectometry (FDR)

Principle

Frequency Domain technique is based on the principle that the electrical capacitance of the capacitor that uses soil as a dielectric depends on volumetric water content. When the capacitor made up of metal rods or plates is embedded in the soil and connected to an oscillator to form an electric circuit, the circuit operating frequency changes with changes in soil moisture. These changes form the basis of frequency domain technique which is used in capacitance sensor or frequency domain Reflectometer sensor.

In capacitance sensor, the apparent dielectric constant, K_a of a soil is determined by measuring the change time of a capacitor made with that soil. In FDR, the oscillator frequency is controlled within a certain range to determine the resonant frequency (the frequency at which the amplitude is the greatest), which is the measure of water content in soil. When probes consisting of two or more metal plates or rods are inserted into the soil and an electrical field is applied, the soil around the electrode forms the dielectric of the capacitor to complete the oscillating circuit.

Equipment and Materials

Soil moisture sensor and read-out device

Procedure

- Check the setting of the reading device.
- One can go with the calibration equation provided usually by the manufacturer for mineral and organic soil or can calibrate soil moisture sensor for specific soil condition before use and for more accurate measurement as the measurement is affected by temperature, salinity, bulk density and clay content.
- To prepare calibration curve, measure the output (in mV) at a given bulk density at various known volumetric soil moisture contents for this specific soil. Draw the calibration curve and develop equation between output and volumetric water content.
- Always ensure a good contact of soil moisture sensor (FDR) with the soil.
- Now, measure the output in mV for that soil at any unknown soil moisture content.
- Repeat the measurement several times.

Calculation

Put the average reading in mV into the calibration equation and determine the volumetric water content of the soil.

6.2 Soil Moisture Constants

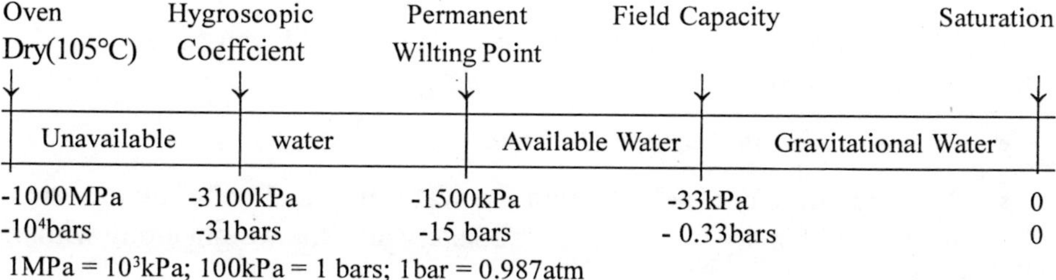

1MPa = 10^3kPa; 100kPa = 1 bars; 1bar = 0.987atm

6.2.1 Determination of Maximum Water Holding Capacity cum Single Value Physical Constants of Soil

In this method alongwith maximum water holding capacity the following other physical constants can be determined simultaneously:

- Bulk density
- Particle density
- Porosity
- Volume expansions per 100 cc soil.

Maximum water holding/retention capacity is defined as the amount of water in the soil when its total pore space - both macro and micro are completely filled with water.

Equipments and Materials

Keen Raczkowski box (5.6 cm internal diameter and 1.6 cm hieght) having perforated bottom (diameter of hole is 0.75 mm and holes are 4 mm apart), slide calipers, spatula, water tray or petridish, watch glass, oven, desiccator.

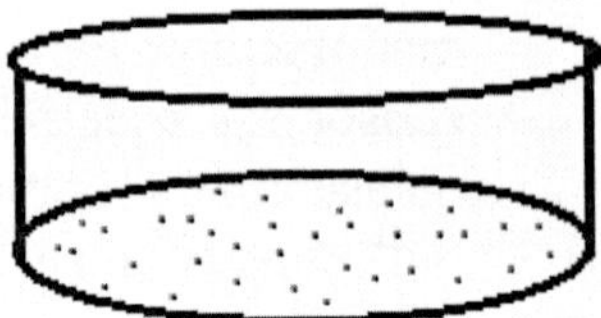

Fig. 6.5: Keen Raczkowski box

Procedure

- Cut a filter paper to fit properly at the bottom of the Keen box.
- Weigh the Keen box fitted with the filter paper.
- Pack the Keen box with air dry soil passed through 2 mm sieve by adding small amount of soil at a time and tapping (20 times at each addition) the Keen box gently on the table.
- When the box is full, slice off the extra amount of soil with sharp spatula. Tap the box, add little more amount of soil, tap on the table and again remove the surplus soil to ensure a level surface.
- Do this for few more times and take weight of the Keen box with air dry soil.
- Place the box in a water tray or petridish and add water to the tray to a depth of 1 cm.
- Leave it overnight to get saturated.
- Next day take out the box from the water tray and wipe the outside of the box with a dry towel and immediately take the weight along with the wet soil.
- Cut the expanded portion of the wet soil above the top of the box carefully and keep it to a previously weighed watch glass.
- Immediately weigh the watch glass with the surplus saturated soil and Keen box with the residual wet soil separately to avoid evaporation loss.
- Dry both the box and watch glass with saturated soil sample in the oven at 105°C for overnight.
- After cooling in the desiccator weigh the box and watch glass with their dry soil separately.

- To apply correction for the amount of water absorbed by the filter paper weigh 10 flitre papers of same size as fitted in the Keen box together. Saturate them with water and roll a glass rod gently over them to squeeze out water uniformly. Weigh and calculate the average amount of water absorbed by one filter paper.
- Measure the internal volume of Keen box with the help of slide calipers.

Calculation

Let,

Weight of Keen box plus filter paper = A g

Weight of Keen box plus filter paper plus air dry soil = B g

Weight of Keen box plus filter paper plus wet soil = C g

Weight of Keen box plus filter paper plus residual wet soil = D g

Weight of watch glass = E g

Weight of watch glass plus surplus wet soil = F g

Weight of Keen box plus filter paper plus residual dry soil = G g

Weight of watch glass plus surplus dry soil = H g

Average weight of water in one wet filter paper = M g

Internal volume of Keen box = $\pi d^2h/4$ = V cc

Where, d and h are the inner diameter and the height of the box, in cm, respectively.

Now, total weight of oven dry soil = (G–A) + (H-E) g = X g

Bulk density of soil (g/cc) $= \frac{X}{V}$

$$\text{Particle density}(g/cc) = \frac{G-A}{[V-(D-G-M)]}$$

$$\text{Maximum water holding capacity}(\%) = \frac{(C-A-M-X)}{X} \times 100$$

$$\text{Porosity}(\%) = \frac{D-G}{V} \times 100 \text{ [considering density of water 1 g / cc]}$$

$$\text{Volume expansion of 100 cc soil (\%)} = \frac{H-E}{G-A} \times 100$$

Interpretation of Results

Rating	Maximum water holding capacity (%)	Volume of Expansion (%)
Low	< 20	<10
Medium	20 - 50	10-30
High	>50	>30

6.2.2 Determination of Field Capacity (Field Method)

According to Veihmeyer and Hendrickson (1931) field capacity is the amount of water held in soil after draining of excess gravitational water from a saturated soil and downward movement of water is practically ceased. Normally, it takes about 2 to 3 days after saturation. It is the upper boundary of plant available water. Water at this condition is present in capillary pores (micro pores). The tension with which the moisture is held ranges from 0.1 bar (pF 2.0) for light soil to 0.33 bar (pF 2.5) for heavy soil.

Equipments and Materials

Black polythene sheet (4m x 4m) or straw mulch, spade, tube auger, balance, aluminium box, oven, desiccator.

Procedure

- Select a uniform area of about 9 m^2 (3m x 3m). Ensure that water table must be more than 2 m below the layer for which field capacity is to be determined.
- Remove weeds, pebbles, etc from the surface and bund the area with the soil from all sides.
- Fill the area with sufficient amount of water to completely saturate the soil to the depth of interest.
- As soon as the surface water disappears cover the area with the polythene sheet or thick (about 40 cm) straw mulch to check evaporation.
- After 24 hours of saturation take soil sample from the middle of the plot from desired depth removing the polythene sheet/straw mulch with the help of tube auger.
- Weigh, dry the soil in oven at 105°C for about 24 hours and again take the weight of dry soil sample after cooling in desiccator.

- After each sampling immediately cover the area with polythene sheet/ straw mulch.
- Take the soil sample daily and determine moisture content until the value of 2 to 3 successive days are nearly same.
- Plot the daily moisture content (%) against days. The lowest moisture content represents the value at field capacity of soil.

Precaution

- Sample should not be taken within 50 cm from the borders of the area.
- Sample should not be taken within 30 cm from the water front.
- Sampling should be done either in the morning or evening hour.

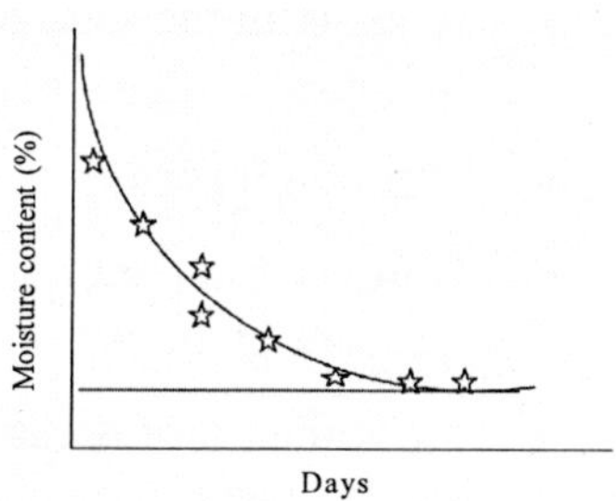

Fig. 6.6: Field capacity curve

6.2.3 Determination of Permanent Wilting Point

Permanent wilting point can be defined as the maximum soil wetness at which the wilted plant fails to recover its turgidity even when it is kept in a saturated atmosphere for 12 hours. The permanent wilting point is characterized as the lower limit of available soil moisture. At permanent wilting point water is held at a tension of 15 bar (pF 4.2).

Equipments and Materials

Pressure plate apparatus, metal rings of 5 cm diameter and 4 cm height, core sampler/soil sampling auger, pipette, moisture box, oven, balance, desiccator.

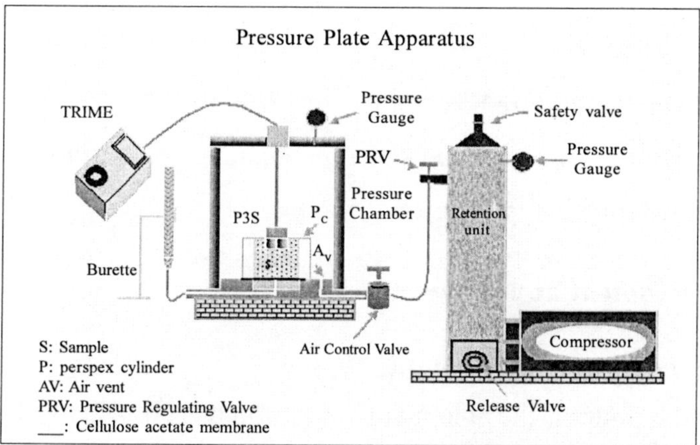

Fig. 6.7: Pressure plate apparatus

Procedure

- In case of disturbed soil sample take requisite amount of air dry soil sample passed through 2 mm sieve in metal rings to maintain the same bulk density as was in the field condition. But in case of undisturbed soil, sample soil in metal rings is collected with the help of core sampler.
- Place the metal rings with soil in the ceramic plate.
- To saturate the soils add water in ceramic plate and allow it to stand for 24 hours.
- Drain out excess water with the help of pipette.
- Transfer the ceramic plate alongwith the saturated soil sample to the pressure chamber.
- Close the lid of pressure chamber by fixing rubber seal.
- Adjust the desired pressure (15 bars) in the chamber by turning the regulator. Water will come out through the outflow tube until an equilibrium between soil water pressure and air pressure attains (It may take 20 to 48 hours).
- After attainment of equilibrium slowly release the pressure by turning off the regulator.
- Open the pressure chamber and transfer the wet soil sample to a previously weighed moisture box and weigh.
- Dry the soil sample in oven at 105°C for 24 hours and weigh again after cooling in desiccator.

Calculation

Let,

Weight of empty moisture box = X g

Weight of moisture box plus wet soil after attaining equilibrium = Y g

Weight of moisture box plus dry soil = Z g

Soil water content at permanent wilting point (%) $= \dfrac{Y-Z}{Z-X} \times 100$

6.2.4 Determination of Available Water

The available water is the amount of water which is utilized by plants for their growth and development.

Gravimetric Available soil water (%) = Moisture at Field capacity (%) – Moisture at Permanent Wilting Point (%)

Volumetric Available Water (cc/cc) = Bulk Density (of the soil layer concerned) × Available soil water (%)

Available water at a particular depth (cm) = Volumetric Available Water (cc/cc) × Depth of that soil layer (cm)

6.2.5 Determination of Hygroscopic Coefficient

Hygroscopic coefficient represents the amount of water taken up by a dry soil kept in an atmosphere saturated with water vapour (Relative humidity, H ≈ 100%) at a given temperature. At hygroscopic coefficient water is held at a tension of 31 bars (pF 4.5). Higher the organic matter or clay content higher will be the hygroscopic coefficient value.

Equipments and Materials

Moisture box, watch glass, desiccator, 2 mm sieve, balance, oven, 3.3% H_2SO_4 (by weight).

Procedure

- Prepare a 3.3% H_2SO_4 (by weight) to get relative humidity of 98% at a vapour tension of 31 atmosphere.
- Take 5g air dry soil sample (<2 mm) on a watch glass and place it in desiccators.
- Close the lid of the desiccator and allow it to be equilibrated for a week.
- After a week open the lid of the desiccator and quickly transfer the soil to a previously weighed moisture box and weigh it.
- Dry the soil in oven at 105°C for 24 hours and again weigh it.

Calculation

Let

Weight of empty moisture box = A g

Weight of moisture box plus equilibrated soil = B g

Weight of moisture box plus dry soil = C g

$$\text{Soil moisture content at Hygroscopic Coefficient (\%)} = \frac{B-C}{C-A} \times 100$$

6.2.6 Determination of Moisture Equivalent

According to Briggs and McLane (1907) moisture equivalent is defined as the amount of water held by a soil of 1 cm thick when subjected to a centrifugal force of 1000 times that of the gravity corresponding to 2400 rpm for 30 seconds in the centrifuge. In medium textured soil its value is very close to that of field capacity, while it is lower for sandy soil and higher for clayey soil than their corresponding field capacity values.

Equipments and Materials

Briggs-McLane centrifuge with sample cups, wire gauged moisture equivalent box, moisture box, oven, balance, spatula, 2 mm sieve, Whatman No. 2 filter paper, water tray, desiccator.

Procedure

- Cut the Whatman No. 2 filter paper and fix properly on the bottom of the wire gauged moisture equivalent box.
- Weigh 30 g air dry soil sample passing through 2 mm sieve and pour in the moisture equivalent boxes.
- Tap the boxes gently on the table and level the surface of the soil with spatula.
- Place the boxes in water tray having about 1 cm water and allow them for 24 hours to get saturated.
- Next day take the boxes out of water, wipe out the outer side of the boxes with dry cloth and put their cover on.
- Arrange the boxes in the centrifuge in opposite direction for proper balancing and centrifuge them for 30 minutes at 2400 rpm.
- After centrifugation transfer the soils quickly to the previously weighed moisture boxes and weigh.
- Dry the soil sample in oven for 24 hours at 105°C and weigh again after cooling in desiccator.

Calculation

Let,

Weight of empty moisture box = A g

Weight of moisture box plus moist soil = B g

Weight of moisture box plus oven dry soil = C g

$$\text{Moisture Equivalent (\%)} = \frac{B-C}{C-A} \times 100$$

6.3 Determination of Soil Moisture Potential

6.3.1 Determination of Soil Suction by Resistance Block

Principle

The principle behind the working of resistance block is the conductance or in other words resistance to the flow of electricity in relation to moisture content. Bouyoucos and Mick (1940) proposed that when two electrodes placed parallel to each other in a porous non-conducting material and current is passed, then the flow of electricity will experience resistance depending upon the moisture content of the porous material. The resistance is inversely proportional to the moisture content of the porous medium i.e. the moisture content of the soil where it is placed. Thus, dry soil offers maximum resistance while wet soil has the minimum.

The use of resistance block is a relatively inexpensive device of providing continuous measure of matric potential. Resistance block made up of plaster of Paris works effectively over a soil moisture range of 0.5 to 15 bars of matric suction; hence it works more satisfactorily in dry than in moist soil. Fibre glass block performs well over the entire range of available soil water (0.3 to 15 bars), while nylon block is most effective in very moist soil (2 atm).

Preparation of Resistance Block

Equipments and Materials

A frame of mould compartment, PVC coated twin cable (1/22 gauge), plaster of Paris, stirrer, 20 mesh stainless steel screen (for electrode).

Procedure

- Solder the wires of a twin cable to two pieces of stainless steel screen which will function as electrodes.

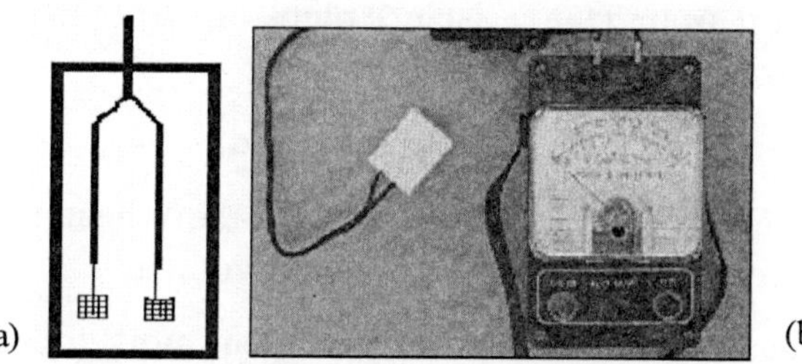

Fig. 6.8 (a) Internal arrangement of parallel soldered wire electrodes in the resistance block; (b) Resistance block with Wheatstone bridge

- Set two electrodes parallel to each other in a frame of mould compartment.
- Hook them to prevent the electrodes from being pulled out of the block.
- Prepare slurry of plaster of Paris with water in 1:1 ratio (by weight), mix thoroughly with the stirring rod.
- Pour the slurry in the compartment in one stroke and fill it to brim.
- Remove air bubbles (if any) by gentle tapping of mould.
- Allow it to hard for at least ½ hour.
- Remove the block from the mould and cure it in shade for about 15 days.

A. Standardization and Calibration of Resistance Block

Equipments and Materials

A bucket of water, Wheatstone bridge of A.C. type (1000 cycles), garden pots, moisture box, oven.

Procedure

- Dip the resistance block in a bucket of water and allow to get saturated for ½ hour.
- Take out the block from water and expose it to the atmosphere for 10 minutes.
- Repeat the process for at least three times to be ensured that the block does not contain any entrapped air.
- Connect the wires of the resistance block with the Wheatstone bridge and record the reading dipping the block in water. Record the reading for all the blocks and calculate the mean reading. Reject the resistance blocks which showed more than 5% deviation from the mean value.
- Install the resistance block in a garden pot containing soil collected from the same field whose matric potential is to be measured periodically.
- Pack the soil around the resistance block in the pot to maintain same bulk density as was in the field condition.
- Irrigate the pot to make the soil saturated.
- Record simultaneously the resistance on the Wheatstone bridge and soil moisture content at the block's depth periodically as the soil dries.
- Plot the data of resistance at different soil moisture content on a graph aper and draw the calibration curve.

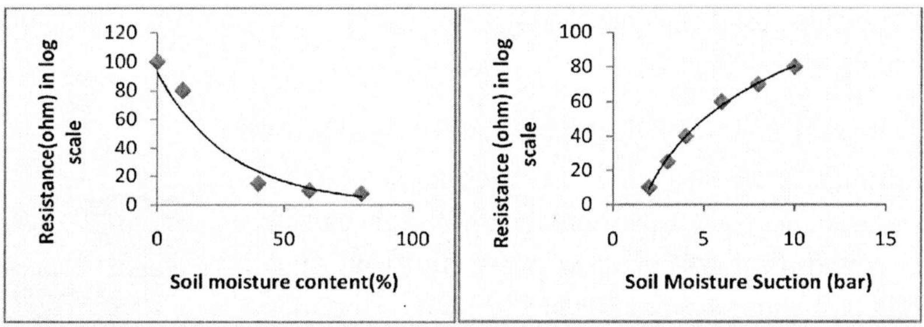

Fig.6.9: Relationship between resistance vs soil moisture content/soil moisture tension

[*Note:* Temperature Correction: As units are temperature sensitive temperature correction of block readings are necessary. Correction is made to 70°F with the following equation]:

$$\textbf{Log R (t-70) = Log R.t [1 + 0.002 (t - 70)]}$$

Where, R is the resistance reading in ohms, t is the temperature in °F

B. Determination of Soil Moisture Content or Soil Moisture Tension

Equipments and Materials

Resistance block unit (Resistance block along with Wheatstone bridge), post hole auger.

Procedure

- Dig a hole to the specified depth for installation of the resistance block with the help of posthole auger.
- Place the water saturated resistance block in the hole at the specified depth.
- Fill the hole with moist soil and ensure good contact of the block with the soil.
- Keep the vertical distance between two blocks more than 30 cm if more than one resistance blocks are placed in the same bore.
- Label the terminal wires of the resistance blocks carefully with their respective depths before installation.
- To avoid water stagnation make a heap with the soil near the surface of the bore spot to about 3 cm.
- Irrigate the field and record the resistance reading on the Wheatstone bridge periodically as soil starts to dry.

- Calculate soil moisture content or soil moisture tension from the calibration curve against the resistance reading.

6.3.2 Determination of Soil Suction by Tensiometer

Tensiometer is widely used for years together to measure soil moisture tension/suction which can be used as the guide for scheduling irrigation. It is relatively cheaper, simple and easy to install. It works well over a range of soil moisture tension of 0 to 0.8 bars.

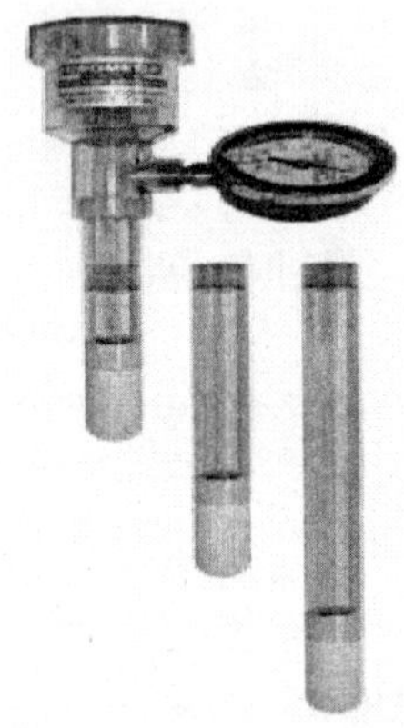

Fig. 6.10: Tensiometer

Tensiometer consists of a fine porous ceramic cup of 7.5 cm long connected with a manometer or vacuum gauge through a plastic tube. The manometer or vacuum gauge indicates vacuum pressure created relative to atmospheric pressure. Tensiometer is filled with water and is made air tight. Now, if the porous cup is placed in relatively dry soil, water will come out through the pores of the ceramic cup to attain equilibrium between the suction on manometer or gauge and the soil. This leads to a drop of hydrostatic pressure which is measured by manometer or vacuum gauge.

Characteristics of a Standard porous cup

The material of the ceramic cup should be such that it will pass through the following standards:

(i) **Air Entry Value**: Air entry value is the minimum air pressure difference required to cause air leak through the pores of saturated cup. In the laboratory it can be tested by connecting the cup with a compressor fitted with a gauge through a plastic tube. Water saturated cup is then dipped in a beaker filled with water. By releasing pressure slowly from the compressor observe the point at which air bubbles start to appear. This is the air entry value which must be in the range of 0.8 to 0.9 bars.

(ii) **Cup Conductivity:** It is the volume of water passing through the cup wall per unit time per unit hydraulic pressure head difference. Its value should be 3.95 ml/ sec/atm.

Testing of Tensiometer

Before installation each tensiometer should be tested whether it passes through standards otherwise reject it.

Procedure

- Immerse the porous cup of the tensiometer into water for 2 to 3 days to get it saturated.
- For tensiometer having manometer add some mercury to the manometer and then fill the tensiometer and manometer with water. In case of tensiometer with vacuum gauge fill the tensiometer with water only.
- Close the opening of the tensiometer with the cap to make it air tight.
- Dry the porous cup in air so that about 0.7 bar suction develops on the tensiometer.
- On dipping the cup in water for good tensiometer the suction goes down to zero within 5 minutes.

Installation

- Make a hole in the field to slightly deeper than the desired depth with the help of screw type/tube auger having almost same outer diameter as that of porous cup.
- Drop some amount of loose excavated soil of the same depth and some amount of water into the hole.
- Insert the tensiometer ensuring close contact of the cup with the field soil.
- Compact the soil around the tube and make a small heap with the soil to avoid water stagnation.
- Fill the tensiometer with deaerated water; tensiometer will show zero reading.
- With time as water from the cup will come out to the soil to attain equilibrium record the tensiometer readings.
- Correlate the tensiometer readings with soil moisture content.

Calculation

Say,

In case of tensiometer with vacuum gauge (Fig. 6.11a):

Matric potential $\mathbf{\Phi m = -(F_g \times R_g) + Z}$

Where, F_g = Gauge factor (suction value for each unit of gauge reading)

R_g = Gauge reading

Z = Vertical distance from the centre of the cup to the centre of dial gauge

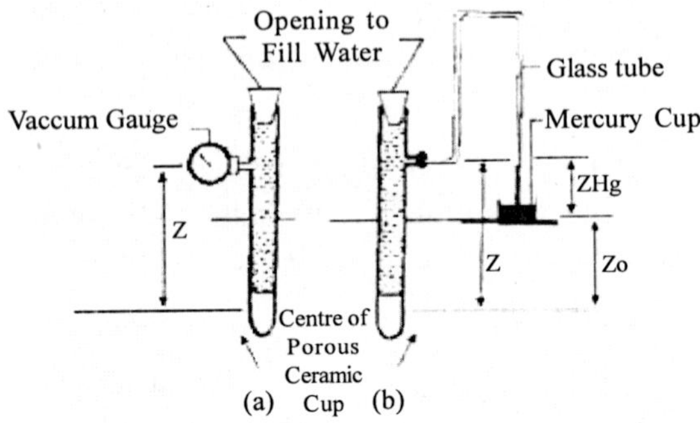

Fig. 6.11: Tensiometer (a) with vacuum gauge and (b) with mercury manometer

In case of tensiometer having manometer (Fig. 6.11b):

$$\textbf{Matric potential } (\Phi m) = -Z_{Hg} \times \frac{\rho m}{\rho w} + Z$$

$$= -13.6Z_{Hg} + Z \left(\text{as } \frac{\rho m}{\rho w} = 13.6\right)$$

$$= -13.6Z_{Hg} + Z_{Hg} + Z_o$$

$$= -12.6Z_{Hg} + Z_o$$

Where, ρm = Density of mercury; ρw = Density of water; Z_{Hg} = Height of mercury column from the centre of reference mercury level

Z = Vertical distance from the centre of the cup to the surface of the mercury column

Z_o = Vertical distance from the centre of the cup to the reference mercury level

6.3.3 Determination of Soil Moisture Suction by Pressure Plate Apparatus

Principle and procedure of the instrument has already been discussed in the section 6.2.3. Here, instead of determining volumetric moisture content at only 15 bars, determine moisture content at different suctions and prepare a curve of soil moisture suction against volumetric soil moisture content.

Note: Soil Moisture Retention Curve can also be prepared by determining soil water content at different suctions say, 0.1, 0.5, 1.0, 5.0, 10.0 bar with the help of Pressure Plate Apparatus or any other means and plotting gravimetric soil water content (%) against soil moisture tension/suction (bar).

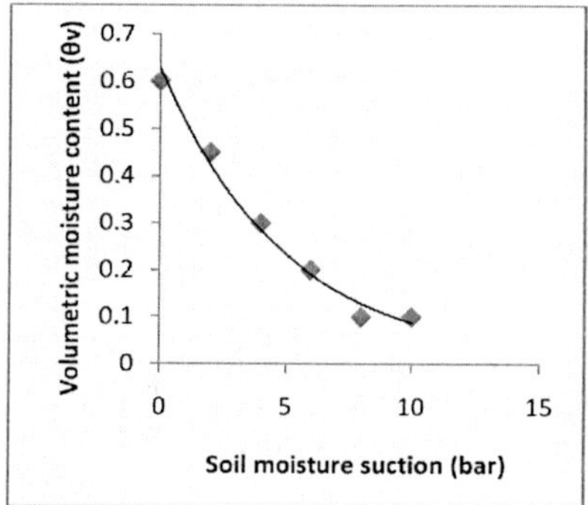

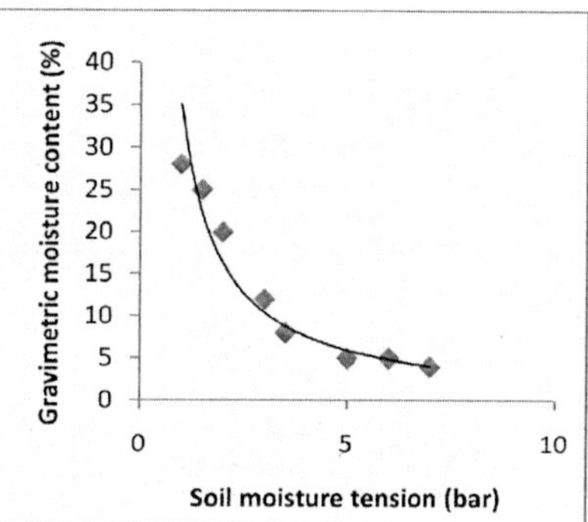

Fig. 6.12: Moisture retension curves

From these curves soil moisture content at any soil moisture tension can be determined.

Alternative Methods

A. Determination of Soil Moisture Suction by Tempe Pressure Cell

Soil moisture retention curve or the moisture characteristic of undisturbed soil core can be determined using a rather simple method involving Tempe Pressure Cell within 0 to 1 bar range. Air pressure is applied to the Tempe Pressure plate containing soil core sample saturated with water to extract moisture under controlled condition. When air pressure inside the Temple Cell is raised above atmospheric pressure, higher pressure forces to drain excess water through microscopic pores of porous ceramic plate. The maximum air pressure (bubbling pressure or air entry value) that any given wetted porous cermic plate can withstand before letting air passes through the pores is determined by the diameter of the pore. The smaller the pore size, the higher the air pressure will be to allow air to pass through. Porous ceramic plates must be used at air pressure extraction below the air entry value for that plate.

Equipments and Materials

1. A Tempe Cell consists of 3 parts: (a) the Top and Base Cap Assembly, (b) the Brass Cylinder and (c) the Porous Cermaic Plate. The Top and Base Cap Assembly is made of plexiglass that permit clear view of the internal parts and activity. The "O" rings in the caps make the pressure seals. The Tempe Cell accepts an undisturbed soil core, contained in the Brass Cylinder directly from Soil Core sampler (Fig. 6.13). Depending upon type of application user should choose the correct Porous Ceramic Plates that vary in bar rating.

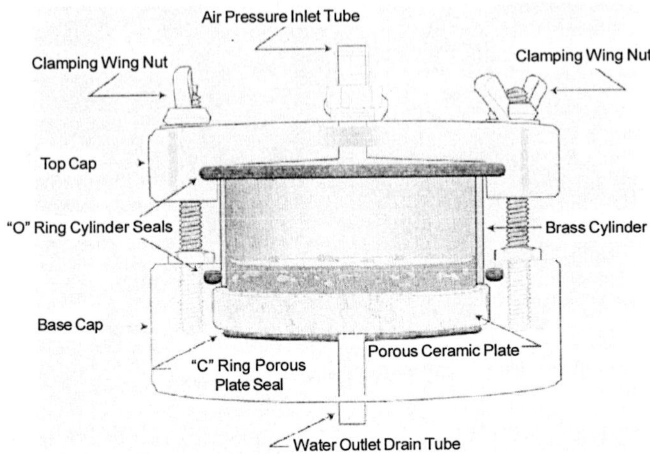

Fig. 6.13: The Tempe Cell with sample

2. A source of regulated gas pressure: compressed air from a compressor or compressed air or nitrogen in high pressure tank can be used for water extraction work. If the Tempe Cell is to be used extensively, compressed air from a compressor is the most satisfactory source of supply.
3. Automatic direct reading balance with a capacity of 800 g and of sensitivity of 0.03g.
4. Hot air oven.

Procedure

- Before operation fully saturate and deaerate the ceramic porous plate of the Tempe Cell. It is best done by saturating the plate in vacuum desiccators with deaerated distilled water. Remove the porous plate from the cell by first removing the "O" ring cylinder seal. The drain tube of the Tempe Cell is then connected to a leveling bulb, and the height of water in the bulb adjusted to the same height as the bottom of the cell.
- Reinstall the saturated porous plate. Once the plate is in place, raise the leveling bulb so that the water level is just at the top of the porous plate. There should be no free water on the plate.
- Place the soil core, in its brass cylinder, on the bottom of the cell. Gently twist the cylinder as it is being pushed down past the "O" ring and then firmly press the core into the porous plate to ensure good contact.
- Saturate the soil core by raising the bulb to the same height as the top of the sample.

- Reattach the top part of the Tempe Cell once the core is saturated and tight the wing nuts.
- Connect the Tempe Cell to the pressure supply and raise the pressure in the cell to the desired level as indicated on the pressure gauge.
- Immediately before weighing the cell for the first time, expel any air entrapped below the plate using a syringe filled with water. To do this, hold the cell upside down and inject water into the drainage tube. This filling procedure may be necessary before each weighing so that weight of water in the bottom of the cell is constant.
- Weigh the cell on an automatic, direct reading balance by wiping off any droplet hanging from the drain tube and inserting a clamp over the drainage tube to prevent air from being pulled through the drain tube into the drainage grooves when the air supply hose is disconnected. Equilibrium is reached when weight remains constant.
- Repeat the weighing process particularly at higher pressure values. Record the change in weight from one soil suction value to another by noting the difference in weight. Dry the sample at 105°C at the end of the run.
- Interpret the changes in weight in terms of moisture content by weight or volume of soil, since the inner volume of brass cylinder is the volume of the soil.
- Plot the moisture content at each equilibrium value against the soil suction (in bars) for that equilibrium value and draw the moisture charcteristic curve (Fig. 6.12).

B. Determination of Soil-Moisture Characteristics Curve and Pore-Size Distribution using Hanging Water Column

Principle

A hanging water column consists of a glass cup fitted with a highly permeable porous ceramic plate at the bottom and connected on its underside to a water column terminating in a reservoir open to the atmosphere. Freshly collected core soil sample is placed on the flat porous plate and the height of water reservoir is maintained at the top of the plate to get saturated. Then the reservoir is lowered to a new height of a distance 'z' below the top of the porous plate. By the equilibrium principle, water will flow from the soil sample through the porous ceramic plate to the outflow until the total water potential at the outflow point and inflow point becomes equal.

At hydraulic equilibrium follows:

Ψ_{tA}(water potential at inflow point) = Ψ_{tB} (water potential at outflow point)

or

$\Psi mA + \Psi apA + \Psi gA + \Psi oA = \Psi mB + \Psi apB + \Psi gB + \Psi oB$

Since, the system is under atmospheric pressure: $\Psi apA = \Psi apB = 0$; considering free diffusion of salt at everywhere of the system: $\Psi oA = \Psi oB$ and reference level at soil sample: $\Psi gA = 0$

$\Psi mA = \Psi gB$ [as the matric potential at outflow, $\Psi mB = 0$]

or, $\rho wgh = \rho wgz$

[h is the distance from the midpoint of the soil core to the water level in the burette]

or $h = z$

Water flows out of the sample until static equilibrium is reached. The water drained out by this suction is measured in a graduated burette attached to the other end of the hanging water column. At equilibrium the moisture content (θ) is determined and one obtains consequently one point of the soil moisture characteristic curve (h_1 vs θ_1).

Equipments and Materials

Hanging water column (Fig. 6.14): This consists of a sintered ceramic disc (porous plate) fitted to the base of a Buchner funnel (sintered glass funnel) and a long flexible transparent tubing attached to the end of the funnel, burette with stand, soil core, hot-air oven, aluminum moisture cans and balance

Procedure

- Put an appropriate size filter paper on the ceramic disc of the Buchner funnel.
- Fill the whole apparatus with deaerated water, without any air bubble below the porous plate and in the tube.
- Place an undisturbed soil core taken from the field on a filter paper in the funnel.
- Saturate the soil core by bringing the water level in the burette at the same level to the surface of the soil core.
- Allow the system to equilibrate for 24 h. Record the height of water column in the burette as initial reading.

- Lower down burette by 10 cm from the midpoint of the core. Again, allow the system to equilibrate for 24 h and record the burette reading. The difference between two readings indicates the volume of water drained out at that suction.
- Repeat the above step by lowering the burette at different levels and record the decrease in the height of burette (h) and burette reading (V).
- At the end of this study remove the soil core from the sintered glass funnel and record its initial fresh weight.

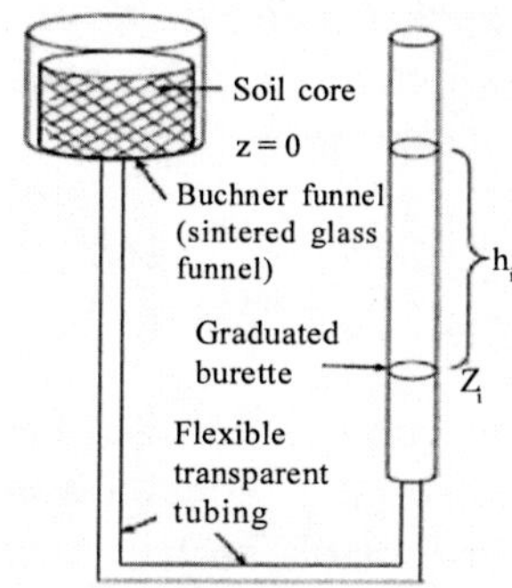

Fig. 6.14 Hanging water column for soil moisture characteristic curve

- Dry the soil core in the oven till a constant weight is achieved and note down the dry weight for determining the soil moisture content.

Calculation

Observation	Burette height (cm)*	Water drained (ml)	Cumulative water drained (ml)	Water content in the soil
1	10	X_1	$X_1=Z_1$	θ_1
2	20	X_2	$Z_1+X_2=Z_2$	θ_2
3	30	X_3	$Z_2+X_3=Z_3$	θ_3
4	40	X_4	$Z_3+X_4=Z_4$	θ_4
5	50	X_5	$Z_4+X_5=Z_5$	θ_5

*Height = distance between the midpoint of the core and the water level in the burette

Z_1, Z_2, Z_3, Z_4 and Z_5 represent the total amount of water drained from the saturated soil sample under 10, 20, 30, 40 and 50 cm water tensions, respectively. The water drained at 50 cm water tension is called as *Aeration porosity*.

Moisture content: Data on drained water can be used to back-calculate the moisture content of the soil at that corresponding tension using the following formula.

$$\text{Water content at a given tension} = \frac{(\text{Water content at the previous tension} + \text{amount of water drained at that tension})}{\text{Oven dry weight of the soil}} \times 100$$

The water content values corresponding to tensions were plotted in a graph produced the *soil moisture characteristic curve*

Pore size distribution: This experiment is also used to determine the pore radii which retain water at that corresponding tension using the following equation of capillarity

$$r = \frac{2\gamma \cos\alpha}{\rho g h} \cong \frac{0.148}{h}$$

where, r = radius of the pore in cm, γ = surface tension of water (72.5 dynes cm^{-1}), α = contact angle (normally taken as 0 for non-hydrophobic soil), h= tension (cm), g = acceleration due to gravity (980 cm s^{-2}), and ρ = density of water (1g cm^{-3}).

If the volume of water drained at a given tension hi (corresponding to radius ri) = Vi

Total volume of water present in the sample at saturation = Vt

Then, % of porosity corresponding to tension hi (or having radius $\geq$ ri) = (Vi/Vt) x 100

Questions and Hints

Q.1 Whether organic matter in soil has any influence on the determination of soil moisture by Neutron Moisture Meter?

Hint: Yes; because the basic principle of determining soil moisture content by Neutron Moisture Meter is to measure the amount of hydrogen nuclei present in the soil. Organic matter also contains sufficient amount of hydrogen nuclei, thus with increase in organic matter content there will be over estimation of soil moisture value.

Q.2 Why some methods of soil measurements are called indirect method?

Hint. Some methods of soil measurements are called indirect because by these methods one cannot measure soil water content directly; instead he measures some other variable with which soil water content can be calculated.

Q.3 What are the advantages of TDR method of soil moisture determination?

Hint: The advantages of TDR method over the others are: (i) simple, nondestructive, and relatively fast method (takes approx. 30 sec per measurement), (ii) superior accuracy to within 1-2% of volumetric water content, (iii) in many cases calibration are not needed, (iv) no radiation hazard as in neutron probe attenuation technique, and (v) capable of providing continuous measurement through automation and multiplexing.

Q.4 Why apparent dielectric constant of soil depends largely on the volumetric water content in soil?

Hint: The dielectric constant of a soil should include the dielectric constants of all the soil constituents in the equation, *i.e.* dielectric constant of air (K_o), water (K_w) and solid (K_s). Therefore, the actual relation should be

$$\sqrt{K_a} = f_a\sqrt{K_o} + f_w\sqrt{K_w} + f_s\sqrt{K_s}$$

where f_o, f_w and f_s are the volume fraction of air, water and solid matter, respectively.

Rearranging the above equation substituting the values of f_o, f_w and f_s, we get

$$\sqrt{K_a} = (n-\theta)\sqrt{K_o} + \theta\sqrt{K_w} + (1-n)\sqrt{K_s}$$

Where n is the porosity of the soil sample.

Since, the dielectric constant of air and solid matter are very small in comparison to that of water, the apparent dielectric constant depends largely on the volumetric water content of soil.

Q.5 What are the drawbacks of TDR method?

Hint: The main drawback of this method is that the instrument is costly. Secondly, Topp *et al.* (1980) equation fails to determine the Ka relationship adequately for water content exceeding 0.5% and for organic soil or mineral soil with high organic matter or clay content. Soils having high clay or organic matter content require soil specific calibration.

Q.6 In crop fields where will you install the resistance block?

Hint: It should be placed in the root zone. For convenience of intercultural operation install it in a row between two plants.

Q.7 What is pF?

Hint: The negative potential of unsaturated soil is expressed in terms of equivalent hydraulic head or the height of water column in cm. To avoid the use of large value pF scale was introduced which is the logarithm of the negative pressure head i.e. the height of water column in cm (h)

$$pF = \log h$$

Q.8 Which soil water potential actually tensiomter measures?

Hint: Matric potential. However with modification osmotic potential can be included.

Q.9 Why matric potential is expressed with negative sign?

Hint: The potential energy of free water is arbitrarily considered as zero. Under unsaturated soil condition water is subject to two forces: one is adsorptive force and other is capillarity. These two forces combine to produce matric potential. These forces attract and bind water on the surfaces of the soil particles and hold water in the capillary which ultimately reduces the free energy of soil water in comparison to pure free water. That's why matric potential is negative. Matric potential with zero value means soil is in saturated condition.

Q.10 What is soil water tension / suction?

Hint: Soil water tension/suction is the soil water potential without negative sign. In unsaturated condition to avoid the use of negative sign of osmotic and matric potential, the term suction/tension is used.

Q.11 Two soils having same texture but have matric potential -0.2 and -0.4 bars, which soil has more water content?

Hint: Higher the absolute value of matric potential (ignoring sign) i.e. higher the matric suction lesser is the water cotent. Thus, water content is less in -0.4 matric potential soil.

Q.12 What are the advantages/disadvantages of determining soil water retention curve using hanging water column method?

Hint: **Advantages**: 1.The hanging water column method is probably the easiest method to determine water retention curves, especially for h >-200 cm, and no external vacuum source and pressure regulators are needed.

2. The hanging water column method always works at atmospheric pressure and therefore, has no effect on the shape of the air-water interface.

Disadvantages: Only one sample can be analyzed at a time, unless more funnels are set up.

Q.13 What are the precautions need to be taken for determining soil water retention curve using hanging water column method?

Hint: 1. Avoid any air leakage in the hanging water column i.e., there should not be any air bubble below the ceramic porous plate and always use deaerated water.

2. One should also be aware that evaporation of water can occur through the tubing. If the measurements are critical, as for basic research, an additional hanging water column should be set up without a soil sample to measure the evaporation through the tubing, so that corrections can be made during the calculations of the water retention curves.

7

Flow of Water in Soil

7.1 Determination of Hydraulic Conductivity of Saturated Soil Laboratory Methods

Hydraulic conductivity of a saturated soil indicates the ease with which water is transmitted through the soil pores. If a constant water head (h) is maintained on a saturated soil column of length (L) and the volume of water (V) percolating per unit time (t) through per unit cross sectional area (A) to the other end, then according to Darcy's Law, the rate of flow of liquid or flux (q) is proportional to the hydraulic gradient (ÄH/L) across the length of soil column.

$$q = \frac{V}{A.t} = K_S \frac{\Delta H}{L} \qquad (7.1)$$

$$\text{or, } K_S = \frac{V.L}{A.\Delta H.t} \qquad (7.2)$$

Where K_S is the proportionality constant or the hydraulic conductivity means the rate of flow at unit hydraulic gradient.

$$\Delta H = H_{\text{inflow}} - H_{\text{outflow}}$$

n case of downward movement of water, water drips out freely from the bottom of the soil column (Fig. 6.1), H_{outflow} is zero, then

$$\Delta H = H_{\text{inflow}} = H_p + H_g$$

Constant water head

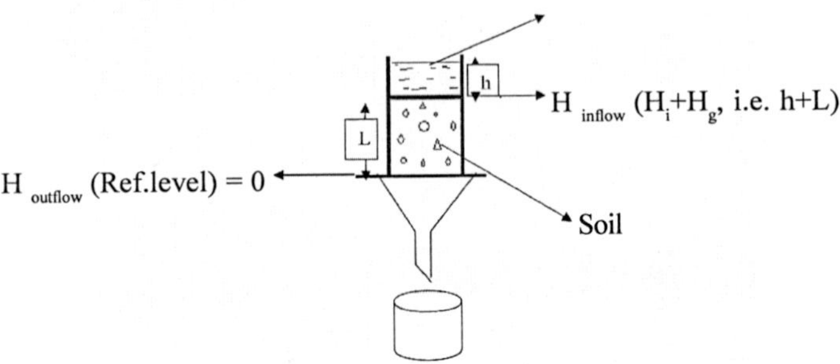

Fig.7.1: Downward movement of water

H_p and H_g are the pressure head and gravitational head, respectively. The equation 6.2 can be rewritten as:

$$K_S = \frac{V.L}{A.t(h+L)}$$

Here, H_p is the height of constant water head (h) in cm and H_g is the length of soil column (L) in cm. But, when water moves vertically upward through the soil column and water drips from the surface (Fig. 7.2), then

$$\Delta H = H_{inflow} - H_{outflow}$$

$$\Delta H = H_i - H_g$$

Where, H_i and H_g are the pressure head and gravitational head, respectively. H_i is the height of water column from the bottom of the soil column (reference level) to the constant water head.

$$K_S = \frac{V.L}{A.t(H_i - L)}$$

7.1.1 Determination of Saturated Hydraulic Conductivity by Constant Head Method

This method is suitable for the very porous soils.

Equipments and Materials

Brass permeameter of 7 cm inside diameter and 15 cm height fitted with inlet and outlet device at the bottom and at the top and a 20 mesh screen fitted above the inlet at bottom (Fig. 7.2), permeameter inside diameter size screen, constant level water reservoir, polythene tube, iron stand, measuring cylinder, stop watch, filter paper of Whatman No. 42

Procedure

- Place a filter paper on the screen of the permeameter.
- Pack about 200 g of air dry soil in the permeameter by gentle tapping on the table through a height of 2.5 cm for 15-20 times.
- Put another filter paper and a screen on the top of the soil to avoid soil loss.
- Saturate the soil by placing the permeameter in a tray containing water for overnight. The water level in the tray should be few cm above the bottom screen height.
- Place the permeameter on a table and fix the constant level water reservoir at a desired height supported by an iron stand.
- Connect the inlet at the bottom of the permeameter with a constant level water reservoir by a polythene tube to have a constant hydraulic head difference for the entire duration of experimentation.
- Start water supply and after some time record discharge rate at least for few replicates to have an idea about the measurement variability.
- When steady flow is reached start collecting discharge in the measuring cylinder and record the time interval.
- Record few such consecutive readings.
- Measure the length of soil column by pushing glass rod vertically in the soil column after removing the top screen and also the height from the bottom screen to the water surface at the constant water reservoir.
- Record the temperature of water.

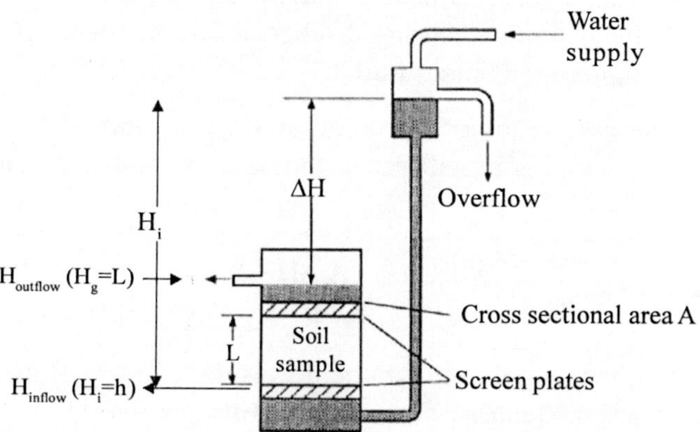

Fig. 7.2: Vertical movement of water from the bottom of the soil column

Calculation

Say

Diameter of the permeameter = d cm

Inner cross sectional area of the permeameter (A) = $\pi d^2/4$ sq. cm

Height of the soil column = L cm

Height from the screen to the water surface of the constant level water reservoir = H_i cm

ÄH = $(H_i - L)$ cm

Volume of discharge water at time t min = V cc

Temperature of water = T°C

$$\text{Hydraulic Conductivity at T°C } (K_{ST}) = \frac{V.L}{A.\Delta H.t}\ \text{cm}/\text{min}$$

Temperature Correction

$$\text{Hydraulic Conductivity at 20°C } (K_{S20}) = K_{ST}\frac{\eta_T}{\eta_{20}}\ \text{cm / min}$$

η_T and η_{20} are the viscosity of water in poise at T°C and 20°C, respectively.

7.1.2 Determination of Saturated Hydraulic Conductivity by Falling Head Method

This method is suitable for soils having very low permeability. Here, instead of using a constant hydraulic head a narrow stand pipe is attached to the soil column and instead of measuring discharge volume the drop of water level in the narrow stand pipe is measured.

If –dH be the change in head from initial (H_0) to final (H_1) in dt time interval and 'a' be the cross sectional area of the stand pipe, then according to Darcy's law

$$\text{The rate of flow } (q) = -\frac{a.dH}{dt} = K_{s.}\frac{A.\Delta H}{L}$$

Where, ΔH/L is the hydraulic gradient, A is the cross sectional area of the soil sample and L is the length of the soil column.

$$\frac{A.K_S}{a.L}dt = -\frac{dH}{H}$$

Integrating both sides between t= 0 to t = t and between H = H_0 to H = $H_{1:}$

$$K_S = \frac{a.L}{A.t}\ln\frac{H_0}{H_1}$$

Equipments and Materials

Brass permeameter of 7 cm inside diameter and 15 cm height fitted with inlet and outlet device at the bottom and at the top and a 20 mesh screen above the inlet at bottom (Fig. 7.3), screen, glass stand pipe of suitable diameter depending upon the permeability of the soil (lesser the permeability narrower the diameter), iron stand, polythene tube, stop watch, filter paper of Whatman No. 42.

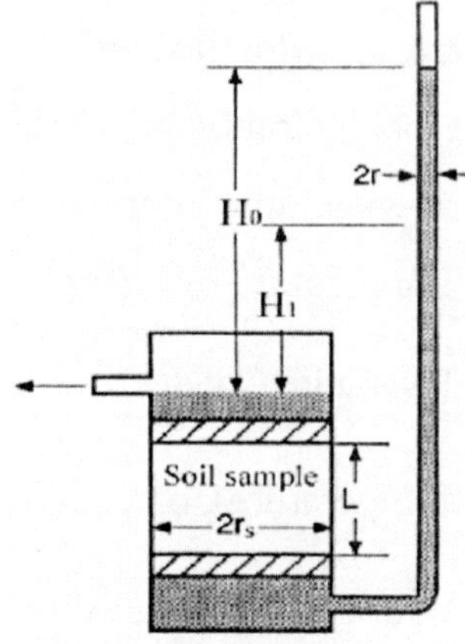

Fig.7.3: Falling head permeameter

Procedure

- Follow same steps as in section 7.1.1 upto saturation of soil of the permeameter in the water tray for overnight.
- Place the permeameter on a table and erect the glass stand pipe with the help of an iron stand.
- Connect the glass stand pipe with the peameameter by polythene tube and add sufficient water to the stand pipe.
- Take a note of initial height of water column in the stand pipe and the time.
- Record the final water level in the stand pipe and time.
- Refill the stand pipe with water to maintain same initial head.
- Take at least 4 such readings.
- Measure the length of soil column and record the temperature of water as the method described in the section 7.1.1.

Calculation

Say

Diameter of the stand pipe = d cm

Area of the stand pipe = $\pi^2/4$ sq. cm = a sq. cm

Diameter of the permeameter = D cm

Inner cross sectional area of the permeameter = $\pi D^2/4$ sq. cm = A sq.cm

Height of the soil column = L cm

Initial hydraulic head = H_0 cm

Final hydraulic head = H_1 cm

Average time interval = t hr

Temperature of water = T°C

$$\text{Hydraulic Conductivity at T°C}\left(K_{ST}\right) = \frac{a.L}{A.t} \ln \frac{H_0}{H_1} \text{ cm / hr}$$

Temperature Correction

$$\text{Hydraulic Conductivity at 20°C } (K_{S20}) = K_{ST} \frac{\eta_T}{\eta_{20}} \text{ cm/hr}$$

η_T and η_{20} are the viscosity of water in poise at T° and 20°C, respectively.

Rating

Permeability Class	Hydraulic conductivity of a sat. soil (cm/hr)
Very slow	< 0.125
Slow	0.125 – 0.5
Moderately slow	0.5 – 2.0
Moderate	2.0 – 6.25
Moderately fast	6.25 – 12.5
Fast	12.5 -25.0
Very fast	> 25.0

7.1.3 Determination of Intrinsic Permeability of Soil

Hydraulic conductivity is a property of a soil which depends on the characteristics of both the porous medium (soil) as well as the fluid used (here, water), while intrinsic permeability of a soil is property which depends only on the property of the soil (pore geometry) nullifying the effect of the fluid. Thus, intrinsic permeability of a soil is always same irrespective of the fluid used.

Intrinsic permeability of a soil in cm² (k/) $= \frac{K_S}{f}$ (7.3)

Where, K_s is the hydraulic conductivity of a soil in cm/sec at a given temperature (say, T°C) with respect to water, f is the fluidity of water.

Fluidity of water (f) $= \frac{\rho_w}{\eta} \cdot g$ (7.4)

Where, η = viscosity of the water in poise at T°C, η_w = density of water in g/cm³ at T°C and g = acceleration due to gravity in cm/sec²

Thus, after calculating hydraulic conductivity of a soil one can easily calculate the intrinsic permeability of soil:

Intrinsic permeability of a soil in cm² (k') $= \frac{K_S}{g} \times \frac{\eta}{\rho_w}$

Rating

Class	Intrinsic permeability (cm² x 10^{-10})
Very slow	< 3
Slow	3 – 15
Moderately slow	15 – 60
Moderate	60 - 170
Moderately fast	170- 350

7.2 Determination of Unsaturated Hydraulic Conductivity by using Campbell Equation

Hydraulic conductivity of any unsaturated soil depends on soil water potential or content. As the relation between soil water potential and water content of an unsaturated soil is affected by hysteresis, hydraulic conductivity of unsaturated [K (θ)] soil is also affected by hysteresis and cannot be measured directly in all ranges of interest. However, hydraulic conductivity of an unsaturated soil can be calculated by equation of Campbell and Oswal.

$K(\theta) = K_s \cdot (\theta/\theta_s)^{2b+3}$ (7.5)

Where, K_s is the saturated hydraulic conductivity, è is the volumetric water content of the unsaturated soil, θ_s is the volumetric water content at saturation and b is the slope of the curve between log (h/h_e) and –log (θ/θ_s); where, h is the suction obtained at different moisture content and h_e is the air entry suction [obtained from the intersect of the curve drawn plotting log h against –log (θ/θ_s)].

Equipments and Materials

Pressure plate apparatus, metal rings of 5 cm diameter and 4 cm height, core sampler / soil sampling auger, pipette, moisture box, oven, balance, desiccator

Procedure

- Follow the same procedure as described in the section 6.3.3 for determination of soil moisture suction by pressure plate apparatus and measure the moisture content (θ) at different pressure (h). Plot log h against–log (θ/θ_s) and determine the value of h_e from the intersection of the curve at Y axis.
- From the graph of log (h/h_e) vs –log (θ/θ_s) calculate the value of b, the gradient of the slope.

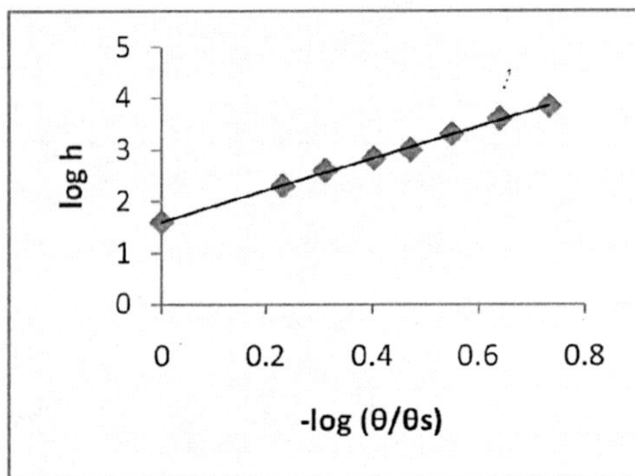

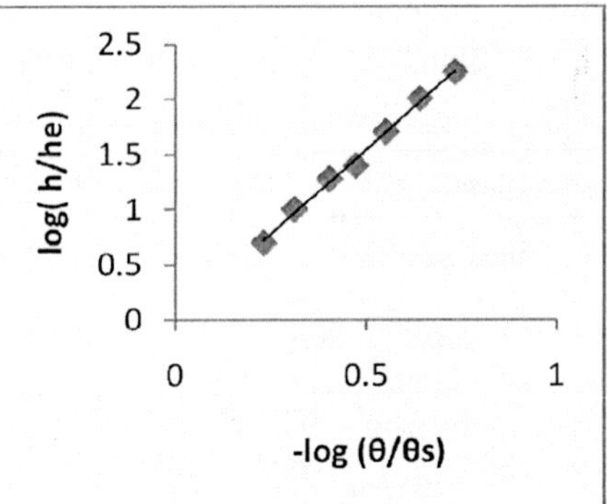

Fig. 7.4: Curves of log h vs –log (θ/θ_s) and log (h/h_e) vs –log (θ/θ_s)

Calculation

Say,

Saturation volumetric water content = θ_s cm³/cm³

Saturation hydraulic conductivity = K_s cm/sec

Now, putting the value of K_s and b in the following equation, the unsaturated hydraulic conductivity value at any moisture content (θ) can be calculated:

$$K(\theta) = K_s . (\theta/\theta_s)^{2b+3}$$

So, from the above equation it is conspicuous that K (θ) values are different at different moisture content (θ).

7.3 Determination of Diffusivity and Sorptivity of Soil

Except rice cultivation under flooded condition most of the time crops are subjected to the condition where water moves under unsaturated soil condition. Neither conductivity nor potentials associated with unsaturated flow are easily measured in all ranges of soil moisture. Though, liquid flow under unsaturated condition is not truly a diffusion process rather a resultant effect of different forces like matric suction of soil for water, surface tension in air-liquid interface and intermolecular attraction of liquid molecules. Still the liquid flow can be written in diffusion form and driving force here is the water content gradient.

In case of one dimension flow in the X direction (horizontal) from the source of zero hydraulic head, the soil water diffusivity [D (θ)] can be expressed according to Bruce and Klute (1956)

$$D(\theta) = -\frac{1}{2t}\frac{dX}{d\theta}\int X.d\theta \quad (7.6)$$

Where, t = the total time of water infiltration, X is the distance of wetting front from the source, è is the volumetric water content and integration is done from initial to the desired volumetric water content.

Sorptivity is another hydraulic property of unsaturated soil which can be measured from the same experiment. Sorptivity which refers to the affinity of the soil for water is closely related with hydraulic conductivity and soil water diffusivity.

According Phillip (1957) sorptivity (S) is the proportionality constant between cumulative infiltration (I) and square root of time (t). If cumulative infiltration (I) is expressed in cm and time (t) in min, then the slope of the curve passing through the origin is the sorptivity expressed in cm/ (min)$^{1/2}$.

Equipments and Materials

A transparent plastic or acrylic column made from 20 transparent rings each having a breadth of 1 cm and inner diameter of 5 to 6 cm joining side by side by cellotape, glass tube of 60 cm length and 4 cm diameter having a small pin pointed aperture on the side of tube at the narrow end, glass funnel fitted with a sintered

(a) (b)

Fig. 7.5: (a) Plastic column with rings (b) a complete setting of instrument

glass filter, plastic plates, packer, moisture box, rubber tube, iron stand to support the glass tube, sharp knife, oven, Whatman No. 1 filter paper.

Procedure

- Tightly fix one end of the transparent plastic column with a hard plastic plate having a small aperture at the middle for exclusion of air during wetting by cellotape.
- Place a filter paper inside the column.
- Carefully pour and pack the calculated amount of air dry soil (< 2 mm) ring by ring leaving the top one with the help of packer to maintain the field bulk density of soil and to avoid gaps during filling.
- Clamp another plastic plate to the other end of column and make it air tight by fixing cellotape.
- Horizontally place the column on a table.
- Place the glass funnel fitted with a sintered glass filter to the soil column and fix a wooden plate with the clamp.
- Connect the glass funnel with the glass tube erected with the help of iron stand by a rubber tube in such a way that zero hydraulic head is maintained.
- Paste a narrow cm graph paper on the glass tube for recording water level.
- Pour water in the glass tube, close the wide mouth of the glass tube with air tight rubber cork and immediately note the initiation time and water level.
- Record the volume of water infiltrated into the soil from water meniscus in the glass tube and the length of wetting front from the source at different cumulative time.
- Disconnect water supply when wetting front reaches about to 15th to 16th rings.
- Note the final time, volume of water infiltrated and the length of wetting front.
- Segment soil column into one cm sections by cutting the cellotape in between rings by sharp knife and take soil from each section separately in previously weighed moisture boxes.
- Weigh them, dry in oven at 105°C for overnight, cool in desiccator and weigh again.
- Calculate the volumetric water content of the soil in each ring.

Calculation

Let,

Breadth of each ring = a cm

Inner diameter of ring = d cm

Volume of soil in each ring = $\pi d^2.a/4$ cc

Weight of soil in each ring for desired bulk density (ρb) = $\rho b.\ \pi d^2.a/4$ g

Distance of wetting front from water source = X_i cm

Volumetric moisture content = θi = Gravimetric moisture content × ρb

Sl. No.	X_i	dX	θi	dθ	X. dθ = Z	“ X. dθ	dX/ dθ
15	X_{15}	-	Θ15	-	-	-	-
14	X_{14}	X_{15} -X_{14}	Θ14	Θ14-Θ15	Z_{14}	Z_{14}	
13	X_{13}	X_{14}- X_{13}	Θ_{13}	Θ13-Θ14	Z_{13}	Z_{14}+ Z_{13}	
-							
-							
3	X_3	$X_4 - X_3$	Θ_3	Θ_3- Θ_4	Z_3	Z_4+ Z_3	
2	X_2	$X_3 - X_2$	Θ_2	Θ_2- Θ_3	Z_2	Z_3+ Z_2	
1	X_1	X_2 - X_1	Θ_1	Θ_1- Θ_2	Z_1	Z_2+ Z_1	

Total time of infiltration = t min

a) Diffusivity at different moisture content can be calculated using the following formula:

$$D(\theta) = -\frac{1}{2t}\frac{dX}{d\theta}\int X.d\theta$$

Plotting D (θ) values against their corresponding volumetric water content (θ) in the graph paper, diffusivity of soil can be determined at any moisture content.

b) Sorptivity of soil can be calculated from the curve drawn by plotting cumulative infiltration against square root of cumulative time “t using the following formula.

$S = I/\sqrt{t}$

Where, cumulative infiltration (I) is the Σ X. Δθ. The gradient or the tangent of the curve is the sorptivity of the soil.

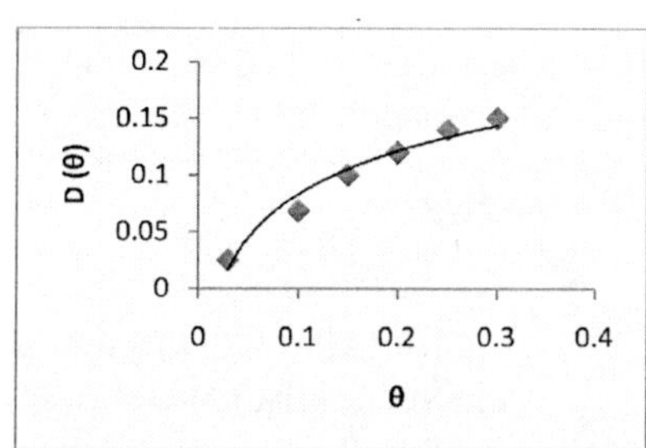

Fig. 7.6: Soil water diffusivity at different moisture content

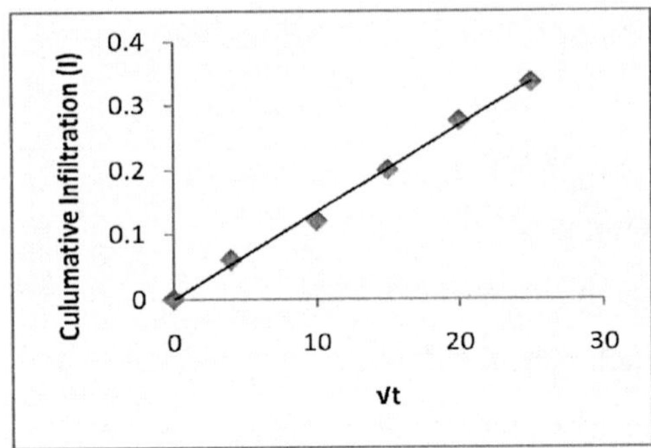

Fig.7.7: Cumulative infiltration vs square root of cumulative time

7.4 Determination of Infiltration Rate of a Soil by Ring Infiltrometer

Infiltration refers to the process of downward entry of water from air to soil. The infiltration rate or capacity is the maximum rate of absorption of rain or irrigation water by a soil when it is present at the soil surface.

Equipments and Materials

Both side smooth galvanized two iron cylinders of 14 to 16 gauge having 30 cm and 45 cm diameter, respectively and 40 to 45 cm height. One end of the cylinder is sharpened from outside for easy penetration, hammer, circular driving cup for each ring, stop watch, hook gauge.

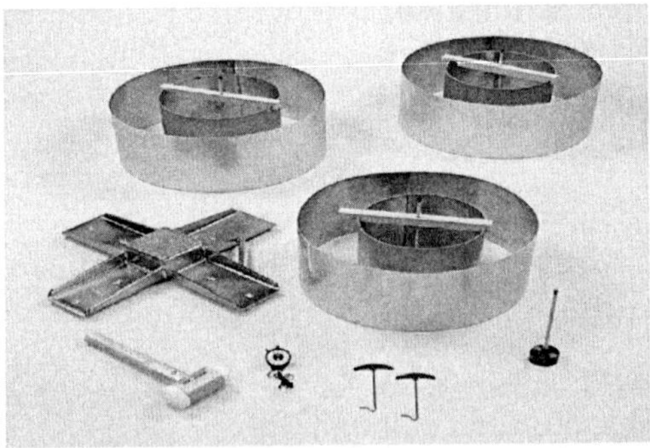

Fig. 7.8: Double ring infiltrometer

Procedure

- By hammering on the central guide rod of the circular cup drive the smaller ring (30 cm diameter) in the level field soil to a depth of 15 – 20 cm vertically. Care should be taken during penetration that the ring moves uniformly downward from all sides.
- Remove the circular cup from the ring.

- Drive the larger ring (45 cm diameter) into the soil concentrically with central smaller ring. The larger outer ring will prevent the divergent flow of water.
- Tap the soil into the space between soil column and the cylinder to bring the soil inside the ring to its natural condition as far as practicable.
- Apply 10–15 cm depth of water inside the central ring and the space in between two rings.
- Place a hook gauge in the central ring at the desired level.
- Record the initial and the final water level against a suitable time interval on the hook gauge.
- If required refill the cylinder to the desired depth and record the initial level.

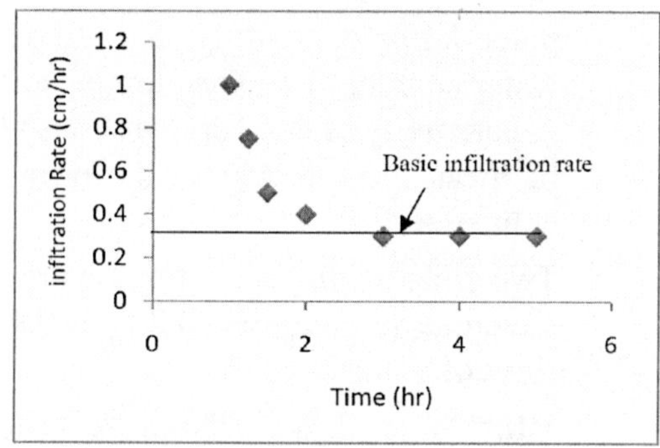

Fig. 7.9: Infiltration rate against time

- Plot infiltration rate (cm/hr) against time. A time will attain when the rate of infiltration will be constant which gives the final or basic infiltration rate expressed in cm/hr.

Basic infiltration rate

Rating

Soil Class	Infiltration rate (cm/hr)
Very slow	< 0.254
Slow	0.254 – 1.27
Moderate	1.27 – 2.54
Fast	> 2.54

Questions and Hints

Q.1 Hydraulic conductivity of saturated soil follows which law?

Hint: Darcy's law. Darcy's Law states that if a constant water head (h) is maintained on a saturated soil column of length (L) and the volume of water (V) percolating per unit time (t) through per unit cross sectional area (A) to the other end, then the rate of flow of liquid or flux (q) is proportional to the hydraulic gradient ($\Delta H/L$) across the length of soil column.

Q.2 Why hydraulic conductivity of soil is affected by soil temperature?

Hint: Hydraulic conductivity of a soil is inversely proportional to the viscosity of soil water and viscosity of water decreases with increase in temperature. Thus, hydraulic conductivity increases with increase in temperature.

Q.3 Why hydraulic conductivity of an unsaturated soil can not be measured in all soil water ranges?

Hint: Hydraulic conductivity of any unsaturated soil depends on soil water potential or content. As the relation between soil water potential and water content of an unsaturated soil is affected by hysteresis, hydraulic conductivity of unsaturated soil is also affected by hysteresis and cannot be measured directly in all ranges of soil water potential or soil water content.

Q.4 Two soils having same texture are in close contact with each other but have matric potential -0.2 and -0.4 bars. What will be the direction of flow of water?

Hint: Water moves from lower matric suction to higher matric suction. Higher absolute value of matric potential (ignoring sign) means higher matric suction. So, water will flow from -0.2 to -0.4 bar matric potential soil.

Q.5 Whether intrinsic permeability of a soil varies with the fluid used?

Hint: No; it is independent on the characteristics of the fluid. Intrinsic permeability is a property which depends on the soil characteristics, better to say the pore geometry.

Q.6 What is the difference between infiltration and permeability?

Hint:

Infiltration	Permeability
1. The process of downward entry of water into the soil	1. The process of movement of water through the soil
2. It depends on the surface characteristics of soil and initial water content	2. It depends on transport property of the soil i.e. texture , structure, compactness, organic matter content and type of exchangeable cation

Q.7 Why sandy soil has higher saturated hydraulic conductivity than the clayey soil?

Hint: According to Poiseuille's equation the volume of flow is proportional to the radius of pores in the order of fourth power. Sandy soil has more number of macropores, while clayey soil contains more number of micropores. That's why sandy soil has higher saturated hydraulic conductivity than the clayey soil.

Q.8 Why does heavy textured soil support plant for a longer period of time than light textured soil under moisture stress condition?

Hint: Under moisture stress condition macropores (prevalent in light textured soil) are mostly remained empty which discontinue the movement of water through the soil. Under water stress condition water remains mostly in the capillaries or the micropores which are prevalent in heavy textured soil. That is why heavy textured soil can supply water better than the light textured soil under water stress condition.

8

Soil Consistency

Soil consistency is a term used to designate the manifestation of different physical forces acting within the soil at various moisture levels. The consistency limits, also known as '**Atterberg Limits**' (Atterberg, 1911; 12) are defined by the water content of the soil to create specified degree of consistency. These limits are – (i) Liquid limit or upper plastic limit, (ii) Plastic limit or lower plastic limit and (iii) Sticky limit.

8.1 Determination of Liquid Limit or Upper Plastic Limit

Equipments and Materials

Liquid limit device (Casagrande, 1932) and accessories, spatula, dish, oven, moisture box, balance, desiccator

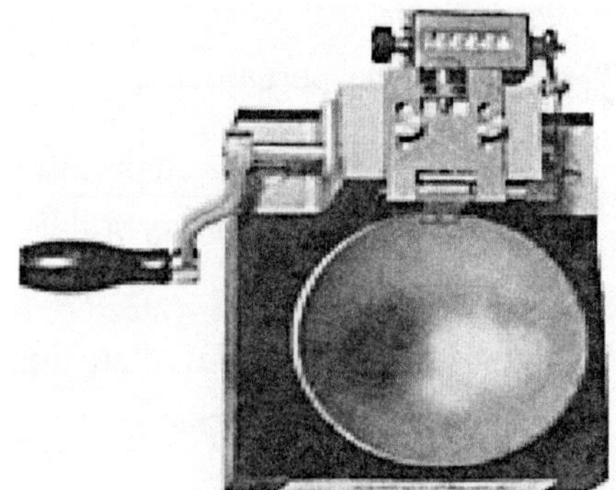

Fig. 8.1: Liquid limit device (Casagrande device)

Procedure

- Clean the liquid limit device so that the cup drops freely.
- Thoroughly mix about 100 g air dry soil in a dish with gradual addition of distilled water to make soft paste.
- Put the soil paste in the cup and smooth the surface with spatula and be assured that the depth of soil in the cup is more than 1 cm at the point of contact.
- With the help of grooving device supplied with the instrument cut a V shaped groove of 1.3 cm wide at the bottom that makes an angle of 60° with the horizontal surface.
- Remove any soil along the wall of groove more than 1 cm above the cup bottom.

Fig. 8.2: Grooving device and soil groove

- Rotate the cam with the handle at the rate of 2 revolutions per second and record the number of taps required to fill the bottom of the groove.
- Remix the soil in the cup and repeat previous three steps. If the number of taps required is ±1 to 2 the number of previous determination and in between 18 and 32, then the determination is valid.
- Take about 10 g of soil paste and weigh it.
- Dry the soil sample in oven at 105°C and reweigh it.

Calculation

Calculate water percentage, W_N by following the relation

$$W_N = \frac{\text{Weight loss on oven drying}}{\text{Weight of oven dry soil}} \times 100$$

Liquid limit (L_L) is that water content of soil which required 25 taps to fill the bottom of the groove. Calculate liquid limit by the following relation

$$L_L = W_N \times \left(\frac{N}{25}\right)^{0.12} \qquad (8.1)$$

Where, W_N = water percentage corresponding to the number of taps, N of the determination.

Table.8.1 Values of $(N / 25)^{0.12}$ corresponding to different values of N

N	$(N / 25)^{0.12}$	N	$(N / 25)^{0.12}$
18	0.961	26	1.005
20	0.974	28	1.014
22	0.985	30	1.022
24	0.995	32	1.030
25	1.000		

8.2 Determination of Plastic Limit or Lower Plastic Limit

Equipments and Materials

Dish, glass plate or ceramic tile, moisture box, oven, desiccator

Procedure

- Make a stiff paste of 10 -15 g air dry soil in a dish with distilled water and form a soil ball.
- Roll the soil ball on a glass plate or ceramic tile with the palm of the hand until a thread of 3 mm diameter is formed and begins to crumble into pieces of 1.25 to 1.50 cm long.

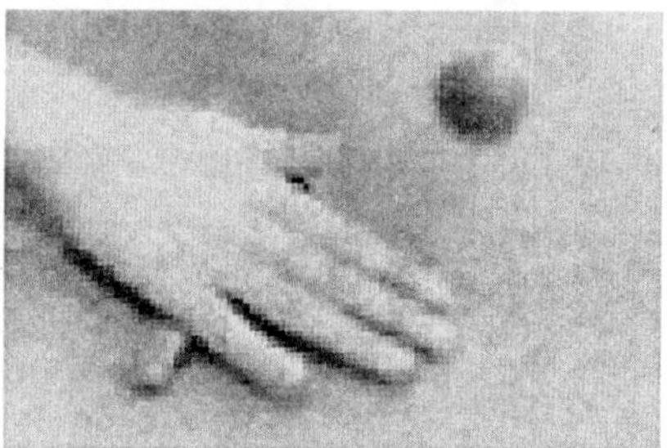

Fig. 8.3: Rolling of soil ball

- Remold the soil into a ball and repeat previous step.
- Compute water percentage of crumbling soil as that of section 8.1.
- Repeat the determination at least for thrice.
- Plastic limit is the average moisture percentage of these determinations.

Calculation

Say,

Weight of empty moisture box = X g

Weight of moisture box + moist soil = Y g

Weight of moisture box + oven dry soil = Z g

$$\text{per cent soil moisture at plastic limit} = \frac{Y-Z}{Z-X} \times 100$$

Average of three such determinations is the plastic limit of soil.

Plasticity Index or Plasticity Number = Upper Plastic Limit – Lower Plastic Limit

8.3 Determination of Sticky Limit

Equipments and Materials

Dish, glass plate or ceramic tile, spatula, moisture box, oven, desiccator

Procedure

- Put about 50 g of stiff soil paste on a glass plate.
- Add little amount of distilled water and thoroughly mix to make it uniform paste with the help of stainless steel spatula.
- Draw a clean spatula across the surface of soil paste exerting a firm pressure on the soil to test for adherence of soil to the spatula.
- If the soil does not adhere to the spatula repeat step 2 and 3 until water content is just sufficient that adherence commences.
- Determine moisture percentage at this point as described in section 8.1.
- Repeat the determination for another 2 to 3 times.
- Sticky limit is the average moisture content of these determinations.

Calculation

Say,

Weight of empty moisture box = X g

Weight of moisture box + moist soil = Y g

Weight of moisture box + oven dry soil = Z g

$$\text{per cent soil moisture at sticky limit} = \frac{Y-Z}{Z-X} \times 100$$

Average of three such determinations is the sticky limit of soil.

Note: (i) The important source of error in all three limits test is lack of uniformity in mixing of soil and water.

(ii) In most of the cases sticky limit and liquid limit are nearly the same.

(iii) Oven dried sample can be employed for these determinations.

Questions and Hints

Q.1 What is soil consistency?

Hint: Soil consistency is usually defined as the term to designate the manifestations of the physical forces of cohesion and adhesion acting within the soil at various moisture contents. These manifestations include (i) the behaviour towards gravity, pressure, thrust and pull; (ii) the tendency of the soil mass to adhere to foreign bodies; and (iii) the sensation which is evidenced as feel by the fingers of the observers.

Q.2 What does adhesion and cohesion refer to?

Hint: Adhesion refers to the attraction of the liquid phase on the surface of the solid phase. Cohesion in soils is the bonding of the particles due to attractive forces between the particles arising from physico-chemical mechanisms. These bonding forces may be: (i) van der Waals force, (ii) electrostatic attraction between oppositely charged surface of the colloids, (iii) linking of the particles through cationic bridge, (iv) the cementation effect of various cementing agents and (v) the surface tension of the curved menisci at the air-water interface.

Q.3 What is the significance of Atterberg limits?

Hint: The plastic limit represents the moisture content of the change from the friable to the plastic consistency. The lower plastic limit signifies the minimum water content at which soil can be puddled. Soils with high plasticity indices are difficult to plough. Soils with expanding lattice clays have high plasticity indices.

Q.4 What is the factors on which Atterberg limits depend on?

Hint: Factors are: (a) clay content- plastic limit and plasticity index increase with clay content, (b) nature of clay- expanding clays have higher values of Atterberg limits, (c) exchangeable cation- Na-saturated clay exhibits the lowest plastic limit and the highest plasticity index, K-saturated clay has the lowest plasticity index and lowest liquid limit, while Ca and Mg-saturated clay generally exhibits higher plastic and liquid limits than that of Na and K-saturated clay.

Q.5 What is plasticity index?

Hint: It is the difference in water content between upper plastic limit and lower plastic limit. It is an indirect measure of the force required to mould the soil.

9

Soil Air

9.1 Determination of Oxygen Diffusion rate (ODR) of Soil

The rate of movement of oxygen from atmosphere to respiring plant root cells is important for root respiration and thereby the growth and development of plant. While diffusion through the air-filled pores maintains the exchange of gases between the atmosphere and soil, diffusion through water films maintain the supply of oxygen and disposal of carbon dioxide from active root tissues which are typically hydrated.

Principle

Lemon and Erickson (1952) introduced a method for measuring oxygen diffusion in soil with the help of platinum electrode. The process measures the extent of chemical reduction of oxygen by the platinum electrode maintaining a constant potential of 0.65V with a reference electrode.

$$O_2 + 2H_2O + 4e \rightarrow 4OH^-$$

The current flowing between two electrodes is proportional to the rate of O_2 diffusion to the platinum electrode. The oxygen rate (ODR) is calculated from the estimated electric current by following equation:

$$\text{ODR} = \frac{MI}{nFA} \tag{9.1}$$

Where ODR = oxygen diffusion rate (g m^{-2} s^{-1}); M= molecular weight of O_2 (32 g mol^{-1}); I = current reading (amperes); n = No. of electrons required for reduction of one molecule of O_2 = 4; F = Faraday constant = 96500 coulombs $mole^{-1}$, and A = exposed surface area of the Pt-electrode (m^2).

Instrument

ODR meter comprised of the following components: (i) Pt-electrode, (ii) Ag-AgCl reference electrode with a KCl salt bridge, (iii) a brass counter electrode,

(iv) read-out unit, (v) the variable resistor, and (vi) the battery. The counter electrode has two functions: the reference electrode is used to measure and monitor the potential difference between the Pt-electrode and the soil, and the brass electrode is used to close the electric circuit for the current needed to reduce the oxygen. The basic circuit is shown in Fig.9.1a & b.

Pt-electrode: The Pt-electrode is a 6 mm length and 1.2 mm diameter of Pt-wire hardened with 10% addition of Ir. The wire is mounted at the tapered end of a 75 cm long rod that facilitates the placement of electrode at various depths of soil. At the top of the rod is a female BNC connector for connection of the cable to the read-out unit.

Reference electrode: An Ag-AgCl electrode is used as the reference for voltage measurement. The Ag-AgCl electrode system is mounted on the top of a porous ceramic cup that filled with saturated KCl solution to serve as salt bridge to the soil.

Brass electrode: The counter electrode in the current circuit is a brass rod. It has an integral coaxial signal cable.

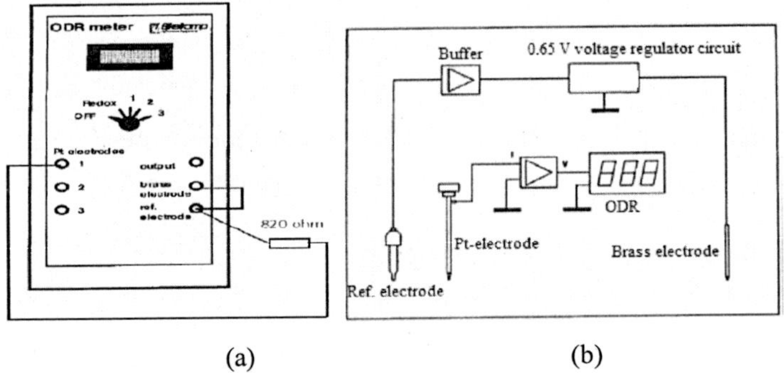

(a) (b)

Fig. 9.1: Schematic diagram of ODR meter

Procedure

- Connect the signal cables from Pt-electrodes to the BNC input connectors on the read-out unit.
- Prepare a place in the soil for the reference electrode and the brass electrode. A distance of about 2 metres between these electrodes and the Pt-electrodes will usually be found satisfactory in the field, though it is not at all mandatory. In horticulture a distance of a few centimetres may be chosen without any problem, depending on the size of the object.
- Connect a signal cable to the reference electrode. Bury the salt bridge and the brass electrode in the top layer of the soil. In dry soils, it is often necessary to water the soil to improve the electrical contact between these electrodes and the soil. Watering should be limited to the immediate

vicinity of the salt bridge and the brass electrode. Its effect must never extend to the Pt-electrodes.

- Connect the signal cables from the reference electrode and the brass electrode to the corresponding inputs on the read-out unit.
- Turn the switch on the read-out unit and maintain voltage between Pt-electrode and reference electrode at a value of 0.65V and read the ODR current at the corresponding electrodes.
- The reading will rise to a high value and then continue to fall for some time, when the already dissolved oxygen is reduced. It takes 2 to 4 minutes to reach equilibrium. Therefore, take a reading after about 3 minutes, wait for another 3 minutes and take a second reading to see if equilibrium has been reached or not. One millivolt in the display corresponds to a current of one microampere.

Calculation

Say,

Output current = Z μ ampere or Z $\times 10^{-6}$ ampere

Length of Pt-electrode = l m = 0.006 m

Electrode radius = r m = 0.0006 m

Therefore, the surface area of Pt-electrode exposed to the soil = $(2\pi rl+\pi r^2)$ m^2 = 0.0000237 m^2 or 23.7 $10^{-6}m^2$

$$\frac{MI}{nFA}=\frac{32\times Z\times 10^{-6}}{4\times 96500\times 23.7\times 10^{-6}}g\ m^{-2}s^{-1}=\frac{8\times Z}{228705}g\ m^{-2}s^{-1}$$

[*Note*: Regardless of temperature, roots do not grow when oxygen diffusion rates are less than 20×0^{-8} g cm^{-2} min^{-1}.]

Questions and Hints

Q.1 Why a constant potential between Pt-electrode and reference electrode is maintained at 0.65V?

Hint: The electro-chemical processes in the soil depend largely on the applied voltage. Reduction of oxygen is the dominant process at a voltage of about 0.428 volt (at 25°C) between the Pt-electrode and the soil. The potential of Ag-AgCl reference electrode is -0.222 volts. Thus, a potential between Pt-electrode and reference electrode is maintained at 0.65 volts. The magnitude of the current will vary significantly with the voltage. It is therefore important to specify the applied voltage and to keep the voltage within narrow tolerances during measurements.

Q.2 What are functions of reference electrode and brass electrode in ODR meter?

Hint: Reference electrode is used to measure and monitor the potential difference between Pt-electrode and the soil, and the brass electrode is used to close the electric circuit for the current needed to reduce oxygen.

10

Soil Tempearture

Temperature is a fundamental property of usually used to characterize the thermal condition of a system. The growth of any biological system is optimal within certain ranges of temperature and inhibited or ceased beyond this boundary. In addition to plant growth, the significance of temperature on agriculture can be ascertained by its role on physical, chemical and biological processes occurring in soil. Therefore, to study these soil properties intensively measurement of soil temperature is essential.

10.1 Measurement of Soil Temperature

Temperature of any system cannot be directly measured. It can only be estimated by its influence on some properties (thermometric properties) which respond to the variation in the intensity of heat of that system. These thermometric properties include volume, pressure, length, electrical resistance, thermal emf of the matter. Instruments built to take the advantage of any of these thermometric properties of the matter are called **thermometer**.

10.1.1 Types of Thermometers Used in Soil Studies

There are numerous types of thermometers that can be used to measure soil temperature. Some commonly used thermometers are: (i) mercury or liquid in glass thermometer, (ii) bimetallic thermometer, (iii) bourdon thermometer, and (iv) electrical resistance thermometer. The choice of thermometer depends on-the degree of precision, availability and accessibility to the location of sensing device.

(i) Mercury or Liquid in Glass Thermometer

In this thermometer volume of expansion of mercury or liquid (alcohol) with increase in temperature is used to measure the temperature of the system. Before use this thermometer it is advisable that it should be checked and calibrated against a standard thermometer.

(ii) Bimetallic Thermometer

This thermometer is commercially made by welding together two bars of two metals having different linear extensibility and rolling the resulting bimetallic bar into a strip. Usually invar and brass or invar and steel are used for this purpose. Because of difference in their linear extensibility of those metals, the strip bends in response to the temperature changes and the angular deflection is calibrated in terms of temperature. Therefore, the extent of angular deflection of the strip can be used to measure the temperature of the system.

(iii) Bourdon Thermometer

This thermometer consists of a curved tube of elliptical cross section connected to a bulb inserted into the soil through capillary tube. The system is totally filled with an organic liquid. With increase in temperature, the organic liquid expands and increases the pressure inside the bulb as well as capillary tube. This causes the curved capillary tube to become little less curved and to change the position of the pointer attached to it.

(iv) Electrical Resistance Thermometer

Resistance thermometer may be of two types. In first type the electrical resistance of the metallic wire increases with increase in temperature, e.g. nichrome, copper, silver, platinum, etc. For example, temperature coefficient of platinum wire at 0° C is about 0.35% per degree centigrade. Resistance changes which are related to the temperature changes can be measured either by using Wheatstone bridge or by using potentiometer.

The second type resistance thermometer is actually a semiconductor called **thermistor**. Thermistors have a high negative temperature coefficient (resistance decreases with increase in temperature) in the order of 4% per degree centigrade.

Among the different thermometers, the liquid (usually mercury) in glass thermometer is used most widely for soil temperature measurement. Therefore, here only the use of mercury in glass thermometer in soil temperature measurement is discussed.

10.1.2 Calibration of Thermometer

There are no single methods for measuring soil temperature that can be recommended to the exclusion of others. However, the user must be sure that the thermometer being used is properly related to the standard scale. A set of fundamental fixed points to which specified values have been assigned by international agreement is used as a basis for calibrating this thermometer. These fixed points are usually the freezing and boiling points of distilled water.

Equipments and Materials

Primary standard resistance platinum or a secondary standard thermometer, Dewar flask or thermos bottle, stirrer, ice made from distilled water, heating element.

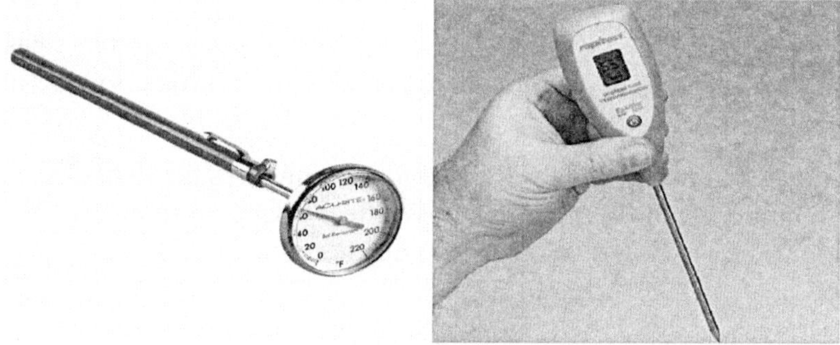

Fig. 10.1: Soil thermometer (a) analog type, (b) digital type

Procedure

- Suspend the standard thermometer as well as the thermometer to be checked or calibrated near the center of a Dewar flask.
- Fill the Dewar flask with crushed ice made from distilled water.
- Add distilled water to the crushed ice to displace all the air, add allow the system to establish thermal equilibrium. The temperature of the system should be 0°C or 32°F.
- Gradually add heat until the ice melts.
- Place the stirrer in the Dewar flask, and stir the solution vigorously.
- Add hot water to the system to raise the temperature.
- When stable readings are attained, compare the reading of the standard thermometer to that of thermometer being checked or calibrated. Maintain the water at the proper level near the top of the Dewar flask by siphoning off the excess water.

10.1.3 Soil Temperature Measurement

Equipments and Materials

Thermometer, soil auger, polythene tube having same outside diameter as that of the soil auger, rubber cork to plug polythene tube.

Procedure

- Sharpen one end of a wooden bar to a point so you can push it into the soil easily. Line the bar up with a ruler and make a mark on the bar for every inch.
- Push the wooden bar into the ground down to the specified depth. Pull the bar back out of the ground to leave a small hole.
- Push the thermometer down into the hole you made with the measuring wooden bar; ensure that it reaches the bottom of the hole for an accurate reading.
- Leave the thermometer in the soil for few minutes for stable reading or, if it's a digital thermometer, until you hear a beeping sound. [If the sun is bright, shade the thermometer with your hand to keep the reading accurate].
- Remove the thermometer and record the reading.
- For periodic measurement make a new hole in the soil each day before inserting the thermometer.

10.2 Determination of Specific Heat of Soil

Soil temperature depends directly upon the heat capacity and specific heat of soil. The heat capacity is the amount of heat required to change the temperature of the substance by one degree. On the other hand, specific heat is the ratio of the amount of heat required to raise the temperature by one degree of a little mass of any substance to the amount of heat required to raise the same temperature of the same mass of water, i.e. specific heat is the heat capacity per unit mass of substance. Therefore, specific heat is an intensive property while heat capacity is the extensive property of any substance. The specific heat of different soils varies due to the variation in the nature and amount of soil minerals, organic matter and moisture content. Specific heat of dry mineral soils varies from 0.17 to 0.29 cal g^{-1} $^{0}C^{-1}$. The information about specific heat of soil is important to evaluate the amount of heat transfer through the soil.

Principle

In an isolated system (no exchange of energy or matter between system and surroundings) when two bodies of different temperature are brought in close contact with each other higher temperature body will lose heat while lower temperature body will gain until an equilibrium is attained. The amount of heat received by lower temperature body will be equal to the amount released by higher temperature body provided no chemical interaction is taking place between two bodies. This unique property can be used to determine the specific heat of any substance.

Let an oven-dry soil of mass M_s, specific heat S_s and initial temperature T_1 be dropped all of a sudden into water of mass M_w, specific heat S_w and initial temperature T_2 ($T_2 > T_1$) contained in a calorimeter having mass M_c and specific heat S_c and if the equilibrium temperature of soil-water mixture is T_3, then according to the principle of calorimetry (Heat gained = Heat lost)

Heat gained by soil = Heat lost by water and calorimeter

$M_s S_s (T_3 - T_1) = M_w S_s (T_2 - T_3) + M_c S_c (T_2 - T_3)$ [where $T_1 < T_3 < T_2$]

Or, $M_s S_s (T_3 - T_1) = (M_w S_s + M_c S_c)(T_2 - T_3)$

$$Ss = \frac{(MwSs + McSc)(T2 - T3)}{Ms(T3 - T1)}$$

Equipment and Materials

Calorimeter (copper) with an insulation box, thermometer, precise balance, tripod stand.

Procedure

- Weigh the empty calorimeter.
- Heat some amount of water (100 ml) to a temperature higher than room temperature and pour about 50-60 ml of hot water to the calorimeter.
- Weigh calorimeter with hot water, keep it in the insulation box and cover it.
- Weigh some amount (20-25 g) of oven-dry soil and read the temperature (T_1) to the nearest of ± 0.01°C.
- Record the temperature of water (T_2) in the calorimeter to the nearest of ± 0.01°C.
- Remove the lid of the insulation box and drop the soil immediately into the calorimeter and cover it again.
- Stir the suspension slowly to bring it at thermal equilibrium. Record the equilibrium temperature of the soil-water suspension (T_3) to the nearest of ± 0.01°C.

Calculation

Weight of the oven-dry soil = M_s g

Weight of the empty calorimeter = M_c g

Weight of the calorimeter + hot water = M_{cw} g

Therefore, weight of hot water = (M_{cw} - M_c) g

Initial temperature of soil = $T_1{}^0C$

Temperature of hot water in calorimeter = $T_2{}^\circ C$

Equilibrium temperature of soil-water suspension/calorimeter = $T_3{}^0C$

Specific heat of water = 1 cal g^{-1} $^0C^{-1}$

Specific heat of copper (calorimeter) = 0.093 cal g^{-1} $^0C^{-1}$

Therefore,

$$\text{specific heat of soil } (S_s) = \frac{(Mw \times 1 + Mc \times 0.093)(T2 - T3)}{Ms(T3 - T1)} \text{ cal g}^{-1}\ {}^0C^{-1}$$

Questions and Hints

Q.1 What are the criteria for selecting metals in a thermocouple?

Hint: The choice of two metals used in a thermocouple depends on: (a) the amount of emf they can produce and linearity of response within the scale of tempearature measurement, and (b) the metals should be stable against chemical contamination, oxidation, reduction and corrosion.

Q.2 What are the principles of functioning of electrical resistance thermometers?

Hint: See section 10.1.1(iv)

Q.3 When soil temperature reading should be taken?

Hint: Take a reading in the morning and late afternoon, and then average the two values.

Q.4 Why specific heats of soils differ?

Hint: The specific heat of different soils varies due to the variation in the nature and amount of soil minerals, amount of organic matter and soil water.

11

Soil Penetrabilty

11.1 Determination of Penetrability of Soil

Penetrability of a soil is usually assessed by the penetration resistance, which a measure of soil strength and thereby the soil compaction. Penetration resistance is the capacity of the soil in its confined state to resist penetration by a rigid object. Besides soil strength, the penetration resistance also depends on the size, shape and orientation of the axis of the penetrating probe. The penetrating probe may be a finger, pencil, stick, root or any specially designed object having a specified geometric shape and a device to measure the resistance as it is pushed into the soil.

Soil penetrability is intimately related to agriculture because of its adverse impact on draught requirement during ploughing, root development and crop yield, to civil engineering due to its relation to settlement, stability and ground water flow.

Soil penetrability is a measure of the ease with which an object can be pushed or driven into the soil. Any device designed to measure the resistance to penetration may be called penetrometer. There are several types of penetrometers: (i) pocket penetrometer, (ii) proctor penetrometer, (iii) cone penetrometer, and (iv) split-spoon penetrometer. Here, we will discuss about two most commonly used penetrometers viz. pocket penetrometer and cone penetrometer used for agricultural purposes.

11.1.1 Pocket Penetrometer

The pocket penetrometer is a portable, hand-operated, calibrated spring penetrometer developed to estimate the engineering consistency of the cohesive fine-textured soil (Fig.11.1). As the piston needle is pushed into the soil deformation of the spring is correlated with the compressive strength of the soil in kg cm^{-2}. Since root growth has correlation with point resistance, Bradfield suggested the conversion of compressed strength scale against a set of known weights.

Equipments and Materials

Direct-reading pocket penetrometer having diameter of 19.1 mm and a piston needle diameter of 6.4 mm.

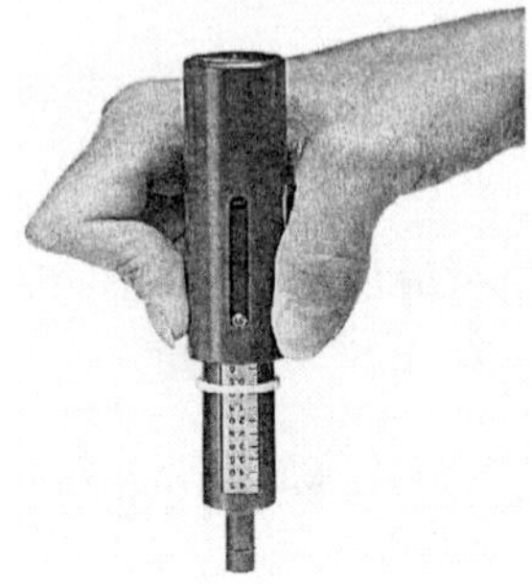

Fig.11.1: Pocket penetrometer

Procedure

- Select the test location and determine soil properties which affect the soil resistance, e.g. moisture content, bulk density and soil structure.
- Move the indicator sleeve to zero on the penetrometer scale.
- Grip the handle and push the piston needle into the soil with a steady rate of penetration until the engraved line 6 mm from the tip is flushed with the surface of the soil.
- Remove the penetrometer from the soil and read the scale. Clean the piston and return the sliding indicator to the zero position.
- Repeat the test several times at different locations to get an average value of penetration resistance.

[*Note*: Penetration scale reading can be converted to total probe resistance by calibrating the scale readings with a set of known weights. Plot the added weight (kg) per tip area (0.322 cm^2) against sleeve reading (Fig.11.2). Convert the compressive strength value to total probe resistance from the calibration curve.]

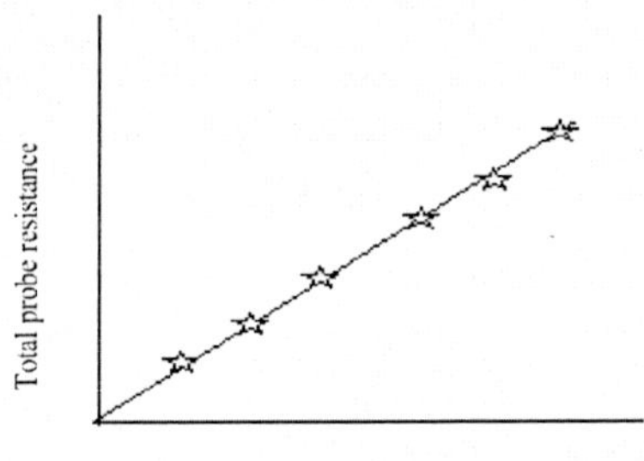

Fig.11.2: Calibration curve of penetrometer reading with total probe resistance

11.1.2 Cone Penetrometer

The cone penetrometer is the most widely used device to measuring soil's mechanical condition. In agricultural research it is extensively used to locate hardpans or traffic compaction area, to correlate soil strength parameters with root growth and crop yield, and to quantify the physical state of soil. The applied force required to press the cone penetrometer into a soil is an index of shear resistance of the soil and called the '*cone index*'. The cone index is defined as the force per unit basal area required to push a cone penetrometer through a specified increment of soil. The recommended insertion time is 2 and 4 seconds for the smaller and larger cones, respectively. Small-diameter friction cone type penetrometer is used to study root elongation, soil structure, and rate of penetration.

The cone penetrometer consists of a shaft or central push-rod used for advancing the cone attached to the end of push-rod into the soil. Surrounding the central push-rod is moveable sleeve with outside diameter equal to the cone base diameter.

This penetrometer is easily adapted to both laboratory and field determinations. In laboratory study, the penetrometer is pushed into the soil with a rate controlled laboraory load machine, but for field studies with penetrometer of less than 5 mm diameter a drive machine is used for this purpose. The weight of machine is used to drive the penetrometer.

Equipments and Materials

1. Friction-sleeve cone penetrometer: The cone has a 60^0 angle and base diameter of 3.74±0.02 mm; the cone is made from brass or stainless steel and has a polished surface. The push-rod is made from stainless steel hypodermic needle, tubing (12-gauge) having outside diameter of 2.76 ± 0.02 mm and length of about 150 mm.

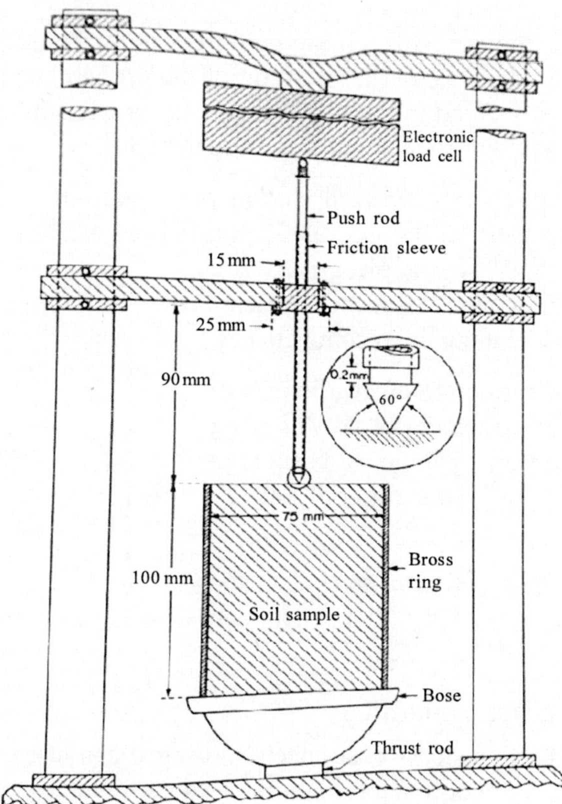

Fig. 11.2: Friction sleeve cone penetrometer

2. Load weighing system: The system consists of a 10 to 50 kg capacity strain gauge load cell, a signal conditioning unit, and a recorder.
3. Penetrometer drive machine: For laboratory tests, any motorized compression or strength-testing machine with variable speed drive can be used.

Procedure

A. For Laboratory Determination

- Attach the friction sleeve penetrometer to the penetrometer drive machine. The friction sleeve of the penetrometer is secured to a metal cross-bar which is then fastened to the compression machine guide rods.
- Position a 4 to 8 mm diameter steel ball between the end top of the load shaft and the base of the load cell.
- Move the cross-bar upwards or downwards until the distance between the cone base and the end of the friction sleeve is about 0.2 mm.
- Position the soil core on the compression machine platen. [The cone diameter should be at least 20 times the probe diameter and should be radially confined in the metal rings; for brittle soil sample the core to probe diameter ratio should be more than 30].
- Advance the core upward so that the penetrometer enters the soil at constant rates of 0.0005, 0.005 and 0.5 cm min^{-1}. [For re-formed soil cores, only two replications are needed; for natural soil cores, at least 3 to 5 replications are recommended].
- Convert the load cell readings (kg) into cone resistance (k Pa) by dividing the load (kg) by the cone basal area (0.1099 cm^2) and multiplying by 98.07.
- Plot cone resistance in kPa on the abscissa against penetration depth on the ordinate (Fig.11.3).

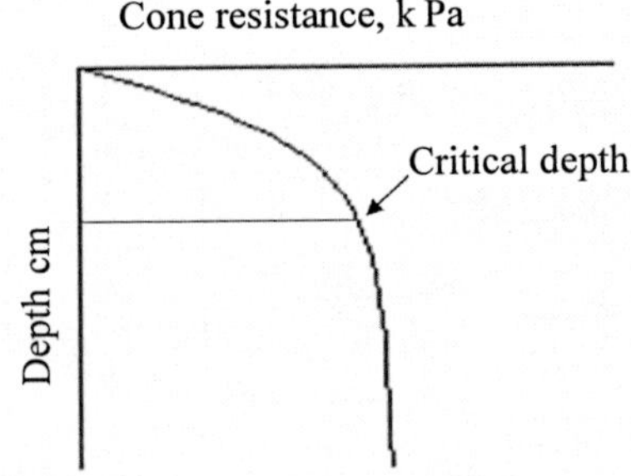

Fig. 11.3: Cone resistance vs depth curve

B. For in situ Determination

- Attach friction-sleeve cone penetrometer to the field drive unit by following the procedure outlined for laboratory determination.

- Position the penetrometer above the test area and advance the penetrometer into the soil at a constant rate of 5.0 cm min^{-1} to a depth of about 10 cm.
- Repeat the test 5-10 times. Remove the soil overburden to the next desired depth and repeat the procedure.
- Continue sequence until the desired depth is reached.
- Plot cone resistance (k Pa) against penetration depth and report soil bulk density, matric potential and soil structural type and class.

Table 11.1: Penetration resistance classes

Class	Penetration resistance (M Pa)
Small	< 0.1
Extremely low	< 0.01
Very low	0.01- 0.1
Intermediate	0.1- 2
Low	0.1 - 1
Moderate	1 - 2
Large	> 2
High	2 - 4
Very hoigh	4 - 8
Extremely high	8

Questions and Hints

Q.1 At what soil condition penetrability test is recommended?

Hint: Penetrabiltiy test in situ is recommended at the field capcity soil water content (approx. 24 hours after a soaking rain/irrigation).

Q.2 What are the soil factors that affect penetrability of a soil?

Hint: Penetrasbilty of a soil depnds on matric potential (i.e., water conetnt), bulk density and soil structure.

Q.3 What is the critical limit of penetrability for root growth?

Hint: There is a critical limit for root growth, but it differes among crops and soil texture. For example, critical limit for cotton root growth at sandy loam soil is 3000 k Pa.

Q.4. What is there any specification regarding core to cone diameter for laboratory study of soil penetrability by cone penetrometer?

Hint: Yes; the core diameter should be at least 20 times the cone diameter and should be rigidly confined within metal ring. For brittle soil the core to cone diameter should be more than 30.

Q.5 What do you mean by critical depth in penetrability measurement?

Hint: Penetration resistance increases with depth to a certain depth, which is dependent on soil hardness and probe diameter, is known as critical depth. Below the critical depth, failure mechanisms changes from shear to one of shear and compression, and the rate of increase of resistance with depth is less or resistance is constant with depth.

Section -II

SOIL MINERALOGICAL ANALYSIS

12
Thermal Analysis

Thermal Analysis

Physical and chemical properties of any soil are governed largely by the mineralogical make up of the soil, and more specifically by the clay fraction. Identification, characterization and an understanding of properties of different minerals are important in evaluating soils in relation to classification, agronomic practices and engineering purposes.

Clay fractions of soils are usually composed of mixture of one or more secondary phyllosilicate minerals along with primary minerals inherited from the parent materials. Identification of minerals species as well as their quantitative estimation in such polycomponent system require application of several qualitative and quantitative analysis.

The methods used for identification and quantification have steadily been improved over the last 50 years. Among the several methods often used for this purpose three are very common, e.g. (i) differential thermal analysis (DTA), (ii) x-ray diffraction analysis (XRD), and (iii) infrared spectroscopy (IR).

Thermal analysis is a term covering a group of analyses that determine the changes of some physical parameters, such as weight, energy, dimension and evolved volatile substances with change in temperature. Many soil constituents undergo various thermal reactions upon heating, which can be utilized as a diagnostic tool for qualitative identification and quantitative estimation of the substances. These reactions may be exothermic or endothermic, and phase change or crystal change may take place during reactions. Whatever be the reactions, if properly measured, they give some valuable information to identify the substance under consideration.

Among the several methods available for thermal analysis, three methods are frequently used in soil research: differential thermal analysis (DTA), differential scan calorimetry (DSC), and thermogravimetry (TG). Among these, DTA and TG will be discussed here.

Water is removed from clay on heating. Water is present in clay in two different forms: OH^- ions and H_2O molecules. Customarily, the OH^- ions which are present as individual ion among the oxygen of tetrahedra or octahedra or as continuous sheet as in 1:1 and 2:2 mineral species, are referred to as crystal lattice water, and their removal from minerals requiring high temperature is termed as dehydroxylation.

$$2OH^- \rightarrow \underset{\text{(gas)}}{H_2O\uparrow} + \underset{\text{(in solid)}}{O^{2-}}$$

The H_2O molecules which are present either on the external surfaces of clay particles (*e.g.*, kaolinite) or both external and internal surface (*e.g*, montmorillonite, vermiculite, hydrated halloysite) are referred to as water of hydration/water of adsorption and their removal done at low temperature is called dehydration/desorption.

$$\text{HOH-Solid} \rightarrow \underset{\text{(gas)}}{H_2O\uparrow} + \text{Solid}$$

Removal of water molecules from clay particles may or may not cause any reversible change in the crystal lattice. For example, upon removal and readsorption of H_2O molecules crystal lattice of kaolinite remains unaltered, but of montmorillonite and vermiculite changes in a reversible manner.

The loss of water upon heating is measured by loss in weight of clay mineral. But, presence of organic matter and carbonates in soil may introduce errors in this measurement. Again, loss of weight due to loss of water may to some extent be counterbalanced by gain of oxygen upon oxidation of reduced elements (say, Fe^{2+}) present in mineral sample. The temperature at which the major amount of crystal lattice water is lost differs among mineral species, thus can be used as a most important parameter for the identification of the mineral species.

12.1 Differential Thermal Analysis

The basis of differential thermal analysis is the determination of energy changes between sample and a reference material when both are heated side by side at a controlled rate. Upon heating if sample undergoes a transformation, the heat effect causes a difference in temperature between the sample and reference material. In DTA, this difference in temperature is measured by means of a set of thermocouples and recorded as a function of temperature. If the temperature of the sample falls below that of the reference material (ΔT is negative) an endothermic peak develops and when the temperature of sample rises above that of the reference material (ΔT is positive) an exothermic peak develops. The portion of the curve for which ΔT is approximately zero is considered to be the base line. Endothermic reactions are attributed to reactions, such as dehydration, dehydroxylation, fusion, evaporation and sublimation. On the other hand, exothermic effects are achieved in oxidation, crystalline structure formation and some decomposition reactions.

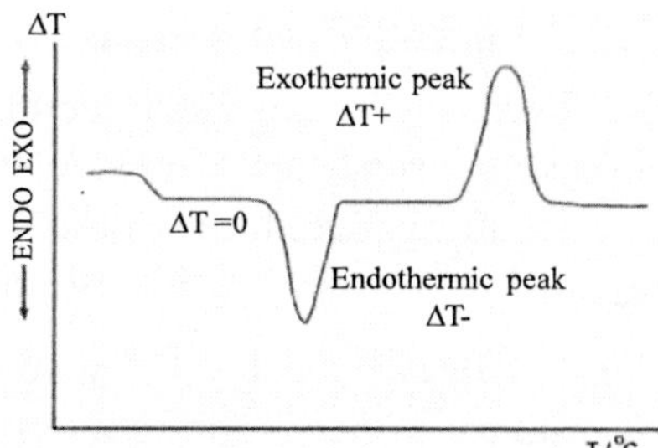

Fig. 12.1: Idealized DTA curve depicting endothermic and exothermic peak

12.1.1 Instrument

All DTA instruments have the following basic components: (i) furnace or heating unit, (ii) a sample holder, and (iii) a temperature regulating and measuring system.

Furnaces used in DTA are operable from 0 to 2800°C with the heating provided by infrared radiation, by high frequency of oscillation, by coiled tubing through which heated gas/liquid is circulated or by resistance elements. Among them heating by resistance elements are very common.

The sample holder consists of either a rectangular or circular nickel or ceramic block. They are either equipped with cylinder holes (well-holder), in which samples and thermocouples are placed, or they are designed to hold ring-type thermocouples upon which a small platinum pan can be seated.

For soil analysis, the thermocouple elements are preferably made from platinum (90%) and rhodium (10%) for their resistance against corrosion. The size of two thermocouples in the DTA circuit must be identical; otherwise the DTA curves tend to drift upward or downward from the base line.

The temperature differences and furnace temperature are recorded by placing either a reflecting galvanometer, which records the variation in EMF on a photographic paper fastened to a rotator drum or a pen and ink recorder in the thermocouple circuit.

12.1.2 Reference Material

The reference or standard material is a known substance which is thermally inert within the temperature range under experimentation. The reference materials include both organic (octanol, benzol, tributyrin) and inorganic (alumina, clay, quartz, NaCl, KCl, AgCl, glass beads and aluminium foil) compounds. Among them, the most commonly used reference material is α- or γ-alumina or calcined (preheated to 1200°C) alumina or clay (e.g., kaolinte). Calcined alumina or clay needs to be replaced after 2 to 3 runs due to development of hygroscopicity on use in DTA.

12.1.3 Sample Preparation

Soil is in fact a mixture of mineral and organic matter widely varying in particle size and in properties. For DTA, soil is fractioned into different particle sizes by fractionation procedure as practiced in other mineralogical analysis, such as x-ray diffraction (see section 14.2.2).

Sand Fraction: Analysis of sand by DTA is only necessary in the investigation of primary mineral and/or Fe-Mn concretion. Since, particle size has little influence on the nature of DTA curve of sand; further fractionation of sand (2.00-0.05/0.02 mm) into various size groups is not required. DTA of sand fraction is featureless other than that of endothermic peak of quartz at 573°C.

Silt Fraction: Like sand, silt fraction (0.05/0.02-0.002 mm) can directly (without further fractionation) be analyzed, though the DTA curve is more complex than that of sand, particularly for such silt size fraction which contains a good amount of sesquioxide and kaolinite yielding endothermic peaks at 270 and 498°C, respectively in addition to the quartz peak at 573°C.

Clay Fraction: Depending upon the objective of study, clay fraction (< 2μm) of soil can be used directly or can be separated into coarser (2.0-0.2 μm) and finer (< 0.2μm) clay fractions. For analysis of amorphous minerals (*allophane*) use of fine clay fraction is suggested. However, for all general purposes, the clay fraction (< 2μm) yields DTA curve that can satisfactorily be used for qualitative and quantitative interpretation.

Organic Fraction: DTA can be used to distinguish between humic acid and fulvic acid extracted and purified from soil organic matter (see section 19.3). Humic acid is characterized with a strong exothermic peak in DTA curve at 400°C, while fulvic acid exhibits a DTA curve with an exothermic peak at 500°C.

Cation Saturation: Saturation of the sample with a known cation prior to DTA has been proposed for clay or material having cation exchange property. Clay saturated with variety of cations (Na^+, H^+, Ca^{2+}, Al^{3+}) affects the shape and size of endothermic and exothermic peaks because of their variation in hydration behaviour, though the temperature of the peaks remains unaffected. For example, the peak intensity for both 530°C and 1000°C peaks of kaolinite increases from Ca-kaolinite to Na-kaolinite to H-kaolinite to Al-kaolinite. Saturation of clay with Na^+, H^+, Ca^{2+}, and Al^{3+} is done by shaking the clay with 0.1M solution of NaCl, HCl, $CaCl_2$, and $AlCl_2$, respectively for an hour and allowing them to stand for overnight. After washing with distilled water, samples are dried at 45°C, ground and stored in desiccators over $CaCl_2$.

Hydration and Solvation: Hydration of samples prior to DTA may result changes in the low temperature endothermic peaks within 0 to 200°C. Soil colloids differ greatly on their moisture absorption capacity at a particular relative humidity. Samples equilibrated at constant relative humidity exhibit low temperature endothermic peaks which can be used for characterization of amorphous and 2:1 type of lattice clay. Equilibration is carried out in a desiccators over a compound [e.g., $Mg(NO_3)_2$. $6H_2O$, $Mg(NO_3)_2$, $CaCl_2$ or H_2SO_4]. Samples should not be kept over desiccating agent for prolonged period.

12.1.4 Sample Size

The amount of sample required for analysis depends on the type of DTA instrument available. Instruments equipped with well-type sample-holder require 100 mg sample to fill the hole. Analysis of large amount of sample produces extremely large peaks which are difficult to record on a normal size paper. In that case, dilution is done with inert materials. The amount for mineral sample is usually in the order of 1 to 10 mg and for humic compounds it may be as little as 0.5 mg.

In qualitative analysis, weighed sample is not required for DTA, although for comparison of DTA curves identical amount of sample should be taken. In quantitative analysis, the sample must be weighed precisely for DTA.

12.1.5 Sample Packing

Packing of sample has an important bearing on the DTA, particularly with well-type sample-holder and in low temperature ranges where heat transfer is mainly done by conduction. In high temperature region where heat transfer is mostly through radiation, packing-effect is negligible on DTA curve. However, aggregate size and bulk density of both sample and reference material should be identical to avoid any base line drift.

Packing is done either by hand tapping of sample or by block and layer packing. In later method, sample is packed as a sandwiched in between two layers of reference material around the thermocouple junction. In this method peak intensity will be one half the intensity observed by filling the whole cavity with sample.

12.1.6 Furnace Atmosphere

It is necessary to control furnace atmosphere, because it affects DTA through one or a combination of reactions: (i) a change in furnace atmosphere results a change in partial pressure of gas atmosphere which may induce a shift of peak temperature, (ii) interaction occurring between the gas supplied and gas evolved may change furnace atmosphere and obscure the appearance of characteristic

DTA peaks, or (iii) peaks are enhanced if the oxidation reaction is eliminated. The methods such as use of vacuum, static air or gas, dynamic gas, gas flow over the sample, self generated gas are tested. Each method has some advantages and disadvantages over the other. For example, in evacuated or N_2 atmosphere removal of organic matter is not necessary for qualitative DTA, since no exothermic reaction associated with oxidation is possible in this atmosphere. However, for most analysis, DTA in presence of self generated gas atmosphere is suitable.

To provide a uniform pressure and thermal condition around sample and reference material, sample holder is covered with loose-fitting lid, which protects both the specimens from direct radiation of the furnace.

12.1.7 Heating Rate

The heating or cooling rate is the rate of change of temperature expressed in degrees centigrade per minute. The heating must be controlled at a uniform and steady rate all through the analysis. Heating rate used in DTA varies from 0.1 to 200°C/min, but for most purposes 5 to 20°C/min heating rate is used. Increase in heating rate increases the peak height of kaolinite gradually and consistently and shifts the peak temperature to a higher side.

12.1.8 Interpretation of Results

A. Qualitative Interpretation: Identification of soil mineral can be done by using the DTA curves as fingerprints and comparing or matching them with the DTA curves of known mineral samples. Table 12.1 and Figure 12.2 depict the endothermic and exothermic peaks of major clay minerals along with the reactions involved in their formation.

Table 12.1: Endo- and exothermic DTA peaks of clay minerals and reactions involved

Mineral	Endothermic peak at (°C)	Main Reaction	Exothermic peak at (°C)	Main Reaction
Kaolinite	500-600	Dehydroxylation	900-1000	γ-alumina formation
Dickite, Nacrite	500-700	Loss of adsorbed water	900-1000	γ-alumina formation
Halloysite	100-200	Dehydroxylation		
Montmorillonite, Beidellite	100-250	Loss of adsorbed water	900-1000	Recrystallization
Nontronite	100-200	Loss of adsorbed water	900-1000	Recrystallization
	500	Dehydroxylation		

Vermiculite	150	Loss of adsorbed water	800-900	Recrystallization
	850	Dehydroxylation		
Illite	100–200	Loss of adsorbed water	920-950	Recrystallization
	600	Dehydroxylation		
	900-920	Dehydroxylation		
Chlorite	500 - 600	Dehydroxylation	800	γ-alumina formation
Gibbsite	250 - 350	Dehydroxylation		
Goethite	300 - 400	Dehydroxylation		
Allophane	50 - 150	Loss of adsorbed water	800 - 900	γ-alumina formation
Quartz	573	α to β inversion		

Source: Tan (2005)

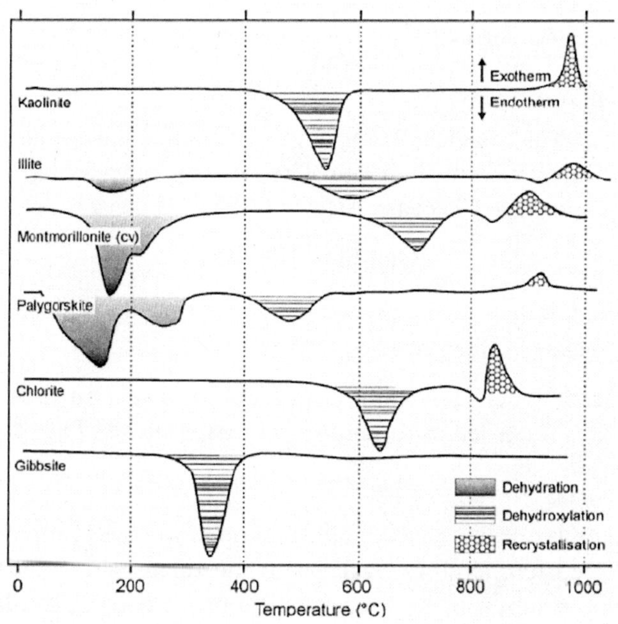

Fig. 12.2: DTA curves of some clay minerals

B. Quantitative Interpretation: Quantitative analysis of soil clays using DTA requires an accurately weighed sample. Aggregate size, packing density of samples and heating rate must be similar to maintain uniformity. Availability of automated and programmed control instruments makes the quantitative analysis with DTA much easier. Since quantitative analysis is based on the height, area, or width of the main endothermic peak, DTA is applicable only to those soil minerals yielding endothermic peaks of sizes large enough to be measured accurately and precisely. 1:1 type of minerals, gibbsite and goethite or their mixture exhibit well-developed endothermic peaks, thus lend themselves for quantitative analysis. On the other hand, montmorillonite and vermiculite have poorly developed endothermic peaks, making quantitative assessment of these minerals very difficult.

For quantitative estimation of clay mineral, first construct a standard curve of that mineral (identified using DTA) by plotting height or area of the endothermic peak against the mineral concentrations (Fig. 12.3).

Now, concentration of the mineral present in the sample can be estimated by comparing the peak height or area of that mineral with the standard curve.

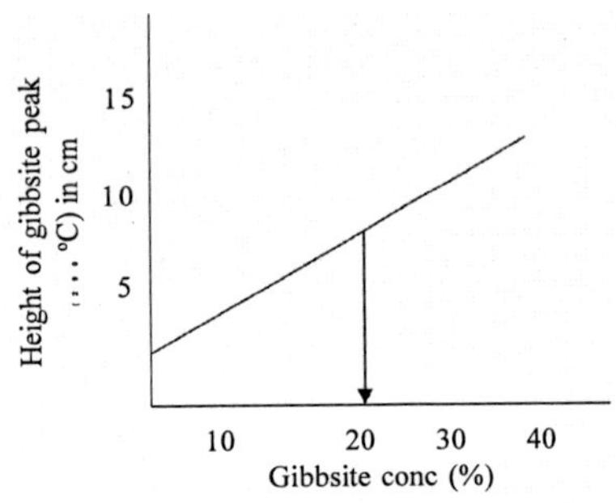

Fig. 12.3: Standard curve of gibbsite

12.2 Thermogravimetry

Thermogravimetry is based on the principle of change in weight of sample with heating or cooling at a controlled rate. The changes in weight are recorded as a function of temperature (T) or time (t):

$$\text{weight} = f\,(T \text{ or } t)$$

The curve obtained by plotting the weight of the sample against T or t is called thermogravimetric (TG) curve.

On the other hand, in derivative thermogravimetry (DTG) time derivative of the weight (dw/dt) is recorded as a function of temperature (T) or time (t):

$$dw/dt = f\,(T \text{ or } t)$$

Change in weight is due to rupture and formation of various physical and chemical bonds at elevated temperature, leading to the release of volatile substances (e.g., H_2O) or formation of heavier reaction product (e.g., oxidation of iron, $FeO \rightarrow Fe_2O_3$)

12.2.1 Instrument

The basic components of thermogravimetric instrument include a furnace, a temperature programmer, and a thermobalance. The thermobalance records the weight of sample continuously during analysis as a function of temperature or time. Presently manually operated automatic recording thermobalances are available. Currently available TG instruments are equipped with sensible solid state automatic recording capability and low mass cooling furnaces have made TG a rapid, accurate and relatively simple analytical method for mineral identification.

Since DTA alone is often not sufficient for analysis of clays, TG and derivative TG can be used as valuable complementary technique. In soil studies TG is used for determination of desorbed water and crystal-lattice water that are removed from different soil minerals at different temperature. The temperature at which dehydroxylation and desorption starts and completes and theoretical weight losses are given in Table 12.2. The theoretical weight loss is due to theoretical amount of crystal-lattice water present in a given pure species. It is estimated by determining the chemical composition, then its formula and molecular weight on water-free basis. The theoretical number of moles of water is then added to the formula, and the composition is recalculated on percentage weight basis, either on formula weight with water or without water.

12.2.2 Sample Preparation, Sample Size and Heating Rate

As with DTA, 10 mg of untreated whole soil sample after grinding of air-dry soil and passing through 140-mesh sieve is used for qualitative and semi-quantitative in TG. Clay fraction obtained by fractionation method (see section 14.2.2) is used for TG and derivative thermogravimetric analysis. Analyses are made at a heating rate of 10°C/min in N_2 gas atmosphere.

12.2.3 Interpretation

Crystal-lattice water of all the important silicate minerals is lost by heating between 150 or 350 and 1000°C and clay minerals can be divided into two distinct groups on the basis of the amount of crystal-lattice water. The total amount of crystal-lattice water for 2:1 minerals such as montmorillonite, vermiculite, and micas lies in the range of 4.2 to 5.1%, while that for 1:1 and 2:2 minerals lies in the range of 13.4 to 16.2% (Table 12.2).

On the basis of the amount of crystal-lattice water, i.e. weight loss by heating between 150 or 350 and 1000°C approximately quantitative estimation of 2:1 minerals and of 1:1 and 2:2 minerals is also possible (Fig. 12.4). If an x-ray analysis and a chemical analysis are available, it is possible to quantify each mineral species present in the sample provided there should not be any mineral species which gives overlapping temperature range of crystal-lattice water removal with others.

Table 12.2: Adsorbed and crystal-lattice water of clay minerals and temperature of desorption and dehydroxylation

Clay mineral		Adsorbed water		Crystal-lattice water		Temperature at completion of desorption		Temperature at start of dehydroxylation	Temperature at completion of dehydroxylation
	Layer water	%*	Cavity water		Layer water	°C	Cavity water		
Ca Montmorillonite	20.16		3.37	5.08	250		370	370	1000
Na Montmorillonite	14.00		0.00	5.05	150		-	150	1000
Ca Vermiculite	20.00		3.08	4.80	250		700	250	1000
Na Vermiculite	10.14		4.66	4.76	150		700	150	1000
Ca Illite	5.16		1.06	4.97	150		370	370	1000
Na Illite	3.45		0.66	5.07	150		370	370	1000
Muscovite	1.00		0.45	4.74	250		350	370	1000
Biotite		0.4-1.5		4.17		350		350	1000
Talc		0.7-1.0		4.99		350		350	1000
Pyrophylite		0.5-1.5		5.26		350		350	1000
Kaolinite		0.2-1.2		16.20		350		350	1000
Halloysite (hydrated)		13.2		16.20		250		350	1000
Chrysotile and antigorite		1.0-2.0		14.96		250		350	1000
Clinochlore		0.2-0.5		14.81		250		370	1000
Gibbsite		0.5-0.6		52.95		100		150	350
Goethite		0.0-0.1		33.80		100		200	370
Brucite		0.5-0.6		44.70		170		200	370

*On the ignited basis

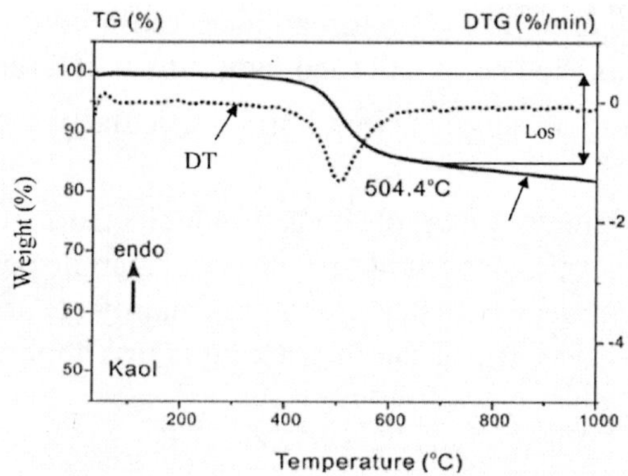

Fig.12.4: TG and derivative TG curves for kaolinite

To differentiate between adsorbed and crystal-lattice water, heating rate should be such that adsorbed water is lost before the loss in crystal-lattice water starts. For this purpose, instead of heating sample (Na-saturated) at a continuously rising temperature, it is more practical to heat the sample at fixed temperatures (150, 370, and 1000°C) for periods between 3 and 12 hours.

Questions and Hints

Q.1 For qualitative information on mineral species present in soil by DTA, whether instead of analyzing only clay, whole soil can be analyzed even without removing organic matter?

Hint: Yes; only if the soil is high in clay, and low in organic matter. Organic matter exhibits exothermic peak at approximately 300°C. Therefore, if any endothermic peak develops between 250 (gibbsite) and 400°C for any clay mineral will be obscured by this strong exothermic peak. However, in soils having low organic matter whole soil can be used without removal of organic matter by H_2O_2 treatment, though the peaks will be less intense. In coarse textured soils, the peak may be obscured because of dilution by high quartz content. For quantitative analysis by DTA, whole soil without removal of organic matter is not suitable.

Q.2 Does absence of a particular thermal peak in DTA curve mean absence of constituent associated with this peak?

Hint: No; if an endothermic reaction of one component and exothermic reaction of other component take place simultaneously with equal heat transfer, in such case no thermal peak is visible in DTA curve.

Q.3 Why calcined alumina or clay needs replacement after 2 to 3 runs in DTA?

Hint: After 2 to 3 runs in DTA the calcined alumina or clay may become hygroscopic. The large difference in thermal conductivity of kaolinite

(0.72×10^{-3} cal-cm $°C^{-1}$ s^{-1}) and calcined kaolinite (1.6×10^{-3} cal-cm $°C^{-1}$ s^{-1}) restricts the use of calcined kaolinite as reference material.

Q.4 Why the packing of sample is important particularly in low temperature regions?

Hint: Packing affects the sample density, which in turn influences thermal conductivity as well as thermal diffusivity of sample particularly in low temperature regions where heat transfer is mainly occurs by conduction. In high temperature, the effect of packing is less apparent.

13

Infrared Spectrometry

The total energy of a molecule consists of translation, vibrational, rotational and electronic energies. Transition among different types of energy levels occurs in different regions of electromagnetic spectrum. Transitions between rotational and vibrational energy levels of the ground state molecules give rise to absorption bands throughout the infrared region of the electromagnetic spectrum. Organic molecules are characterized by possessing bonds and groups that vibrate independently of each other. Thus, initially infrared spectrometry was used for identification of functional groups of organic molecules. There is lot of ambiguity in the interpretation of infrared spectra of organic matter preparations due to complexity and diversity of the components of soil organic matter. But if used carefully it can provide information on: (i) nature, reactivity and structural arrangement of oxygen-containing functional groups, (ii) presence of protein and carbohydrate constituents, (iii) presence or absence of inorganic impurities (metallic ions, clays, etc in humic fractions), (iv) the amount of different soil organic matter components, and (v) interaction of organic components with pesticides, herbicides, fertilizers, etc.

On the contrary to organic molecules, minerals tend to have few isolated vibrating groups. It can be used for mineralogical analysis when used in conjunction with x-ray diffraction and other similar techniques.

Infrared spectrometry is used in the identification of inorganic compounds including minerals which have well-defined absorption bands in determining- (i) whether layer silicate is dioctahedral or trioctahedral in composition, (ii) in studying isomorphous substitution, (iii) hydration of minerals, and (iv) in quantitative analysis. Infrared spectrometer provides information about nature and identify of inorganic compounds that are amorphous in x-ray diffraction analysis.

13.1 Infrared Radiation

The region of radiant energy in electromagnetic spectrum can be specified by wavelength (λ), in units of micrometers (μm) and frequency or wavenumber (υ), in units of number of waves per centimeter (cm^{-1}). The relationship between wavelength and wavenumber may be expressed as follows:

$$\upsilon\ (cm^{-1}) = \frac{1}{\lambda(cm)} = \frac{10^4}{\lambda(\mu m)} \quad (13.1)$$

True frequency (υ′) is related to so-called frequency or wavenumber (υ) in the following manner:

$$\upsilon'(sec^{-1}) = \upsilon c \quad (13.2)$$

where c is velocity of light (3 x 10^{10} cm/sec).

The region of infrared radiation is of slightly greater wavelength than that of visible light (Table 13.1). The total infrared region can be divided into three sub-regions: (a) near infrared, 12,800 to 3,330 cm^{-1} (0.78-3.0 μm), (b) mid-infrared or functional, 3,330 to 350 cm^{-1} (3.0 -29 μm), and (c) far-infrared, 350 to 30cm^{-1} (29-350 μm). However, for soil component analysis, the primary interest lies in region of infrared spectrum between wavenumbers of 4,000 to 200 cm^{-1} (2.5-50 μm).

Table 13.1: Spectrum range of electromagnetic radiation

Region	Wavenumber (cm^{-1})	Wavelength (nm)
Ultraviolet		
far	1 x 10^4 - 50,000	10 – 200
near	50,000-26,000	200 – 380
Visible	26,000 – 12, 800	380 – 780
Infrared		(μm)
near	12,800 – 3,330	0.78 – 3.0
mid	3,330 - 350	3.0 – 29
far	350 – 30	29 – 350
Microwave	30 – 0.01	350 – 1,000,000

The frequency of oscillation of atoms and molecules about their equilibrium position is 10^{13} to 10^{14} cycles/sec or Hz. Since, the frequency of infrared radiation is also in this range, it promotes the transitions in a molecule between rotational and vibrational energy levels of the ground state.

There are two types of bond vibration in simple molecules stretching and bending. In stretching mode, the bond between two atoms A and B in a diatomic molecule stretch periodically along bond axis, while in bending mode the displacements occur at right angles to the bond axis of A-B. These vibrations produce displacement of atoms with respect to one another, causing a periodic change in interatomic distance. When the vibrations (bending and asymmetric stretching) are accompanied by a change in dipole moment, causing net change in displacement of atoms, they give rise to absorption of infrared radiation.

Figure 13.1 depicts the stretching and bending of triatomic molecule CO_2. In stretching O atoms move toward and away from each other along the line joining their centers, either symmetrically or asymmetrically, and in bending the angle between three atoms alternately increases and decreases.

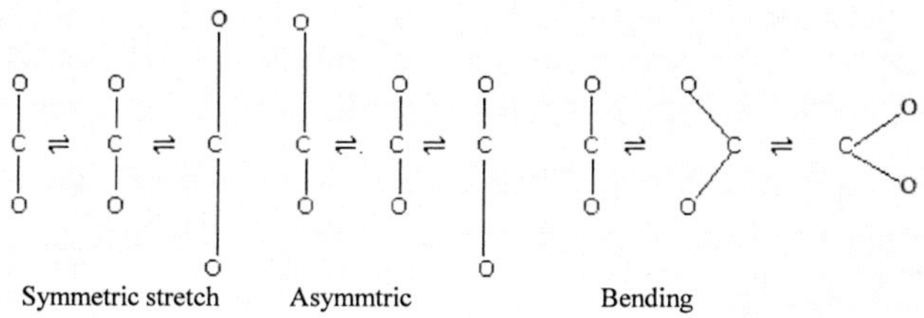

Fig. 13.1: Various vibrational mode of CO_2 molecule

13.2 Infrared Spectrometers

Commercially available infrared spectrometers are of two types: (a) the sequential dispersive spectrometers and (b) multiplex nondispersive interferometer spectrometer. The major components of the sequential dispersive spectrometers consist of a source of radiation, monochromator and filter, and detector. In this instrument, the source of infrared radiation either the Globar (a bonded silicon carbide rod operated at 1200°C) or the Nernst Glower (a mixture of rare earth metal oxide e.g., Zr, Y and Th operated at 1900°C) provides a continuous high energy radiant energy in infrared region (4000 to 100 cm^{-1}). After passing through the sample cell, the radiation is dispersed by the system of gratings and filters (the components of monochromator). The energy of the narrow frequency range passing through the slit of monochromator is measured by the detector, which transforms the received energy into electrical signal. The detectors used to measure the amplified electrical signal include thermocouple, bolometer, Golay or pneumatic detector, and photon detector. Spectra are plotted as absorbance as a function of wavenumber. Development of microprocessor based computerized infrared spectrometer has increased the sensitivity of technique by many folds.

In contrast to sequential dispersive spectrometer, the multiplex nondispersive spectrometer, typified by Fourier transform infrared spectrometer (FT-IR) can utilize the infrared energy more efficiently than the sequential dispersive infrared spectrometer and use all frequencies from the source simultaneously rather than individually. An FT-IR spectrometer basically consists of two parts: an optical system that uses an interferometer, and a computer. The computer controls optical components collects, stores and analyses data and displays spectra. Rapid scanning capability of complete spectral region makes it well-adapted for study of transient systems. The performance of FT-IR has enhanced further by interfacing it with instruments like gas chromatograph and HPLC.

13.3 Organic Spectra

The absorption spectrum of most organic molecules is simplified due to the fact that certain groups of atoms vibrate with same frequency irrespective of molecule to which they are attached. This relative consistency of group frequency is because of uniformity in bond strength from molecule to molecule. These absorption spectra can be used for characterization and identification of functional groups in organic molecules (Table 13.2).

Table 13.2 Characteristic infrared absorption frequencies of humic substances

Frequency (cm^{-1})	Assignments
3400-3300	O-H stretching, N-H stretching (trace)
2940-2900	Aliphatic C-H stretching
1725-1720	C=O stretching of COOH and ketones (trace)
1660-1630	C=O stretching of amide groups (amide I band), quinone C=O and/or C=O of H-bonded conjugated ketones
1620-1600	Aromatic C=C, strongly H-bonded C=O of conjugated ketones (?)
1590-1517	COO^- symmetric stretching, N-H deformation + C=N stretching (amide II-band)
1460-1450	Aliphatic C-H
1400-1390	OH deformation and C-O stretching of phenolic OH, C-H deformation of CH_2 and CH_3 groups, COO^- antisymmetric stretching
1280-1200	C-O stretching and OH deformation of COOH, C-O stretching of aryl ethers
1170-950	C-O stretching of polysaccharide or polysaccharide-like substances, Si-O of silicate impurities

(*Source*: Stevenson, 1982)

13.4 Inorganic Spectra

In contrast to organic molecules, use of infrared spectrometer in soil mineral identification is somewhat restricted due to their order-disorder effects and compositional variations in these complex systems. However, careful studies of well-characterized families of minerals have provided the sound basis for relating structural and compositional variation to spectral characters. Table 13.3 summarizes the assignment of absorption frequency band to functional groups common in soil minerals and other selected inorganic compounds.

Table 13.3: Characteristic infrared group absorption frequencies found in minerals and other inorganic compounds

Frequency (cm^{-1})	Assignments
	O-H vibrations
3700	free O-H stretch
3675-3540	O-H stretch
3390-2500	bonded O-H stretch

1700-1610	H-O-H bending
	O-H librations (dioctahedral)
950-915	Al_2OH
-890	$Fe^{3+}AlOH$
-840	MgAlOH
-800	$MgFe^{3+}OH$
-800	$Fe^{2+}Fe^{3+}OH$
	Si-O vibrations
1100-970	Si-O-Si antisysmmetric stretch
800-600	Si-O-Si symmetric stretch
540-400	Si-O + misc. vibrations
	NH_4^+ vibrations
3330-3030	NH-stretching
1485-1390	NH-deformation
	CO_3^{2-} vibrations
1490-1410	asymmetric stretch
1085-1050	symmetric stretch
875-860	out-of-plane bend
750-680	in-plane bend
	SO_4^{2-} vibrations
1180-1100	stretching
680-580	bending
	PO_4^{3-} vibrations
1100-1000	antisymmetric stretch
635-500	bending

The particle size has a pronounced effect on getting satisfactory infrared spectra, because the range of wavelength usually used (2.5 to 50 μm) for this study can only accommodate clay size fraction (<2 μm). Silt and sand fractions need to be reduced in particle size by careful grinding.

Clay minerals identification is based on the ratio of silica tetrahedral sheet to alumina octahedral sheet (whether 1:1 or 2:1) and occupancy level of octahedral cations (dioctahedral or trioctahedral). In 2:1 minerals (e.g., smectites, mica) the hydroxyls associated with octahedral cations are located only at the base of hole in the tetrahedral surface of clay (known as inner hydroxyl). On the other hand, in 1:1 minerals (kaolinite, serpentine) hydroxyls are located on the surface of crystal in addition to those common to the 2:1 minerals. Thus, 2:1 minerals are characterized by the strong absorption band in 3600 cm^{-1} region, while 1:1 minerals give rise to four distinct absorption bands.

Again, in trioctahedral minerals the orientation of O-H bond axis is normal to 001 plane of clay layer and with decrease in the angle of oriented sample with respect to the incident beam the hydroxyl absorption increases. But in dioctahedral minerals the O-H band axis is at a low angle to the 001 plane and thus exhibits no change in hydroxyl absorption (3620 cm^{-1}) with change in the angle of oriented sample.

13.5 Sample Pretreatment

13.5.1 Humic Substances

The major components of soil organic matter are humic acid, fulvic acid, and humin. Humic acid and fulvic acid both are soluble in dilute alkali and thus dilute NaOH is a logical choice as an extractant. Though water is a natural choice of solvent for many soluble organic and inorganic phases in soils, use of water as an infrared solvent is limited because of the following reasons:

(i) the absorption spectrum of water covers the major part of normal infrared region, 4000 cm^{-1} to 200 cm^{-1}, and

(ii) the common cell window materials, such as NaCl and KBr are extremely soluble in water and hence readily attacked by aqueous solution.

Thus, humic materials after separation and purification are usually freeze-dried for infrared analysis.

13.5.2 Mineral and Inorganic Phases

For x-ray diffraction analysis soils are usually subjected to pretreatment to remove organic matter, oxides of Fe and Al and other cementing agents to enhance distribution, and to get improved appearance and resolution of x-ray diffraction patterns. These pretreatments may create artifacts as well as the change of properties of original mineral and may thwart the capabilities of infrared spectrometry to detect the amorphous and poorly crystalline phases. Therefore, infrared spectrometry may be used for preliminary examination of clay fraction or whole soil after proper grinding before removal of organic matter and other cementing agents for dispersion. For infrared analysis larger size soil particles (> 2μm) requires grinding to reduce their size in order to avoid the distortion of absorption bands due to light scattering by mineral surface (*Christiansen effect*).

Particle size reduction can be done by three methods: (i) filing, (ii) grinding, and (iii) sonification. Filing is accomplished by using a small glass file and making a filing motion perpendicular to the planar surface of layer silicate minerals. Various size particles are then separated out by sedimentation method. Filing causes minimum structural modification in layer silicates.

There are two techniques of grinding: dry grinding and wet grinding. Dry grinding done either by mortar-pestle or by ball-milling is not suitable for particle size reduction in infrared analysis since it causes destruction of mineral structure as well as modification of physical and chemical properties. In wet grinding, 10-15 mg of sample is placed in agate mortar, to which 10-15 drops of ethanol or acetone is added to reduce the friction and pressure during grinding. The sample is then ground with a vigorous rotary motion until ethanol/acetone evaporates completely. Do not continue grinding after the sample becomes dry. Grinding process may require several times repetition to reduce particle size to < 2 μm.

Particle size reduction may also be achieved by sonification - treating the suspension of mineral particles with ultrasonic vibration. The frequency and time period for treatment are adjusted through experimentation to minimize any structural modification.

13.6 Sample Presentation

13.6.1 Dispersion Method

The commonly employed technique for infrared study involves absorption of infrared radiation by the sample evenly dispersed in a continuous optical medium placed directly in the infrared beam. The use of optical medium such as an oil having refractive index near to that of sample (Nujol or Fluorolube) or KBr minimizes the problem of light-scattering at particle-air interfaces. In one of the common methods used for sample presentation, i.e. in mull method, nonvolatile, weakly absorbing Nujol or Fluorolube is mixed with the solid sample. Since, Nujol absorbs strongly in the 3000 to 2800 cm^{-1} and in 1460 to 1375 cm^{-1} region during analysis of organic components these interferences must be taken into consideration. In KBr-pellet technique, finely ground solid sample is mixed with spectroscopically pure KBr and pressed in an evacuable die to make it transparent disk.

13.6.1.1 Alkali Halide Pressed-Pellet Method

Reagents

- Spectroscopic grade potassium bromide (KBr)
- Acetone (CH_3COCH_3) or carbon tetrachloride (CCl_4)
- Agate mortar and pestle (50 or 65 mm outer diameter)
- Die for making pellets. The die consists of a ram and anvil enclosed in such a way that when they are pressed together using hydraulic press, sample kept between them compressed into a pellet or disk. Evacuation of die by vacuum pump removes entrapped air and moisture to yield a better pellet.

- Laboratory hydraulic press for creating pressure on the sample equivalent to 9,000-11,000 kg total force on a ram of a 13 mm diameter pellet-pressing die.
- Vacuum pump

Procedure

- To prepare a pellet of 1 mm thickness and 13 mm diameter, take 0.5 to 3 mg of sample and 300 mg of dry KBr.
- Add small amount (5 to 10 mg) of dry KBr (dried at 105°C for 12 hours) and mix with sample in the mortar by gentle rubbing by pestle, but don't grind the KBr.
- Again, add 15 to 30 mg of KBr and mix as before. Continue addition and mixing of KBr until all the KBr is added.
- Transfer the mixture to the die for pressing the pellet.
- Place small lower plunger of the die at the bottom of the die body. Pour sufficient amount of mixture (sample plus KBr) into the bore of die and evenly distribute the powder by lightly shaking the die.
- To complete distribution, slowly insert large upper plunger or ram and rotate it for few times using light finger pressure to hold it against the powder. Withdraw the plunger very slowly.
- Place the small upper plunger on the top of the compacted powder, insert large upper plunger and place the die in a hydraulic press.
- Apply very slight pressure on the die so as to produce a good air seal.
- Attach the vacuum pump and remove air from the die for 5-10 minutes.
- Now, apply pressure equivalent to 9,000 to 11,000 kg total force to press the compacted powder and maintain for 10-15 minutes.
- First, slowly release the vacuum and then the pressure on the hydraulic press.
- Remove the pellet from the die by forcing the plunger to extrude the pellet from the bottom of the die.
- Don't touch the pellets with the hands. Transfer the pellet in a suitable sample-holder by using tweezers and introduce it in the infrared beam for analysis. If necessary, pellets may be stored in desiccators over a zeolite desiccant.

- Finally, clean the die assembly first with water, and then with acetone or CCl_4.

13.6.1.2 Mull Method

Materials

- Nujol, Fluorolube LG 160 or other mulling medium
- Carbon tetrachloride (CCl_4)
- Agate mortar (50 or 65 mm outer diameter) and pestle

Procedure

- Take about 10 to 15 mg of the powdered sample in the agate mortar.
- Add a small drop of the mulling agent to the mortar.
- Grind the sample with vigorous rotary motion to bring all the material in the suspension. If required, add another small drop of mulling agent to make the consistency of the final mixture like vaseline.
- Transfer the mull from the mortar to an infrared-transparent window (ZnS, AgCl, NaCl, etc) by using rubber policeman or microspatula.
- Place a second window on top of the first and distribute the sample evenly between the plates. Adjust the concentration of the mull by gentle squeezing out the excess mull. The film between the windows should appear slightly translucent and free from air bubble (examine the film against the light).
- Finally, clean the windows, mortar and pestle with CCl_4.

13.6.2 Film method

In cases where successive observations (after adsorption and desorption) are required on a single sample film method is preferred to mull or pressed-pellet method. Film method may be of two types: *sedimented film*, and *self-supporting film*.

In sedimented film method finely ground sample (< 2 µm) is dispersed in water or volatile organic liquid (e.g. isoproponol). An amount of suspension sufficient to give a concentration of 1 to 2 mg/cm^2 is placed on infrared transparent windows, such as AgCl, NaCl or ZnS and allowed to dry. Microscope glass slides and cover glass may be used as window materials for exploratory scanning in region of most OH-stretching frequencies (4,000 to 3,000 cm^{-1}). To compensate the OH absorption bands of glass, use duplicate glass slide or cover glass from the same lot; place the pair of matched slides in the sample and reference beam and record the spectrum. Light scattering effect arising from particle-air interfaces can be minimized by putting a drop of Nujol on the irradiated sample.

Self-supporting films of many smectites, vermiculites and fibrous clays may be formed by evaporating aqueous suspension on foil or plastic films and separating the mineral film by pulling the foil or plastic sheet over a sharp edge. The mineral films should have the concentration of 1 to 2 mg/cm^2. Total absorption of the infrared beam may occur in the region of wavenumber 1,100 to 970 cm^{-1} due to strong Si-O absorption bands.

Self-supporting films of amorphous alumino-silicate gels can be prepared by pressing the powdered sample between polished steel faces at a pressure of about 10,000 psi (68.9 MPa). These films are very fragile, so must be handled with care.

13.7 Identification of Functional Group and Qualitative Analysis of Organic Compounds

Reagents

See section 13.6.1.1 page no 129

Procedure

- Extract, fractionate and purify the desired organic components following the procedure as described in section 19.3, freeze-dry and store the sample in vacuum desiccators over activated Na-A type zeolite.
- Prepare sample by alkali halide pressed-pellet method (13.6.1.1) or by mull method (13.6.1.2). Use of split mull in which fluorolube is used for 4,000 to 1340 cm^{-1} region and Nujol for frequencies lower than 1340 cm^{-1} produce a combined spectrum without interference from the mulling liquids.
- Dispersive spectrometer with computer processing capability and FT-IR instruments are capable of recording the spectra of aqueous solution of humic substances through computer subtraction of water bands.

Interpretation of Spectra

- Divide the spectrum into the characteristic functional group region (4,000-1500 cm^{-1}) and the fingerprint region (below 1500 cm^{-1}).
- Examine the strongest absorptions first followed by medium ones. Refer to Table 12.2 for information on major groups.

Stevenson and Goh (1971) classified the infrared spectra of humic substances into three general groups:

Type I. Humic acids: Strong bands are present at 3400, 2900, 1720, 1600, and 1200 cm^{-1}. Intensities of 1600 and 1720 cm^{-1} are approximately equal.

Type II. Low molecular weight fulvic acids: the intensity of 1720 cm^{-1} absorption band is much stronger than 1600 cm^{-1} and the latter is centered near 1640 cm^{-1}.

Type III: *Bands for protein and carbohydrates*: Strong bands near 1540 cm^{-1} are present in addition to major absorption features typical of Types I and II. Absorption bands near 2900 cm^{-1} is also more pronounced.

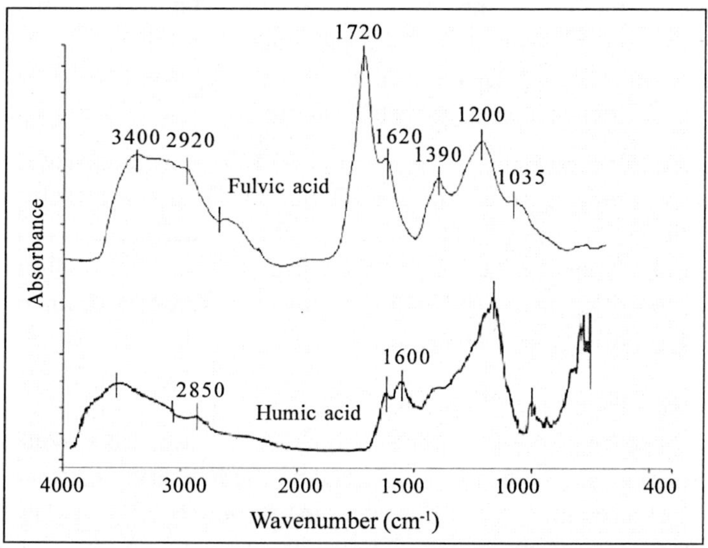

Fig. 13.1: Infrared spectra of humic acid and fulvic acid

Using the above criteria and examining the location and relative intensities of absorption bands in the spectrum of the sample one can determine the types of humic substance present. However, for more definitive information chemical treatments such as acid hydrolysis, complexation, and chemical modification such as methylation, acetylation and saponification are required. Not only infrared spectrometry is used for identification and characterization of isolated organic compounds, it can also be used for identification of functional groups involved in the adsorption of organic compounds on clay. Adsorption reactions perturb the absorption bands of both the clay (adsorbent) and organic molecules (adsorbate). The spectrum of the adsorbed organic molecules can be obtained by placing an amount of clay equal to that in the clay-organic complex in the reference beam, while the clay-organic mixture in the sample beam.

13.8 Identification and Characterization of Amorphous and Crystalline Inorganic Phases

Reagents

See sections 13.6.1.1 and 13.6.1.2 (Page no 129 and 131)

Procedure

See sections 13.6.1.1 and 13.6.1.2 (Page no 129 and 131)

13.8.1 Inorganic Amorphous Phases

Infrared spectra are sensitive to assemblages of groups in which atoms are in specific environment (e.g., the protons in various OH group of gibbsite). Since, infrared spectrometer can detect the presence of such environment in the structure; it can successfully be used in the study of amorphous materials for which x-ray diffraction technique is unsuitable.

The presence of broad band in the region of 3400 cm^{-1} along with a relatively strong band at about 1640 cm^{-1} (H_2O bending) in the spectrum are the characteristics of amorphous materials having high specific surface. Introduction of FT-IR and dispersive instruments having spectral subtraction and data-handling capabilities makes the identification and characterization of inorganic amorphous soil minerals relatively easy.

13.8.2 Inorganic Crystalline Phases

Vibrations of layer silicates can be separated into those of the hydroxyl group, the silicate anion, the octahedral cations, and the interlayer cations. The high-frequency OH-stretching vibrations occur in the 3400 to 3750 cm^{-1} region, while OH-bending vibrations occur in the 600 to 950 cm^{-1} region. Again, Si-O stretching frequencies lying in the 700 to 1200 cm^{-1} region are only weakly coupled with other vibrations of structure; but, Si-O bending vibrations occurring between 150 and 600 cm^{-1} are strongly coupled with vibrations of octahedral cations as well as with translatory vibrations of OH groups. However, interlayer cations have vibrations between 70 and 150 cm^{-1}.

Identification of minerals in the sample may be done by referring to Table 13.3 to check for the presence of specific minerals. The spectrum is then compared to the spectra of the reference clay minerals thought to be present (Fig 13.2).

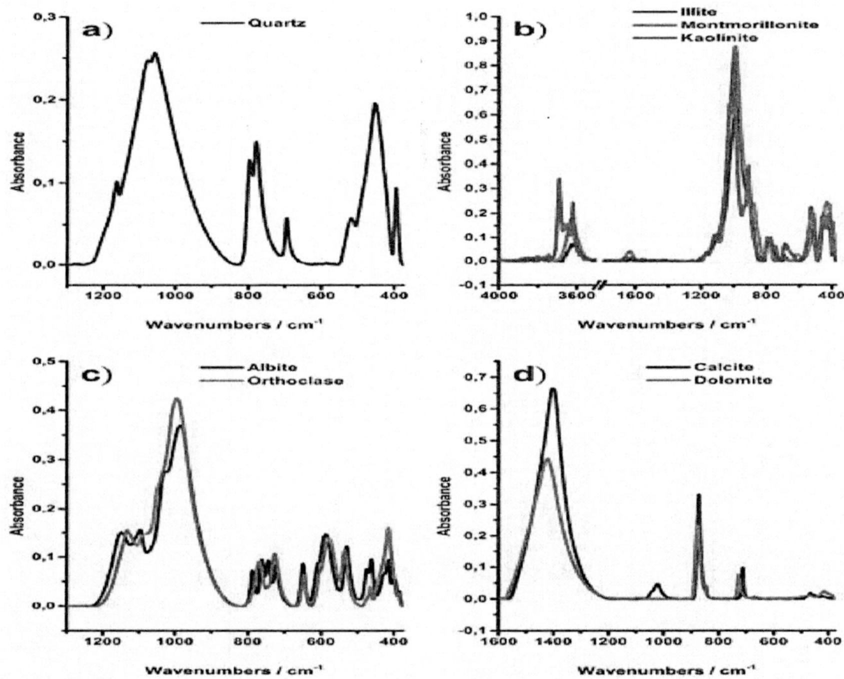

Fig. 13.2: Infrared spectra of some soil minerals

To determine the orientation of structural OH groups in crystalline clay minerals, prepare a sample in the form of film (see section 13.6.2). First record the spectrum of OH-stretching with the film in normal orientation, i.e. perpendicular to the incident infrared beam. Now, rotate the film on its vertical axis so that the angle between the plane of film and the incident beam is about 45° and record the spectrum again. An increase in the intensities of OH-stretching frequencies upon the change in angle signifies that the orientation of OH-bond axis is normal to 001 plane of the clay layer. This variation in the intensity of OH-stretching frequency with increasing angle of incidence for the clay minerals of 2:1 groups indicates the trioctahedral character. The 3620 cm^{-1} band intensity in 1:1 or 2:1 clays is independent of orientation suggesting that the OH groups are directed towards vacant octahedral position which is the character of dioctahedral minerals. However, the hydroxyls of the hydroxylic surface of 1:1 type minerals are normal to the plane of the tetrahedral sheet, and thus the band intensity at 3695 cm^{-1} increase with increasing angle of incidence.

For samples in which the major component is 2:1 type clay, inference regarding octahedral composition can be drawn by comparing the observed OH-stretching and OH-bending frequencies (Table 13.1 & 13.2).

Table 13.3: Infrared absorption bands for selected soil minerals

Mineral	Wavenumber (cm^{-1})																					
Kaolinite	3695	3670	3650	3620							1108		1038	1012	940	915		700		540		472	
Halloysite	3695			3620		3400†		1640‡			1100		1040	1020		918		695		545		474	
Montmorillonite				3620		3400†		1640‡			1100		1040	1020		915				520		470	
Nontronite					3560	3400†		1640‡			1130		1050			827						490	430
Muscovite				3628							1120			1020	928	828	750			535		480	
Biotite			3658		3550									1000			760	690				465	445
Vermiculite					3550		3380†	1640‡							985	812			670			480	
Chlorite(Al rich)			3620	3520		3340							1004		940	825	692		528		475		
Gibbsite				3610	3525	3445	3395				1102		1030		975	800	745		670	560	540		
Quartz										1172		1084				800	780	697		512		462	
Microcline											1110		1030	1000			769	727	647				
Calcite									1435							877		712					

†OH-stretching frequency of water molecules
‡OH-bending frequency of water molecules

Questions and Hints

Q.1 Why for infrared spectrometry sand and silt fractions need to be reduced in particle size?

Hint: For obtaining satisfactory absorption spectra, the size of particle should be less than the range wavelength used for the study. As the infrared spectra used range from 2.5 to 50 μm, sand and silt particles need to grind to reduce their size.

Q.2 Why in alkali halide pressed-pellet method KBr is not ground for mixing with the sample?

Hint: Grinding of KBr will lead to adsorption of water.

Q.3 Why minerals and organic components of soil cannot be quantitatively analyzed with precision?

Hint: (i) For any precise quantitative analysis homogeneity in sample is essential. Solids, more specifically minerals exhibit extreme heterogeneity that restricts its quantitative estimation by IR.

(ii) Light scattering effect of mineral particle limits the application of quantitative relationship between concentration of mineral and infrared radiation intensity.

(iii) Lack of standard reference minerals having same structure, particle size distribution and spectral feature further limit the quantitative determination with infrared spectrometer.

14

X-Ray Diffraction Analysis

14.1 Principles of X-ray Diffraction

Any crystalline structure is characterized by a regular and systematic arrangement of atoms (or ions) in a three-dimensional array. Since crystals are composed of regularly spaced atoms or ions, each crystal contains planes of atoms or ions separated by a constant distance which is the characteristic of that crystalline species.

X-rays are essentially electrostatic and electromagnetic fields oscillating in planes perpendicular to each other and to the direction of propagation in periodic cycles. X-rays having wavelengths in the order of 10^{-3} to 10^{1} nm, are produced within a vacuum x-ray tube by bombardment of a metal anode with high-velocity electrons. High-velocity electrons transfer their energy to the electrons of metal anode causing their elevation to higher energy level (excited state). Excited electrons of metal atom soon come back to the vacant shell and each electron transfer from higher to lower energy state results release of a quantum of energy (x-ray photon) equivalent to the difference in energy between these two levels. The energy and the wavelength of emitted x-ray photons are finite and characteristic of the particular atoms of the metal anode used in x-ray tube. Usually, photons generated due to electron transfer from L-shell to K-shell are used for x-ray diffraction analysis.

Diffraction of x-ray involves scattering of x-rays by planes of atoms of the crystal separated by definite distance and reinforcement of scattered rays in definite direction away from the crystal. Reinforcement of scattered rays is quantitatively related to the distance of atomic planes in a crystal as defined by Bragg's equation:

$$n\lambda = 2d \text{ Sin } \theta \quad (14.1)$$

When a collimated beam of monochromatic x-ray of wavelength λ strikes a crystal, the rays penetrate and are partially scattered from many successive crystal planes. For a given interplanar spacing, there will be a critical angle, θ, at which scattered rays emanating from successive crystal planes will be in phase, *i.e.* will reinforce (magnify) one another as they leave the crystal.

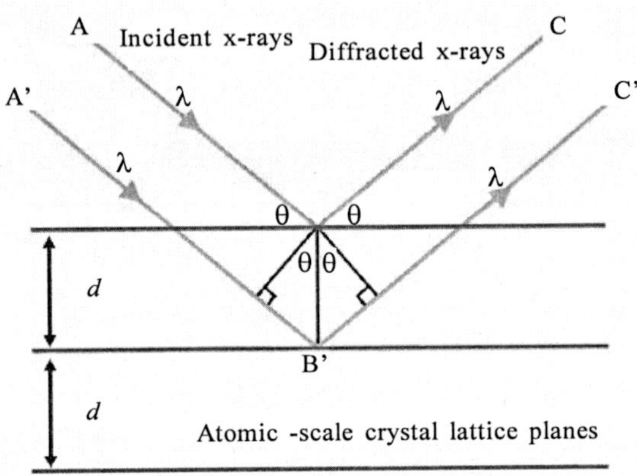

Fig. 14.1 X-ray reflection from crystal planes

The incident x-ray wave front AA' is reflected from the successive layers. Let at the angle of incidence of x-ray, θ (glancing angle) the beam reflected from 1st and 2nd lattice planes will reinforce each other when the reflected beams are in phase i.e., when their path difference, (A'B' + B'C') – (AB + BC) or the difference between A'B'C' and ABC is an integral multiple of λ.

Thus, the conditions of the reflected waves are in phase and interfere constructively, when θ satisfies the Bragg's equation, $n\lambda = 2d \sin\theta$.

Where n = 1, 2, 3———, *etc*, are called first order, second order, third order diffraction of x-ray and so on. Although d remains the same, the second order, third order diffraction will occur from planes of d/2, d/3 spacing, respectively. The wavelength of radiation is characteristic and constant for particular x-ray tube used (*i.e.* metal anode). Therefore, the angle of incidence of primary radiation, θ, can be varied with the crystal planes.

With an x-ray spectrophotometer, the angle of incidence is varied by rotating sample in the path of primary x-ray beam. A suitable detector is also moved in such a way so that it maintains the angle with the sample which is equal to the angle of incidence of primary beam at all positions. The detector (Geiger, proportional, or scintillation counter) intercepts and measures the diffracted rays. The value of 2θ (with reference to primary beam) is available directly from the chart of a direct recording x-ray spectrophotometer. Since no two minerals have exactly the same interlattice distance in three dimensions, the angle at which diffraction occurs is unique for a particular mineral.

Efficiency of diffraction from a mineral sample depends on the wavelength of radiation used and physical and chemical nature of the sample. Therefore, wavelength of radiation should be selected properly and sample should be prepared carefully to get maximum diffraction intensity and quality.

14.2 Preparation of Sample

For identification of mineral species, sample may have to be analyzed several times with various types of pretreatment. Again, the number of analyses, type of pretreatment and manner of feeding of sample to the x-ray diffractometer depend upon the particular mineral combination present in the sample. However, to ensure detection of mineral species in soil or clay sample, fractionation of samples according to particle size is necessary. Segregation of size fraction in the polycomponent mineral sample is accomplished by dispersion mostly through removal of flocculating and aggregate-cementing agents.

14.2.1 Dispersion of Sample

A. Removal of Carbonates and Soluble Salts

Reagents

- Sodium acetate ($CH_3COONa.\ 3H_2O$), 0.5 M: Dissolve 68 g of sodium acetate in 1 litre of distilled water and adjust to pH 5.0 with acetic acid.

Procedure

- Place a suitable amount of dried processed (<2 mm) soil sample into a dialysis chamber, one end of which is tied with a rubber band.
- Add several hundred ml of Na-acetate buffer solution.
- Tie the top of the dialysis membrane around a glass 'breather' tube (approx. 10 cm long) and hang the sample in a reservoir of Na-acetate buffer solution contained in a plastic or glass container (6 L of buffer solution serves well for 100 g sample).
- Knead the membrane after several days when bubble of CO_2 will no longer be evident on kneading (if carbonates are still being dissolved, bubbles of CO_2 will be released).
- Open the dialysis membrane and check some of the coarser particles with strong acid for presence of carbonates.
- Transfer the sample (still in the dialysis membrane) to another container, and desalt it against tap water flowing continuously through the container.

- Check the salt concentration inside the membrane by conductivity measurement on small volume of the supernatant liquid poured out through the 'breather' tube.
- Continue dialysis until salt concentration drops below 10 meq/L.
- Remove excess water in the membrane with a filter candle.
- Transfer the sample without drying and with a minimum of distilled water to a beaker for removal of organic matter.

B. Removal of Organic Matter

Reagents

- Hydrogen peroxide (H_2O_2, 30%)
- Magnesium chloride ($MgCl_2 \cdot 6H_2O$), 0.5M: Dissolve 102 g magnesium chloride hexahydrate in 1 litre of distilled water.

Procedure

- Cover the beaker containing carbonate and salt-free soil suspension with a ribbed watch glass and place it on a hot plate to reduce the volume to soil: water ratio of 1:1 or 1:2. Remove the beaker and allow it to cool. [If the sample requires no treatment prior to removal of organic matter, place the dried, processed soil sample directly into a beaker and add water to give a soil: water ratio 1:1 or1:2 and cover the beaker with ribbed watch glass].
- If necessary, make the sample acidic to litmus with a few drops of 1 N HCl.
- Add 30% H_2O_2 in increments of 5-10 ml, stir the suspension, and allow time for any strong effervescence or frothing to subside.
- Control the reaction that is too vigorous by cooling the beaker in a water bath. Continue adding H_2O_2 in small amounts until sample ceases to froth, then transfer the beaker to a hot plate at low heat (65-70°C) and observe for 10-20 minutes or until danger of any strong reaction has passed.
- Add additional H_2O_2 in amounts to make approximately a 10% solution. Evaporate excess liquid between additions of H_2O_2 to maintain soil: water ratio of 1:1 or 1:2.
- The reaction of soil with H_2O_2 is completed when soil loses dark colour or effervescence.

- Transfer the sample with an aid of a powder funnel and wash bottle to a centrifuge tubes and centrifuge at 1600-2200 rpm for 10-15 minutes and discard the supernatant.
- If clay remains suspended, add few drops of 0.5 M $MgCl_2$, fill the tube approximately 1/3 or less with distilled water, stopper it. Jar the soil loose from the walls of the tube by striking the bottom of the tube on a large rubber stopper.
- Shake for 5 minutes in reciprocating shaker. Wash adhering particles from the wall of the tube with a fine jet of distilled water and centrifuge for 10-15 minutes at 1600-2200 rpm.
- Floccule the suspended clay by addition of few drops of $MgCl_2$. Now, the sample is ready for removal of free iron oxide, if any.

C. Removal of Free Iron Oxide

Reagents

- Trisodium-citrate dihydrate ($C_6H_5O_7Na_3 . 2H_2O$), 0.3 M: Dissolve 88 g of sodium citrate dihydrate in 1 litre of distilled water.
- Sodium bicarbonate ($NaHCO_3$), 0.5 M: Dissolve 42 g of sodium bicarbonate in 1 litre of distilled water.
- Sodium dithionite ($Na_2S_2O_4$)
- Sodium chloride (NaCl), saturated
- Acetone (reagent grade)

Procedure

- Transfer a suitable amount of sample (4 g soil or 1 g clay) to a 100 ml centrifuge tube that has already been treated to remove organic matter, soluble salt and carbonates.
- Add 40 ml of 0.3 M Na-citrate solution and 5 ml of 0.5 M Na-bicarbonate to the sample. Warm the suspension to 80°C in a water bath, and then add 1 g of solid Na-dithionite.
- Stir the suspension for 1 minute continuously and then occasionally for 15 minutes. Avoid heating above 80°C, otherwise FeS forms.
- Following 15 minutes digestion, add 10 ml of saturated sodium chloride. If suspension fails to floccule with NaCl, add 10 ml of acetone.
- Mix the suspension, warm it on a water bath to expedite flocculation.

- Centrifuge the tubes for 10-15 minutes at 1600-2200 rpm.
- Wash the sample finally with Na-citrate solution (with NaCl and acetone, if necessary for flocculation). The sample is now ready for fractionation.

14.2.2 Separation of Particle-size Fractions

For an x-ray diffraction analysis soil clay samples are customarily sub-fractioned into: (i) coarse clay (2-0.2 μm), medium clay (0.2-0.08μm), and fine clay (< 0.08 μm). However, for most of the practical purposes clay samples are separated into two sub-fractions (2-0.2 μm and < 0.2 μm). Further separation of fine clay (< 0.08 μm) from medium clay (0.2-0.08μm) is required if the soil contains very small amount of smectite.

The intensity of diffraction from a particular mineral species is influenced by number of factors including crystal size, crystal perfection, besides its concentration in soil. Since, finer fractions of clay always yield weaker diffraction and greater crystal imperfection than larger particle; each fraction should be analyzed separately. Subdivision of clay fraction is done by differential sedimentation of particles under centrifugal force. Time required for sedimentation of different particle size fractions is calculated using integrated form of Stokes' law as proposed by Svedberg and Nichols (1923).

$$t_s = \frac{\left[\eta \log_{10} \frac{R}{S}\right]}{\left[3.81 N^2 r^2 (\Delta S)\right]} \tag{14.2}$$

Where t_s is the sedimentation time in seconds, R is the radius in cm of rotation of the top of the sediment in the tube (Fig. 14.2), S is the radius in cm of rotation of the surface of the suspension in the tube, N is the revolution per second, r is the particle radius in cm, η is the viscosity in poises at the existing temperature, and ΔS is the difference in specific gravity between solvated particle and the suspension liquid.

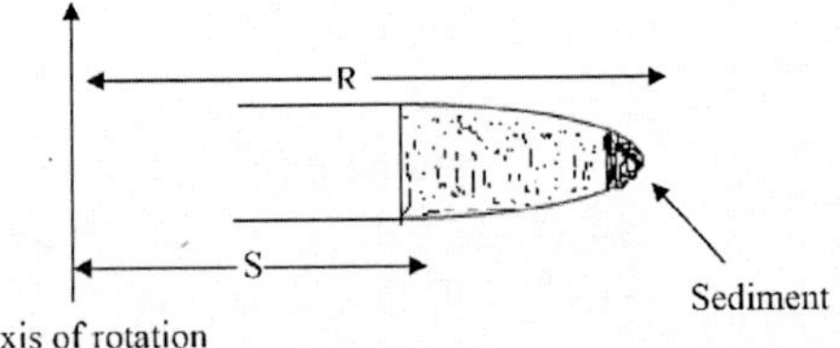

Fig. 14.2

For convenience unit of time is expressed as minutes, t_{min}, N_m as rpm, and particle diameter in microns, D_u, thus the equation (13.2) will be transformed into:

$$t_{min} = \frac{\left[63.0\times10^{8}\eta\log_{10}\frac{R}{S}\right]}{\left[N_m^{\,2}D_u^{\,2}(\Delta S)\right]} \tag{14.3}$$

Table 14.1: Sedimentation time at various centrifuge speeds and temperatures of particles in International No.2 centrifuge with No.240 head and 100 ml tubes

Particle limiting diameter (μm)	Specific gravity of particle sedimented	Centrifuge speed (rpm)	Sedimentation of time* (min) required at suspension temp. of			
			20°C	25°C	30°C	35°C
5	2.65	300	3.3	2.9	2.6	2.4
2	2.65	750	3.3	2.9	2.6	2.4
0.2	2.50	2400	35.4	31.4	28.3	25.4

* Calculated with 1 cm sediment, R= 25 cm and S = 16 cm (10 cm is suspension depth, therefore 9 cm is the net depth of fall)

Equipments and Materials

International Centrifuge No. 2 with speed indicator, timer and a No. 240 head, super centrifuge designed for high speed (25,000-50,000 rpm) equipped with cellulose acetate liner, 100 ml centrifuge tubes fitted with rubber stoppers, glass rod with rubber ball plunger, glass rod with rubber policeman.

Reagents

- Sodium carbonate (Na_2CO_3) solution, pH 9.5: Dissolve 2 g Na_2CO_3 in 18 litres of water.
- Sodium chloride (NaCl)
- Methanol, 95%

Procedure

(i) Separation of Silt plus Clay Particles (< 20 μm) from Soil Suspension

- Transfer the pretreated (after removal of organic matter, soluble salts and free iron oxide) soil suspension to a 250 ml tall beaker (for large quantity of soil sample distribute it in several beakers to facilitate separation by sedimentation and decantation).
- Stir the suspension and allow it to stand for at least 2 minutes for each 5 cm depth.
- Mark on the side of the beaker at the point 5 cm above the top of the sediment in the bottom of the beaker with wax pencil.

- Decant the supernatant liquid (containing particles < 20 µm in diameter) into an appropriate container.
- Add Na_2CO_3 solution to the beaker to just below 5 cm mark. Stir the suspension with glass rod with rubber policeman. When the suspension is well mixed, remove the glass rod and wash with water and make the suspension to 5 cm mark.
- Wait for about 2 minutes (determine the exact time at the laboratory temperature by calculation from Stokes' law or from the nomograph given by Jackson, 1969).
- Decant the supernatant liquid into the container marked for < 20 µm particles.
- Again, add Na_2CO_3 solution, stir the suspension and decant the supernatant liquid.
- Repeat the whole process until supernatant is practically free from suspended materials (usually 5-8 repetitions are required). Sediment of > 20 µm particles (sand) may be discarded if only clay fractions are required.

(ii) Separation of Clay Particles (< 2µm) from Silt plus Clay (< 20 µm) Suspension

- Stir the suspension containing < 20 µm particles and allow it to stand for at least 8 hours for each 10 cm depth.
- Decant the supernatant liquid containing suspended <2 µm particles into an appropriate container. The sediment contains all >2µm and some <2 µm particles.
- Add Na_2CO_3 solution to the sediment, stir and transfer the suspension in 100 ml centrifuge tubes maintaining depth exactly 10 cm.
- To facilitate subsequent separation maintain the depth of sediment in each centrifuge tube (as determined after 1st centrifugation) to approximately 1 cm.
- Centrifuge the suspension at an appropriate speed and time to settle down all > 2 µm particles at the bottom of the tubes [use Eq. 14.2 or nomograph given by Jackson (1969)].
- Decant the supernatant liquid into the container marked for <2µm particles.
- Resuspend the sediment in Na_2CO_3 solution, stir to mix the suspension thoroughly, centrifuge it at the earlier speed and time.

- Decant the supernatant liquid into the container for <2μm particles.
- Repeat the whole process until complete separation of <2 μm particles from the sediment is done, as evident from clear supernatant liquid (usually 5-8 repetitions are required).

(iii) Separation of coarse clay (2-0.2 μm) from Suspension of < 2 μm Particles

- Transfer the suspension containing <2μm particles to 100 ml centrifuge tubes maintaining the depth of suspension to exactly 10 cm.
- Centrifuge the suspension at the appropriate speed and time to settle down > 0.2 μm particles at the bottom of the tubes [use Eq. 14.2 or nomograph given by Jackson (1969)].
- Decant the supernatant liquid into the container marked for <0.2μm particles.
- Resuspend the sediment in Na_2CO_3 solution maintaining exact depth of suspension, stir to mix the suspension thoroughly, centrifuge it at the earlier speed and time.
- Decant the supernatant liquid into the container for <0.2μm particles.
- Repeat the whole process until completion of separation of <0.2 μm particles from the sediment, as evident from clear supernatant liquid (usually 5-8 repetitions are required).
- After last centrifugation, transfer the sediment containing coarse clay (2-0.2 μm) in the bottom of the tube to a small flask and save for x-ray analysis.

(iv) Separation of medium clay (0.2-0.08 μm) from Suspension of <0.2 μm Particles

- Transfer the suspension containing <0.2 μm particles in the feed reservoir of super centrifuge.
- Fit a wetted cellulose acetate liner closely around the clarifier bowl interior and install in the super centrifuge.
- Place 6 liters flask under the upper delivery spout to collect effluent.
- Start centrifugation and when the desired speed is attained, start the flow of suspension through the bowl at proper rate (follow instruction manual of the manufacturer) until feed reservoir is nearly empty.

- Just before the suspension is exhausted from the reservoir, wash down the walls of the reservoir with 1 percent Na_2CO_3 solution from a wash bottle.
- When the wash solution is nearly drained away, pour 300 ml of 1 percent Na_2CO_3 solution into the feed reservoir and allow it to flush out the suspension remaining in the bowl.
- When the effluent flow ceases, allow the centrifuge to come to rest freely. Remove and open the bowl and extract the cellulose acetate liner adhering the sedimented particles.
- Transfer the sedimented particles from the bowl liner into a blender by means of spatula, scrub the residue with a policeman and flush off with Na_2CO_3 solution from the wash bottle.
- Resuspend it with Na_2CO_3 solution in a rotary mixer for 1 to 2 minutes and dilute the suspension to a concentration of $\leq$1 percent.
- Repeat the entire process of super centrifugation. Usually 4-5 repetitions (resuspension and centrifugation) are required for complete separation of 0.08 μm particles.
- After last centrifugation, collect the sedimented particles of medium clay (0.2-0.08 μm) and save for x-ray analysis.
- To the effluent containing particles of < 0.08 μm particles add sufficient NaCl (to make it approximately 1 M with respect to NaCl).
- Allow the suspension to stand until clay flocculates and settles down.
- Discard the clear supernatant liquid with a siphon.
- Transfer the sediments in 100 ml centrifuge tubes and centrifuge them at 1500 rpm.
- Decant the supernatant liquid and wash the sediment five times with 95 percent methanol to remove excess NaCl. Save the fine clay (< 0.08 μm) for x-ray diffraction analysis.

14.2.3 Saturation of Exchange Complex

Depending upon the nature of interlayer exchangeable cation, the expanding phyllosilicates can retain variable amount of interlayer water. Again, to have a uniform hydration cum expansion within all crystals of a species, the clay sample is made homoionic and selection of cation for that purpose is based on the fact that interlayer adsorption of water by air-dry sample will not change with the fluctuation in relative humidity. While magnesium provides relatively uniform interlayer adsorption of water, potassium restricts interlayer water adsorption by vermiculite.

Mg-saturated members of smectite and vermiculite groups exhibit approximately 1.4 nm inter-atomic spacing between (001) planes. On the other hand, 2:1 non expanding phyllosilicates (chlorite) and vermiculites both exhibiting 1.4 nm interlayer spacing can be differentiated from each other by saturating with K as K-saturated vermiculites will collapse to 1.0 nm, while chlorite remains unaffected.

Presence of interlayer hydroxy complexes complicates the differentiation of smectites from chlorites. Heating K-saturated sample at 550^{0}C for 2 hours destroys discontinuous hydroxy complex resulting reduction of inter-atomic spacing to 1.0 nm, but chlorite remains unaffected on heating.

Reagents

- Hydrochloric acid (HCl), 0.1 M: Dilute 86 ml of concentrated HCl to 1 liter.
- Magnesium chloride ($MgCl_2$), 0.5: Dissolve 102g $MgCl_2.6H_2O$ in 1litre to prepare 0.5 M $MgCl_2$ solution.
- Magnesium acetate [$Mg(CH_3COO)_2$], 5 M and 0.5 M: Dissolve 710g magnesium acetate in 1 liter to prepare 5M Mg $(OAc)_2$. Dilute 100 ml of 5M Mg $(OAc)_2$ to 1 liter to get 0.5 M $Mg(OAc)_2$.
- Potassium chloride (KCl), 1 M: Dissolve 74.6g KCl in 1 liter.
- Silver nitrate ($AgNO_3$), 0.1 M: Dissolve 1.7 g $AgNO_3$ in 100 ml water.
- Methanol, 50 and 95%
- Acetone, 95 and 100%
- Bromophenol blue indicator, 0.04%: Dissolve 4 mg bromophenol blue indicator in 100 ml ethanol.

Procedure

Mg-saturation of Clay

- Transfer an aliquot of dispersed clay suspension containing approximately 25 mg of clay to a beaker and add 0.1 M HCl dropwise to bring the pH of the suspension between 3.5 and 4.0 (as indicated by yellow colour in a spot-plate test with bromophenol blue).
- Add sufficient 5 M $Mg(OAc)_2$ to make the suspension approximately 0.5 M with respect to Mg.

- Transfer the suspension in 15 ml centrifuge tube and centrifuge for 5 minutes at 1500 rpm, decant and discard the supernatant.
- Complete exchange saturation with Mg by washing the sample twice with 0.5 M $Mg(OAc)_2$ and twice with 0.5 M $MgCl_2$. For each washing add 10 ml of salt solution, mix the suspension, centrifuge and decant supernatant liquid.
- After complete exchange saturation, remove excess salt from the sample by washing (centrifugation and decantation) once with 50 percent, once with 95 percent methanol, and finally with 95 percent acetone until supernatant liquid becomes free from chloride (negative test with 0.1 M $AgNO_3$).
- Do not allow to dry the sample if it is to be prepared as oriented aggregates. If it is analyzed as random powder, transfer the sample to a large watch glass with acetone and allow it to dry.

K-saturation of clay

- Transfer aliquot of the dispersed clay suspension containing approximately 25 mg of clay to a 15 ml centrifuge tube and add sufficient 1 M KCl solution to flocculate the clay.
- Centrifuge at 1500 rpm for 5 minutes the suspension and discard the supernatant liquid.
- Wash the sample (centrifugation and decantation) four times with 1 M KCl solution to complete exchange-saturation with K.
- Remove excess salt from the sample by washing (centrifugation and decantation) once with 50 percent, once with 95 percent methanol, and finally with 95 percent acetone until supernatant liquid becomes free from chloride (negative test with 0.1 M $AgNO_3$).
- Do not allow to dry the sample if it is to be prepared as oriented aggregates. If it is analyzed as random powder, transfer the sample to a large watch glass with acetone and allow it to dry.

14.2.4 Solvation with Glycerol

Species of both the groups of Mg-saturated vermiculite and smectite exhibit similar basal spacing of 1.4 nm, therefore further treatment is needed to distinguish between these two groups of minerals. While smectites adsorb double sheets of glycerol between adjacent layers to yield a basal (001) spacing of approximately 1.77 nm, vermiculites almost remain unchanged on solvation with glycerol.

Reagents

- Benzene
- Glycerol
- Benzene-ethanol, 10:1 (v/v) and 200:1(v/v)
- Benzene-ethanol-glycerol, 1000:100:4.5 (v/v)

Procedure

Glycerol-solvated Random Powder Sample

- Add 10 ml of 10:1 benzene-ethanol to an Mg-saturated sample as prepared in section 14.2.3.
- Mix the suspension thoroughly, centrifuge for 5 minutes at 1500 rpm and decant the supernatant liquid.
- Repeat benzene-ethanol washing (centrifugation and decantation) once.
- Wash the sample three times with 10 ml portions of benzene-ethanol-glycerol mixture to complete solvation with glycerol.
- Wash the sample once with 200:1 benzene-ethanol solution to remove excess glycerol.
- Resuspend the sample in benzene and transfer to a large watch glass.
- After evaporation of benzene, the sample is ready for x-ray diffraction analysis.

Glycerol-solvated Oriented Aggregate Sample

- Add approximately 2 ml of distilled water and 2 drops of glycerol to the Mg-saturated sample as prepared in section 14.2.3.

  [The correct amount of glycerol to be added depends upon the properties of expanding phyllosilicates and particle size range of the mineral particles. In most cases, the recommended amount is 2 drops per 25 mg of clay. However, the correct amount may be determined by trial and error method].

- Mix the suspension thoroughly to ensure complete dispersion. The sample is ready for mounting as an oriented aggregate (section 14.2.5).

14.2.5 Mounting

Two methods of sample mounting are most commonly used in x-ray diffraction analysis of clay sample: (i) sample either mounted as random powder or (ii) as oriented aggregate. Again, there are several ways in which sample may be mounted as random powder and as oriented aggregate. Both the methods of sample mounting have advantages for particular purposes.

In random powder sample, in which crystals lie in all possible directions, thus sufficient number of crystals oriented properly will yield diffraction maxima from all atomic planes. Therefore, random powder sample enables one to get all possible diffraction spacings from minerals present in the sample. Relative intensities of diffraction maxima obtained from random powder sample are more nearly proportional to the number of crystal present than those obtained from oriented aggregate. On the other hand, oriented aggregate sample results in enhancement of basal diffraction maxima of phyllosilicates which often permits detection of small quantities of phyllosilicate species present in the sample.

A. Rod Method for Random Orientation

Random powder samples are commonly formed into a rod of 0.3 to 0.5 mm diameter. Thin rod of desired diameter may be formed either from dried glycerol-solvated clay sample without addition of binding material or from dry clay sample with addition of $1/5^{th}$ to $1/10^{th}$ volume of binding agent like gum tragacanth and water by rolling it back and forth between two glass slides.

B. Wedge Method for Random Orientation

Powder sample can be mounted in a specially designed metal wedge. Place a small amount of clay sample on a clean glass slide and slowly push it into the recess of the wedge with the help of spatula or with the edge of another glass slide. Carefully smooth the surface of the mounted sample, so that it forms a sharp edge flush with the tip of the wedge.

C. Glass Plate Method for Oriented Aggregates

Oriented aggregate specimens may be prepared directly from Mg-saturated, K-saturated or Mg-saturated glycerol solvated sample. Add sufficient water to the clay sample to make a suspension of about 2 ml. Thoroughly mix the suspension, pipette and carefully transfer to a glass microscope slide resting on a level surface. Add as much suspension as the slide can hold by surface tension. The amount of clay in slide should be between 15 and 25 mg. Dry the suspension on the slide before being analyzed.

D. Porous Ceramic Plate Method for Oriented Aggregates

Unglazed tile of desired dimension (2.6 x 4.6 x 0.5 cm) is commonly used as porous plate. The surface is smoothened by wet-grinding on a glass plate using fine grade corundum abrasive. A dilute suspension containing between 15 and 25 mg clay is transferred to the porous ceramic plate. The liquid is excluded from the clay by applied suction or by centrifugation. After removal of water, moist oriented clay remained on the surface of porous plate is smoothened with a spatula or glass slide to improve the basal (001) x-ray diffraction maxima of phyllosilicates.

14.2.6 X-Ray Diffraction Analysis of Samples

Clay samples prepared for x-ray diffraction analysis are examined in the following order for proper identification and differentiation of mineral species: (i) Mg-saturated, air dried sample, (ii) Mg-saturated, glycerol solvated sample, (iii) K-saturated, air dried sample; and (iv) K-saturated, heated sample (550°C).

Equipments and Materials

Commercial direct reading x-ray diffractometer equipped with a high-voltage generator, an x-ray source, an x-ray beam collimating system and recording system, clay samples prepared for x-ray analyses.

Procedure

- Place the sample in the sample holder provided on the instrument and position the goniometer for its angular scan.
- [In most of the cases, lower limit for 2θ is 2° and upper limit is within 180°. The upper limit for 2θ is dictated by the geometry of the goniometer system and the mineral association in the sample].
- Start goniometer in synchronism with the chart recorder, and scan through the desired angular range.

14.2.7 Qualitative Interpretation of Results

Soils almost always contain a number of mineral species. Qualitative interpretation of results of x-ray diffraction analysis involves identification of crystalline species from the array of diffraction maxima obtained from the sample. In any particular diffraction analysis, the angle of diffraction from the diffracting planes is directly dependent upon the interplanar distances. Diffraction angles are determined directly in terms of 2° from a direct recording spectrometer pattern; and from standard conversion table one may obtain the corresponding diffraction spacing. Commonly identification may be accomplished by measurement of diffraction spacing of Mg-saturated, glycerol-solvated, K-saturated and heated samples and comparison of these spacing with spacing of known standard minerals (Table 14.2).

Table 14.2 X-ray diffraction spacing (001) in nm of some common phyllosilicates

Minerals	Mg-saturated, air dried	Mg-saturated, glycerol solvated	K-saturated, air dried	K-saturated, heated (550°C)
Vermiculite	1.4-1.5	1.4-1.5	1.4-1.5	0.99-1.01
Chlorite	1.4-1.5 (0.715)*	1.4-1.5 (0.715)	1.4-1.5 (0.715)	1.4 (0.715)
Smectite	1.4-1.5	1.77-1.80	1.24-1.28	0.99-1.01
Illite	0.99-1.01	0.99-1.01	0.99-1.01	0.99-1.01
Halloysite	0.99-1.01	1.08	0.99-1.01	-
Kaolinite	0.715	0.715	0.715	-
Serpentine	0.71-0.73	0.71-0.73	0.71-0.73	0.71-0.73

*Figure in parentheses indicate the 2nd order diffraction spacing in nm

Results of x-ray diffraction analysis indicate that diffraction spacing (001) of approximately 1.4 nm obtained from Mg-saturated, air dried sample may be attributed to either by smectite, vermiculite or chlorite or their mixture. Solvation of Mg-saturated sample with glycerol helps one to distinguish smectite from others. Heating of K-saturated sample causes collapse of interlayer hydroxy complexes of vermiculite and thus allows separation of vermiculite from smectite. Heating also destroys kaolin minerals and thus helps to separate and identify chlorite which gives a second order maximum at about same position as the first order maximum of kaolinite (0.71 nm) or serpentine (0.71-0.73 nm). Saturation with K similarly allows separation of vermiculite from chlorite, which does not collapse on heating. Illite remains unaffected with all the treatments and always gives a first order maximum at approximately 1.0 nm.

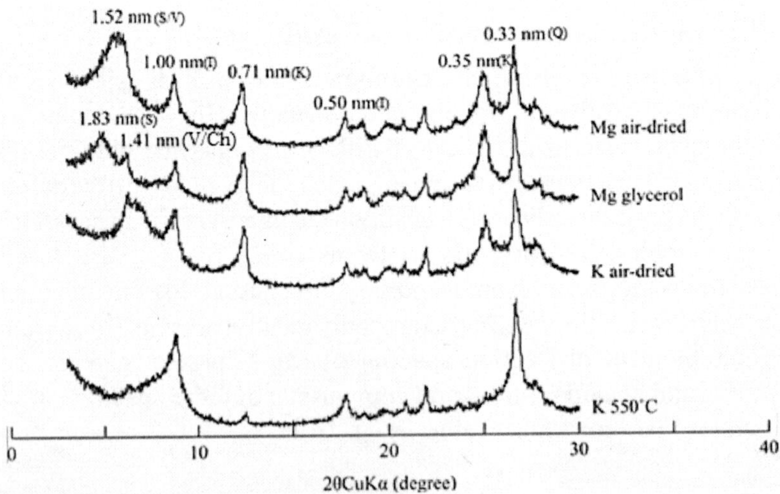

Fig. 14.2: XRD patterns of smectite (s), vermiculite (V), chlorite (Ch), iliite (I), kaolinite (K), and quartz (Q)

14.2.8 Quantitative Interpretation of Results

Direct recording X-ray diffractometer records both spacing and intensity of each maximum. Intensities of maxima recorded as counts per second, are related to the number of corresponding diffraction planes in a sample. Therefore, the relative intensities of maxima provide the basis for estimation of concentrations of mineral species present in the sample. The intensities of diffraction maxima are not only affected by the concentration of mineral species, but by the particle size, crystal perfection, chemical composition, variations in sample packing, crystal orientation and presence of amorphous substances. Some of these factors are difficult to control between reference and test sample, and makes the analysis less accurate. Therefore, the analysis is somehow semi-quantitative rather than quantitative one. For the most reliable and accurate estimation, it is advisable to use of x-ray diffraction analysis in conjunction with other methods such as DTA, infrared spectrometry, surface area, elemental analysis, and other species-specific chemical methods.

Questions and Hints

Q.1 Why separation of fine clay from medium clay is required in soils having small quantities of smectites?

Hint: Presence of specific mineral species within a specific-size fraction is a function of their resistance to weathering and intensity of weathering. Resistant mineral like smectite usually tends to persist in greater quantities in the finer clay fraction, while weathering susceptible minerals, like feldspars and micas disappear before their reach in finer clay dimension.

Q.2 Why acidification of clay is necessary before addition of Mg-solution to prepare Mg-saturated clay?

Hint: In alkaline dispersant Mg will precipitate as $Mg(OH)_2$.

Q.3 Why glycerol solvation is preferred, though smectites form complexes with other polar solvents also?

Hint: Although, smectites form complexes with monohydric, polyhydric alcohols, benzene and ethers, glycerol is preferred because of the following reasons:

(a) It gives maximum first order basal spacing (1.77 nm) with smectites, which is well separated from other maxima likely to occur in x-ray diffraction.

(b) Spacing and intensity of glycerol-smectite maxima remain unaffected by wide variation in water content.

(c) Due to less volatility glycerol-smectite complexes are extremely stable.

Q.4 What is the role of a collimator in x-ray diffractometer?

Hint: X-ray diffraction analysis requires a narrow pencil of x-rays which are obtained by passing x-rays through two slits or pinholes of a collimator. The rays which are nearly parallel-oriented can pass through both the slits and pinholes placed wide apart.

Q.5 What are the criteria for selecting wavelength of x-ray for diffraction analysis?

Hint: The choice of wavelength for diffraction analysis depends on: (i) the nature of minerals present in the sample, (ii) amount of its absorption in air, (iii) elemental composition of the sample, and (iv) the resolution required.

Q.6 What are the metals used in x-ray tube as anode target and which one is most suitable for diffraction analysis?

Hint: The metals commonly used for x-ray diffraction analyses include Mo, Cu, Ni, Co, Fe and Cr.

Among the above metals Cu (K_α, 0.154 nm) is almost universally used for this purpose because of its low air absorption factor. It is satisfactory for most soil analyses, except in soils with high concentration of Fe. However, this problem can be alleviated by removal of free iron oxides before x-ray analysis.

Chromium (K_α, 0.228 nm) suffers from relatively high absorption coefficient in air and elements like Al, Fe and Ca (usually present in relatively high concentration in most soils). The shortcoming with Mo radiation (K_α, 0.071 nm) is the poor resolution which has significance in identification of maxima at low angles of diffraction.

Section -III

SOIL CHEMICAL ANALYSIS

15

Basics of Quantitative Chemical Analysis

15.1 Law of Mass Action

Gulberg and Waage, the two Norwegian chemists in the year 1864 stated the law of mass action in this form: 'The velocity of any chemical reaction at any instant at a given temperature is proportional to the active mass of each of the reactants at that instant present in the system, the active mass in a homogeneous system is defined as the number of gram molecule of the substance present per unit volume i.e., the molar volume'.

Mathematical formulation of the law of mass action

Let us take a simple case of a reversible reaction at constant temperature represented by the equation:

$$A + B \rightleftharpoons C + D$$

The velocity with which A and B react is proportional to their concentration

$$v_1 \propto [A] \times [B]$$

$$= k_1 \times [A] \times [B] \qquad (15.1)$$

Where, v_1 is the velocity of forward reaction, k_1 is a constant known as velocity coefficient of forward reaction and [A], [B], represent the molar concentration of A and B, respectively.

Similarly, the velocity with which the reverse reaction occurs is given by:

$$v_2 = k_2 \; [C] \times [D] \qquad (15.2)$$

Where, v_2 is the velocity of backward (reverse) reaction, k_2 is the velocity coefficient of backward reaction and [C], [D] are the molar concentration of C and D, respectively.

At equilibrium, the velocity of forward and backward reaction is equal (though the equilibrium is a dynamic and not static one). Therefore, $v_1 = v_2$

or, $k_1[A] \times [B] = k_2[C] \times [D]$

$$\frac{([C]\times[D])}{([A]\times[B])} = \frac{k_1}{k_2} = K \qquad (15.3)$$

Where, K is called the equilibrium constant or mass law constant at a given temperature.

The generalized expression of a reversible reaction may be represented as:

$$aA + bB + \text{———} \rightleftharpoons gG + hH + \text{———}$$

Where, small letters are the number of molecule of the reacting substances.

The condition for the equilibrium of the above reaction is given by:

$$K = \frac{G^g \times H^h \times}{A^a \times B^b \times} \qquad (15.4)$$

This equation (8.4) which is the most generalized mathematical expression for the law of mass action can be stated as follows:

'At equilibrium the product of the concentrations of the substances formed in a reversible reaction, each raised to a power which is the coefficient of its formula in the chemical reaction divided by the product of the concentrations of the reacting substances, each raised to the power which is the coefficient of its formula in the chemical reaction is constant'.

15.2 Activity and Activity Coefficient

The most widely used and modern method of quantitatively expressing the deviation of strong electrolyte from the ideal solution law is the use of activity coefficient. In the above deviation of the law of mass action it was assumed that the effective concentration or active masses of the components could be expressed by the stoichiometric concentration. But according to modern thermodynamics this is not strictly true. Any ion in solution at a finite concentration behaves as if it was an effective concentration different from its actual concentration. The effective concentration of any ion is called its ion activity. The activity of an ion varies with the concentration and can be defined that activity of any ion becomes equal to the concentration at infinite dilution.

The equilibrium equation for a binary electrolyte can be presented as:

$$AB \rightleftharpoons A^+ + B^-$$

According to law of mass action,

$$K = (a_A^{\ +} \times a_B^{\ -}) / a_{AB} \tag{15.5}$$

Where, $a_A^{\ +}$, $a_B^{\ -}$ and a_{AB} are the activities of A^+, B^- and AB , respectively and K is the true dissociation constant. The activity is related to the concentration by the factor, activity coefficient.

$$\text{Activity} = \text{Concentration} \times \text{Activity Coefficient}$$

Thus, the above equation (15.5) in concentration term can be written as:

$$\frac{\left(\left[A^+\right]\times f_{A+}\times\left[B^-\right]\times f_{B-}\right)}{\left[AB\right]\times f_{AB}} = \left\{\frac{\left[A^+\right]\times\left[B^-\right]}{\left[AB\right]}\right\}\times\left[\frac{f_{A+}\times f_{B-}}{f_{AB}}\right] \tag{15.6}$$

Where, f_{A+}, $f_B^{\ -}$, f_{AB} are the activity coefficients of A, B and AB, respectively and [ ] represents the concentration.

15.3 The Hydrogen Ion Exponent or pH

It is very cumbersome to express always the hydrogen or hydroxyl ion concentration of solution into gram ion per litre or gram equivalent per litre particularly of solution having very low concentration of hydrogen or hydroxyl ion. Sorensen (1909) introduced the term hydrogen ion exponent or pH which can be expressed as:

$$pH = -\log_{10}[H^+] \text{ or } \log_{10}\frac{1}{\left[H^+\right]} \tag{15.7}$$

Thus, pH is the logarithm (base 10) of the reciprocal of hydrogen ion concentration or the logarithm of hydrogen ion concentration with negative sign; the hydrogen ion concentration expressed in gram ion per litre. At all states of acidity and alkalinity, the pH of the solution can be expressed by a series of positive number between 0 and 14. The neutral solution with $[H^+] = 10^{-7}$ gram ion per litre has the pH of 7. An acidic solution is one having pH less than 7 and an alkaline solution is one having pH more than 7. Alternately, in neutral solution hydrogen and hydroxyl ion concentration are equal i.e. 10^{-7} gram ion per litre. In acid solution hydrogen ion concentration ranges from 10^{-7} to 10^0, while in alkaline solution hydrogen ion concentration ranges from 10^{-14} to 10^{-7} gram ion per litre. At 25°C the product of molar concentration of hydrogen and hydroxyl ion is equal to 10^{-14}.

At 25°C, $[H^+] \times [OH^-] = K_w = 10^{-14}$ (15.8)

or, $[OH^-] = 10^{-14} / [H^+]$

or, $pOH = -\log_{10}10^{-14} - pH$

or, $pOH + pH = 14$

The pictorial presentation of the relation between $[H^+]$, $[OH^-]$, pH and pOH

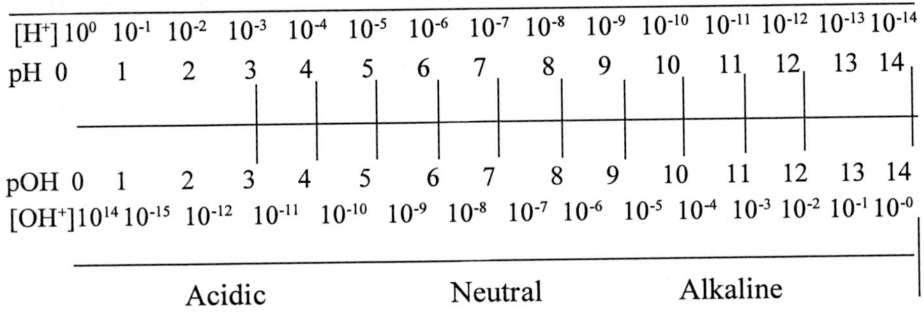

15.4 Volumetric (Titrimetric) Analysis

The term volumetric (titrimetric) used in these analyses to be discussed in this chapter, is because actually the volume of a solution of known concentration (**standard solution**) is determined that reacts quantitatively with the solution of the substance of unknown concentration by the process of titration. Then, the concentration of unknown solution can be determined from the well established law of chemical equivalence:

$$V_1S_1 = V_2S_2$$

Where, V_1 = Volume of solution of unknown concentration, S_1 = Concentration of solution to be determined, V_2 = Volume of standard solution and S_2 = Concentration of standard solution.

The reagent of known concentration is called the titrant and the concentration of the substance to be determined is termed as titrate. Usually, the titrant is taken in the burette and added drop by drop to the titrate until the reaction between these two completes. The point at which the reaction completes is known as equivalence point or end point. Besides titrant and titrate, the auxiliary reagent used in titration to detect the accurate end point of reaction is called indicator.

15.4.1 Criteria of reaction for Volumetric analysis

1. The reaction must be simple and can be expressed in the form of chemical equation.
2. The titrate must react with titrant in equivalent proportion.
3. The reaction must proceed with very high speed.
4. At the end point there must be a marked change in physical or chemical properties of the solution.
5. The indicator should be available which, by change in physical properties (colour or formation of precipitate) helps to detect the sharp end point. If no visible indicator or mixture of indicator is available, the end point can be detected by the change in potential or electrical conductance or optical density of the solution.

15.4.2 Classification of Reaction in Volumetric Analysis

Mainly two types of reaction involved in volumetric analysis

a) Reaction in which no change in valance occurs.

b) Reaction in which change in valence or transfer of electron occurs e.g., oxidation-reduction reaction.

But here for convenience of discussion the reactions are grouped into four classes.

A. Neutralisation reaction or acidimetry and alkalimetry

In acidimetry, titrant is acid and titrate is free base or that formed from salt of weak acid on hydrolysis. In alkalimetry, titrant is base and titrate is free acid or that formed from salt of weak base on hydrolysis. These reactions involve formation of water molecules through combination of hydrogen and hydroxyl ions.

$$HCl + NaOH \rightleftharpoons NaCl + H_2O$$

B. Complex formation reaction

These reactions involve the combination of ions, other than hydrogen or hydroxide ions to form a soluble, slightly dissociated ion or compound.

$$2Cl^- + Hg^{2+} \rightleftharpoons HgCl_2$$

Chelometric (involving chelates like EDTA, ethylene diamine tetra acetic acid) titrations also belong to this group.

C. Precipitation formation reaction

This reaction depends upon the combination of ions other than hydrogen or hydroxide ions to form a simple precipitate.No change in oxidation state occurs.

$Ag^+ + Cl^- = AgCl\downarrow$

D. Oxidation – reduction reaction

In this type of reaction a change in oxidation number or transfer of electron occurs. The reacting substances are: one is oxidizing agent and other is a reducing agent.

15.4.3 Equivalent Weight

A **standard solution** is a solution which contains a known amount of substance dissolved in a definite volume, alternately whose concentration or strength is known. A **molar solution** is one which contains 1 gram molecular weight of any substance in one litre of solution. A **molal solution** is one in which 1 gram molecular weight of any substance is dissolved per litre of solvent. A **normal solution** is one which contains 1 gram equivalent weight of any substance per litre of it.

The equivalent weight of an **acid** is that weight of it which contains 1 gram atom of replaceable hydrogen ion or the molecular weight divided by basicity of the acid. For monobasic acid (HCl, HNO_3, CH_3COOH) equivalent weight is equal to its molecular weight. But for dibasic acid (H_2SO_4) or tribasic (H_3PO_4) acid the equivalent weights are ½ and 1/3 of its molecular weight, respectively.

The equivalent weight of a **base** is its molecular weight divided by acidity or that weight which contains one replaceable hydroxyl group (OH^-). The equivalent weight of NaOH or KOH is equal to its molecular weight and of $Ca(OH)_2$ or $Mg(OH)_2$ is ½ of its molecular weight.

The equivalent weight of a **salt** is that weight of the salt which contains one equivalent of an active element or metal present in it or the equivalent weight of a salt is that weight which contains or react with 1 gram atom of monovalent cation (Na^+, K^+), ½ gram atom of bivalent cation (Ca^{2+}, Mg^{2+}), 1/3 gram atom of trivalent cation (Al^{3+}, B^{3+}), etc.

The equivalent weight of the salt is the molecular weight of the salt divided by the total valence (number of metal atoms in a molecule multiplied by the valence of the metal) of the metal atoms present in a molecule. The equivalent weight of $CaSO_4$ is its molecular weight divided by 2 [1 (No. of Ca atom in $CaSO_4$) 2 (valence of Ca)].

The equivalent weight of a cation is its atomic weight divided by valence. The equivalent weight of an oxiding (oxidant) or reducing (reductant) agent is that weight of reagent which reacts with or contains 1.008 g of available hydrogen or 8.00 g of available oxygen. By 'available' is meant capable of being utilized in oxidation or reduction. But more general view is obtained by considering the amount of electrons involved in the partial ionic equation representing the reaction. In any oxidation-reduction reaction, the oxidation and reduction go simultaneously, alternately one reagent is oxidized and other is undergone reduced, as the two reactions are complementary to each other. The oxidizing agent or oxidant undergoes reduced by gaining electron and reducing agent or reductant undergoes oxidized by loosing electron. So, all redox reactions can be divided into two partial ionic equations, one for oxidation and other for reduction.

For example: Reaction between ferric chloride ($FeCl_3$) and stannous chloride ($SnCl_2$) in aqueous solution. The actual equation is:

$$2\,FeCl_3 + SnCl_2 = 2FeCl_2 + SnCl_4$$

If it is divided into two partial equations, then the partial equation for reduction is

$$Fe^{3+} \rightarrow Fe^{2+} \qquad (15.9)$$

And for oxidation is

$$Sn^{2+} \rightarrow Sn^{4+} \qquad (15.10)$$

The equation must be balanced not only by the number and kind of atom but also electrically i.e., the net charge on left and right hand sides of the equation will be equal.

Equation (15.9) and (15.10) can be balanced in this way:

$$Fe^{3+} + e \rightarrow Fe^{2+} \qquad (15.9.1)$$

$$Sn^{2+} \rightarrow Sn^{4+} + 2e \qquad (15.10.1)$$

As the charge at both sides of the equation is equal, the gain or loss of electron in the redox process must be same. Thus, equation (15.9.1) would be multiplied by 2 and we have:

$$2Fe^{3+} + 2e \rightarrow 2Fe^{2}$$

$$Sn^{2+} \rightarrow Sn^{4+} + 2e$$

— — — — — — — — — — — — —

$$\text{Thus, } 2Fe^{3+} + Sn^{2+} \rightarrow 2Fe^{2+} + Sn^{4+}$$

So here, for calculating the equivalent weight of $FeCl_3$, the molecular weight will be divided by 1, as one $FeCl_3$ molecule gains only one electron. However, equivalent weight of $SnCl_2$ will be calculated by dividing its molecular weight by 2 as its single molecule liberates two electrons.

Another example: Reaction between potassium dichromate ($K_2Cr_2O_7$) and potassium iodide (KI) in presence of dilute sulphuric acid.

$K_2Cr_2O_7 + KI + 7H_2SO_4 = Cr_2(SO_4)_3 + 3I_2 + 4K_2SO_4 + 7H_2O$

One partial equation,

$Cr_2O_7^{2-} \rightarrow 2Cr^{3+}$

$$Cr_2O_7^{2-} + 14H^+ \rightarrow 2Cr^{3+} + 7H_2O$$

To balance electrically, 6e is added to the left hand side

$$Cr_2O_7^{2-} + 14H^+ + 6e \rightarrow 2Cr^{3+} + 7H_2O$$

In other partial equation,

$$2I^- \rightarrow I_2$$

To balance electrically, $2I^- \rightarrow I_2 + 2e$ (15.11)

One dichromate ion receives 6e and two iodide ions liberate 2e. So, to balance the loss and gain of electron, equation (15.11) will be multiplied by 3 and we get

$$Cr_2O_7^{2-} + 14H^+ + 6e \rightarrow 2Cr^{3+} + 7H_2O$$

$$6I^- \rightarrow 3I_2 + 6e$$

Adding, $Cr_2O_7^{2-} + 14H^+ + 6I^- \rightarrow 2Cr^{3+} + 3I_2 + 7H_2O$

Thus, the equivalent weight of $K_2Cr_2O_7$ is its molecular weight divided by 6 and that of KI is equal to its molecular weight.

15.4.4 Preparation of Standard Solution

If a reagent is available in pure state i.e., reagent of primary standard, a solution of any normality or molarity can be prepared just by dissolving a definite fraction or multiple of its equivalent weight or molecular weight, respwctively in the solvent. For example, to prepare 3**N**, 0.5**N** and 0.001**N** solution of $K_2Cr_2O_7$ 3 x 49.04, 0.5 x 49.04 and 0.001 x 49.04 g of $K_2Cr_2O_7$, respectively will remain dissolved per litre of solution (49.04 is equivalent weight of $K_2Cr_2O_7$).

Sometimes, the concentration of the solution is expressed in terms of parts per million (ppm), percentage strength by weight or volume. Parts per million means the weight of solute present per million (10^6) parts (volume or weight) of solution. A solution of X ppm (w/w) means X g of solute is present per 10^6 g of solution or X mg of solute is present per 10^6 mg of solution or X μg of solute is present per 10^6 μg or per g of solution. Similarly, a solution of Y ppm (w/v) means Y g of solute is present per 10^6 ml (10^3 litre) of solution or Y mg solute is present per litre of solution or Y μg of solute is present per ml of solution. If the solvent is water then the above two strengths are same (considering density of water = 1g /cc). When the strength of the solution is expressed in terms of weight of solute per unit weight of solution multiplied by 100 or in simplier term grams of solute per 100 g of solution is the percentage strength by weight (w/w). Say, a 5 % NaCl solution is prepared by dissolving 5 g of NaCl in 95 g i.e. 95 ml of water (considering density of water = 1g /cc). Thus, if the solvent is water then 5% NaCl solution means 5 g NaCl remains dissolved per 100 g solution (w/w) or per 100 ml solution (w/v). The concentration or the strength of solution is also expressed in terms of volume of solute present per unit volume of solution multiplied by 100 or in simplier term ml of solute present per 100 ml of solution is known as percentage strength by volume (v/v). For example a 60% ethanol solution means 60 ml of absolute ethanol is diluted to 100 ml with water which means 60 ml of absolute ethanol is mixed with 40 ml of water.

It is not always essential to weigh out exactly a gram equivalent weight or mole (or a multiple or sub multiple thereof). In practice it is convenient to prepare a solution a little more concentrated than is ultimately required, and then dilute it with the solvent until the desired strength is achieved following the formula $V_1S_1 = V_2S_2$, where V_1 is the volume after dilution, S_1 is the desired strength (molarity or normality), V_2 is the original volume and S_2 is the strength (molarity or normality) originally obtained.

The following substances are available in a state of high purity and therefore suitable for the preparation of standard solution: Sodium carbonate, Na_2CO_3; Borax, $Na_2B_4O_7$; Potassium hydrogen phthalate, $KH(C_8H_4O_4)$; Potassium bi-iodate, $KH(IO_3)_2$; Succinic acid, $H_2C_4H_4O_4$ and Benzoic acid, $H(C_7H_5O_2)$; Silver nitrate, $AgNO_3$; Sodium chloride, NaCl; Potassium chloride, KCl; Disodium ethylene diamine tetra acetate dehydrate (Na-EDTA); Potassium dichromate, $K_2Cr_2O_7$; Potassium bromate, $KBrO_3$; Potassium iodate, KIO_3; Iodine, I_2.

When the reagent is not available in pure form as in cases of most alkali hydroxides, some inorganic acids and various deliquescent substances, solutions corresponding approximately to the desired strength are prepared first, then standardized by titration against a solution of a pure substance of known concentration. The substances in this category for example are: Sodium hydroxide, (NaOH); Potassium hydroxide, (KOH); $_p$otassium permanganate, $KMnO_4$; Sodium thiosulphate, $(Na_2S_2O_3)$; Potassium thiocyanate, (KSCN), etc.

15.4.5 Primary and Secondary Standards

A primary standard is a compound of sufficient purity from which a standard solution can be prepared by direct weighing of a quantity of it, followed by dilution to give a defined volume of solution.

Criteria of a substance to be considered as primary standard:

a) It must be easy to obtain in pure state.

b) It should be unaltered in air during weighing i.e., it should not be hygroscopic and easily oxidized in air.

c) It should have a high equivalent weight, so that weighing error can be minimized.

d) The substance should be readily soluble in desired solvent required for the experiment.

e) The reaction with standard solution should be instantaneous to avoid titration error.

Some Primary Standard Reagents

For acid-base reaction:

Sodium carbonate, Na_2CO_3; Borax, $Na_2B_4O_7$; Potassium hydrogen phthalate, $KH(C_8H_4O_4)$; Potassium bi-iodate, $KH(IO_3)_2$; Succinic acid, $H_2C_4H_4O_4$ and Benzoic acid, $H(C_7H_5O_2)$.

For complex formation reaction:

Silver nitrate, $AgNO_3$; Sodium chloride, NaCl; Potassium chloride, KCl; Disodium ethylene diamine tetra acetate dihydrate (Na-EDTA)

For redox reaction:

Potassium dichromate, $K_2Cr_2O_7$; Potassium bromate $KBrO_3$; Potassium iodate, KIO_3; Iodine, I_2.

A secondary standard is a substance which may be used for standardization, and whose content of the active substance has been found by comparison against a primary standard. It follows that a secondary standard solution is a solution in which the concentration of dissolved solute can be determined not from the weight of the compound dissolved in it but by titration of a volume of solution against a measured volume of a primary standard solution.

15.4.6 Acidimetry and Alkalimetry

15.4.6.1 Indicators

The objective of titration say, an alkaline solution with a standard acid, is the determination of volume of acid which is exactly equivalent chemically to the amount of base present. A large number of substances are available which have the property to change their colour according to hydrogen ion concentration of the solution. Any acid – base titration is characterized by definite hydrogen ion concentration at its equivalence point which is dependent upon the strength of acid and base and also their nature. For example, if the acid and base are strong electrolyte then the resultant solution at equivalence point will be neutral (pH 7.0), but if any of the reactant is weak electrolyte then the salt produced upon the reaction will be hydrolyzed to some degree and the resultant solution will be either acidic or basic.

A considerable number of indicators have so far been brought to light but owing to the following considerations few have been placed in the laboratory:

a) Colour change should be sharp but not sudden i.e., change should happen over a short range of H^+ concentration.

b) The colour should be stable and brilliant. Acid colour and basic colour of the indicator should be contrasting to each other.

c) The colour change should occur at the end point of the reaction being studied.

The position of the colour change interval in the pH scale varies widely with different indicators. Therefore, we have to select correct indicator for a particular titration.

W. Oswald (1891) first introduced the theory of indicator action. As the indicators which are actually weak organic acid or base have different colours in dissociated and undissociated form. The behavior of an indicator in aqueous solution can be represented as follows:

Acid indicator: $HIn \rightleftharpoons H^+ + In^-$

Base indicator: $InOH \rightleftharpoons OH^- + In^+$

(Undissociated colour) (Dissociated colour)

In case of acid indicator (HIn) undissociated form dominates over the dissociated form in acid medium as H^+ of acid solution depressed dissociation (common ion effect) and colour of the undissociated form exhibits. Reverse is true if medium is alkaline. Now, applying law of mass action we get:

$([H^+] \times [In^-]) / [HIn] = K_{ina}$ [assuming activity coefficient to be unity; K_{ina} is the dissociation constant of acid indicator]

or, $[H^+] = (K_{ina} \times [HIn]) / [In^-]$

or, $pH = \log([In^-] / [HIn]) + pK_{ina}$

Similary, in case of base indicator:

$[OH^-] = (K_{inb} \times [InOH]) / [In^+]$ (K_{inb} is the dissociation constant of base indicator)

or, $[H^+] = ([In^+] / [InOH]) \times (K_w / K_{inb})$ [where, $K_w = [H^+] \times [OH^-]$

or, $pH = \log ([InOH] / [In^+]) - pK_{inb} + pK_w$

The Bronsted concept of acid (proton donar) and base (proton acceptor) makes it unnecessary to distinguish between acid and base indicators, but emphasis is placed upon charge types of acidic and alkaline form of indicator. The equilibrium between acidic form In_A and the basic form In_B may be expressed as:

$$In_A \rightleftharpoons In_B + H^+$$

Then, according to law of mass action

$$[H^+] = (K_{in} \times [In_A]) / [In_B]$$

$$pH = \log ([In_B] / [In_A]) + pK_{in} \quad (15.12)$$

Both forms of indicator are present at any hydrogen ion concentration and human eye has limited ability to detect the either of two colours when one of them predominates. Observation says that acid colour will be detected when the ratio of $[In_A] / [In_B]$ is above 10 which mean acid colour will be visible when the corresponding pH of the solution will be:

$$pH = \log 10^{-1} + pK_{in} \text{ [from Eq. (15.12)]}$$

$$= pK_{in} - 1$$

Similarly, basic colour will be visible when $[In_B] / [In_A] > 10$ and the corresponding pH will be:

$$pH = \log 10 + pK_{in}$$

$$= pK_{in} + 1$$

Thus, the colour change interval of the indicator is about two pH units and the change of colour will be gradual as it depends upon the ratio of concentration of the two coloured forms (acid form and basic form) of the indicator. Table 15.1 enlists some indicators suitable for volumetric analysis.

Table 15.1: Colour change and pH range of some indicators

Indicator	pH range	Colour in acid solution	Colour in alkaline solution	pK_{in}
Thymol blue (acid)	1.2- 2.8	Red	Yellow	1.7
Bromo phenol blue	3.0 - 4.6	Yellow	Blue	4.1
Methyl yellow	2.9 - 4.0	Red	Yellow	3.3
Methyl orange	3.1- 4.4	Red	Orange	3.7

Bromo cresol green	3.8 -5.4	Yellow	Blue	4.7
Methyl red	4.2 - 6.3	Red	Yellow	5.0
Azolitmus (litmus)	5.0 – 8.0	Red	Blue	-
Thymol blue (base)	8.0 – 9.6	Yellow	Blue	8.9
Phenolphthalein	8.3 – 10.0	Colourless	Red	9.6
Thymolphthalein	8.3 – 10.5	Colourless	Blue	9.3

15.4.6.2 Mixed Indicator

It is seen that the colour change occurs between two pH units, but in some cases it is desirable to get this change over a very narrow pH range (less than 2 units). This can be achieved only through suitable mixture of indicators. Matching indicators are so selected that their pK_{in} values are close together and the overlapping colours are complementary at an intermediate pH value.

Table 15.2: Some mixed indicators

Components of mixed indicator	Composition	Ratio of the components (v/v)	pH	Colour change
1. Bromocresol	0.1% (Na)* in water green + Methyl orange 0.2% in water	1:1	4.3	Orange → blue → green
2. Bromothymol + Neutral red	0.1% in ethano 10.1% in ethanol	1:1	7.2	Rose pink → green
3.Thymol blue +Cresol red	0.1% (Na) in water 0.2% (Na) in water	3:1	8.3	Yellow violet
4.Thymolphthalein + Phenolphthalein	0.1% in ethano 10.1% in water	1:1	9.9	Colourless → violet

(Na) * means Na salt

15.4.6.3 Universal or Multiple Range Indicator

This is a mixture of certain indicators so as to get colour change over a considerable range of pH. These are not suitable for quantitative analysis. One recipe of such a universal indicator is: 0.1g phenolphthalein, 0.2g methyl red, 0.3g methyl yellow, 0.4g bromo thymol blue and 0.5g thymol blue in 500 ml of absolute alcohol and sufficient sodium hydroxide until the colour is yellow. The colour change will be available from pH 2 to 10,e.g., red at pH 2, orange at pH 4, yellow at pH 6, green at pH 8 and blue at pH 10.

15.4.6.4 Neutralization Process

In neutralization processes hydrogen ion concentration of the titrate changes during the course of titration. Therefore, neutralization curve can be prepared by plotting pH as ordinate against number of ml of titrant addition as abscissa.

Further, it is important to know the changes of hydrogen ion concentration or pH during the course of neutralization reaction particularly at the close vicinity of equivalence point to select the correct indicator to avoid or minimize titration error.

15.4.6.4a Neutralization of a Strong Acid with a Strong Base

Here, assumption is that acid and base are completely dissociated during the course of neutralization reaction and activity coefficients of the ions are unity i.e., during the course of neutralization to calculate pH we may consider the concentration of hydrogen ion in place of activity. Here, we shall discuss the neutralization of 100 ml 1M HCl by 1M NaOH. At any point of titration we can calculate the $[H^+]$ of HCl solution by using the simple formula:

$[H^+]$ = Vol. of unneutralized acid x (Normality of acid / Total vol. of solution)

When no NaOH is added then $[H^+]$ of HCl

$[H^+] = 1$ g mol/litre, pH = 0

When 50 ml of 1 M NaOH is added then,

$[H^+] = (100 - 50) \times [1 / (100 + 50)] = 3.33 \times 10^{-1}$; pH = 0.48

When 90 ml of 1 M NaOH is added then,

$[H^+] = (100 - 90) \times [1 / (100 + 90)] = 5.27 \times 10^{-2}$; pH = 1.3

When 99.9 ml of 1 M NaOH is added then,

$[H^+] = (100 - 99.9) \times [1 / (100 + 99.9)] = 5.1 \times 10^{-4}$; pH = 3.3

Upon addition of 100 ml of 1 M NaOH i.e., at equivalence point $[H^+]$ and $[OH^-]$ will be equal i.e., 10^{-7} and pH = 7.0.

After equivalence point, $[OH^-]$ can be calculated by using similar formula.

$[OH^-]$ = Vol. of unneutralized base x (Normality of base / Total vol. of solution)

So, upon addition of 100.1 ml of 1 M NaOH

$[OH^-] = (100.1 - 100) \times [1/(100 + 100.1)] = 5 \times 10^{-4}$; pOH = 3.3, so pH = 10.7

Thus, upon addition of each ml of base the pH of the solution increases but between 99.9 and 100.1 ml of base addition pH jumps from 3.3 to 10.7. So, to avoid any titration error we have to select that indicator whose colour change interval lies between pH 3 and 10.5 and it is found that except thymol blue (acid) all the indicators listed in table 8.1 satisfy this condition.

In similar way, the pH changes in neutralization reaction of 0.1 M acid and base can be calculated and colour change interval of ideal indicator should lie between pH 4.5 and 9.5. Indicators in this group are bromo cresol green, methyl red and thymol blue (base). Use of methyl orange and phenolphthalein will introduce some titration error though negligibly small.

The pH range of ideal indicator for neutralization of 0.01 **M** acid and base is further narrow and lies between 5.5 and 8.5 and suitable indicators are methyl red, bromo thymol blue or phenol red.

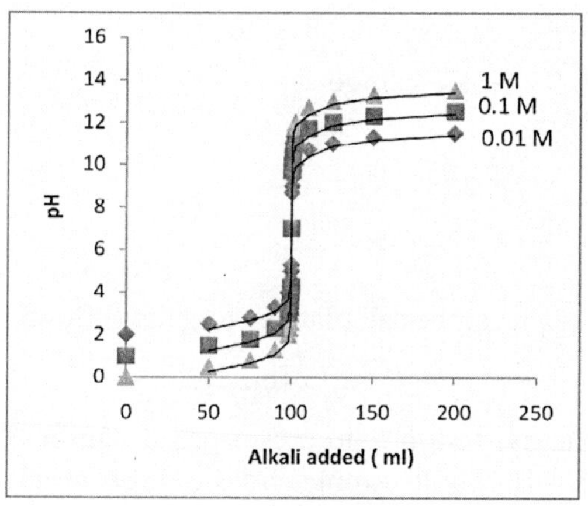

Fig. 15.1: Neutralization curve of strong acid with strong base

15.4.6.4b Neutralization of a Weak Acid with a Strong Base

Let us discuss the neutralization of 100 ml 0.1 **N** acetic acid with a strong base 0.1 **N** NaOH solution. The pH of the solution at the equivalence point can be calculated using the following equation.

$$pH = \frac{1}{2}pK_w + \frac{1}{2}pK_a + \frac{1}{2}\log c$$

Where, K_a = dissociation constant of acid, c = concentration of salt at equivalence point.

So, pH at equivalence point is:

$pH = 7 - \frac{1}{2}\log K_a + \frac{1}{2}\log(100 \times 0.1 / 200)$ [when 100 ml NaOH is added]

$= 7 - \frac{1}{2}\log (1.82 \times 10^{-5}) + \frac{1}{2}\log (0.05)$ [say, K_a of acetic acid is 1.82×10^{-5}]

$= 8.72$

The pH at any other points can be calculated by the equation

$$pH = \log([salt] / [acid]) + pK_a$$

Thus, for the titration of weak acid ($K_a > 5 \times 10^{-6}$) it is necessary to use an indicator with a pH range around 8.7 e.g., phenolphthalein, thymolphthalein or thymol blue (base). However, titration error is less in using thymolphthalein rather than phenolphthalein.

15.4.6.4c Neutralization of a Weak Base with a Strong Acid

As an example we shall discuss the titration of 100 ml of 0.1 N aqueous ammonia ($K_b = 1.8 \times 10^{-5}$) with 0.1 N HCl. The pH of the solution at the equivalence point can be calculated by the following equation:

$$pH = \frac{1}{2}pK_w - \frac{1}{2}pK_b - \frac{1}{2}\log c$$

$$= 7 - 2.37 - \frac{1}{2}(\bar{2}.70)$$

$$= 5.28$$

The pH at other points can be calculated using the following formula:

$$pH = pK_w - pK_b - \log([\text{salt}] / [\text{base}])$$

Methyl orange, methyl red, bromo cresol green can be used for titration of weak base ($K_b > 5 \times 10^{-6}$) with strong acid. For more weak base ($K_b > 1 \times 10^{-7}$) only bromo phenol blue or methyl red is suitable.

15.4.6.4d Neutralization of a Weak Acid with a Weak Base

Titration of 100 ml 0.1 **N** acetic acid ($K_a = 1.8 \times 10^{-5}$) with 0.1 **N** aqueous ammonia solution ($K_b = 1.8 \times 10^{-5}$) is illustrated. The pH of the equivalence point is given by the equation:

$$pH = \frac{1}{2}pK_w + \frac{1}{2}pK_a - \frac{1}{2}pK_b$$

$$= 7.0 + 2.37 - 2.37 = 7.0$$

The difference between the two titrations, one is between strong acid with strong base and other is between weak acid and weak base, is at both the case pH at equivalence point is 7.0, but in later case there is no sharp pH change at equivalence point like the former. So, no single indicator is suitable to exhibit sharp end point colour change. Sometimes mixed indicator is used in this case with success. But for practical purposes it is advisable to avoid this titration.

15.4.7 Precipitation Reactions

The most important and popular precipitation reactions in volumetric analysis utilize silver nitrate as the reagent (argentimetric process). Therefore, the present discussion will be confined to argentimetric processes and the same principle can be extrapolated to other precipitation reactions.

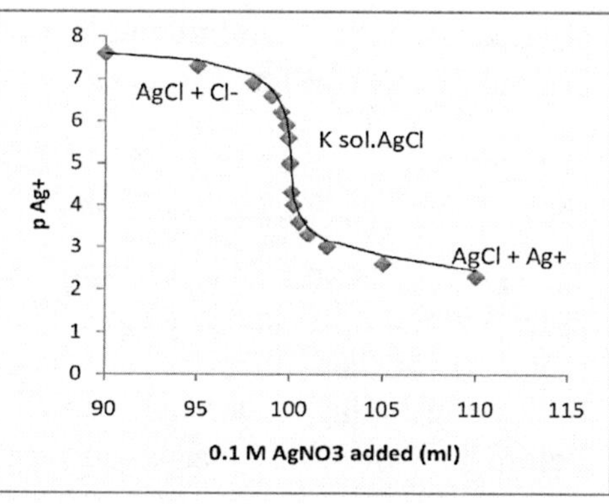

Fig. 15.2: Titration curve of 100 ml 0.1 N NaCl with 0.1 N $AgNO_3$

Let us discuss the changes in ionic concentration which occur during the titration of 100 ml 0.1 **N** NaCl with 0.1 **N** $AgNO_3$. The initial concentration of chloride ion, $[Cl^-] = 0.1$ gram equivalent per litre or $pCl^- = 1$. When 50 ml of 0.1 **N** $AgNO_3$ is added, 50 ml of 0.1 **N** NaCl remains in the total volume of 150 ml; thus $[Cl^-] = (50 \times 0.1) / 150 = 3.33 \times 10^{-2}$ or $pCl^- = 1.48$. Similarly, on addition of 90 ml $AgNO_3$ solution, $[Cl^-] = (10 \times 0.1) / 190 = 5.3 \times 10^{-3}$ or $pCl^- = 2.28$. Up to equivalence point the concentration of $[Cl^-]$ can be computed by this equation.

Again, S_{AgCl} [Solubility product of AgCl] $= [Ag^+]\,[Cl^-] = 1.2\ 10^{-10}$

$$\text{or, } pAg^+ + pCl^- = 9.92$$

At equivalence point $[Ag^+] = [Cl^-]$

$$\text{So, } pAg^+ \text{ or } pCl^- = 9.92 / 2 = 4.96$$

After equivalence point, say on addition 100.1 ml of $AgNO_3$ solution, $[Ag^+] = (0.1 \times 0.1) / 200.1 = 5 \times 10^{-5}$ or $pAg^+ = 4.30$.

Thus, there is a marked change in silver ion concentration in the neighbourhood of equivalence point which can be inspected by the values of silver ion exponents.

Determination of End Point

There are several methods to detect the end point; some of them are discussed below:

a) Formation of coloured precipitate

A simple example of this type of reaction is the titration of NaCl with $AgNO_3$ in presence of small quantity of potassium dichromate solution. At the end point the excess silver ions combine with chromate ions to form sparingly soluble red coloured silver chromate.

b) Formation of soluble coloured compound

The titration of silver with standard potassium or ammonium thiocyanate in presence of nitric acid using ferric nitrate as indicator is a typical example of this group. Thiocyanate ion produces the precipitate of silver thiocyanate which after completion of neutralization reacts with ferric ion to form soluble reddish brown complex of ferricyanate.

$$Ag^{+} + SCN^{-} \rightleftharpoons AgSCN$$
$$Fe^{3+} + SCN^{-} \rightleftharpoons [FeSCN]^{2+}]$$

This principle can be adopted for the estimation of chloride or bromide. First, known amount of standard $AgNO_3$ solution is added to form precipitation of silver halides and then the excess silver ion is back titrated with standard thiocyanate solution. It is advisable to filter off the precipitate of silver halide before back titration otherwise silver halide may react with thiocyante ion.

c) Use of adsorption indicator

The action of this indicator is due to the fact that at the equivalence point the indicator is adsorbed on the surface of the precipitate and due to adsorption a change occurs in the indicator which results in the development of different colour of the substance. The indicator is either acid dye such as those of fluorescein series or basic dye such as those of rhodamine series.

A precipitate has a tendency to adsorb its own ion. For example the precipitate of AgCl during titration of NaCl with $AgNO_3$ will adsorb chloride ions on its surface (primary adsorption) and primary adsorbed layer will again adsorb a layer of oppositely charged ion present in the medium, Na^{+}.

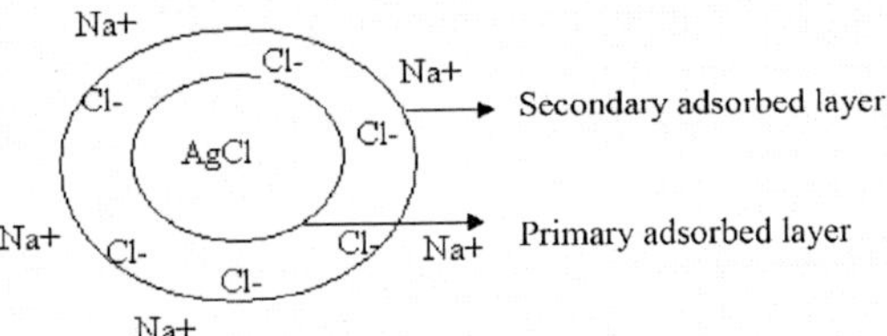

Fig. 15.3: AgCl precipitation in presence of excess Cl^{-}

But immediately after the equivalence point the solution is enriched with $AgNO_3$ and thereby the primary adsorbed layer will be with the silver ions and secondary layer will be with the nitrate ions.

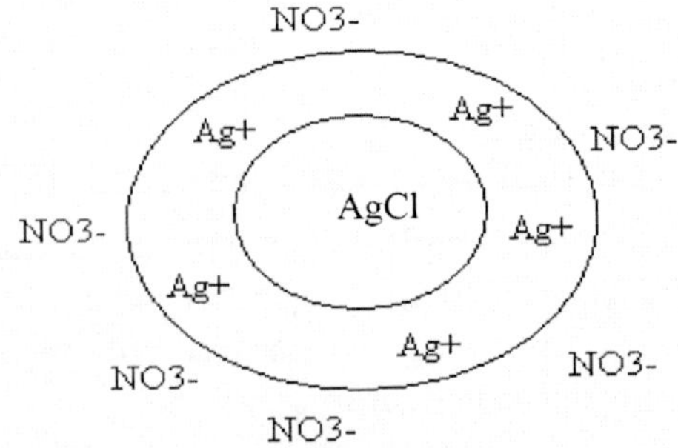

Fig. 15.4: Precipitation of AgCl in presence of excess Ag^{+}

However, in presence of fluorescein (yellowish green) the nitrate ions will be replaced by the negatively charged fluorescein which will be reflected by the change of colour owing to the formation of pink colour complex of silver.

Table 15.3: Some selected adsorption indicators and uses

Indicator	Used in titration of	Colour change at E.P	Further information of interest
1. Fluorescein	Cl^-, Br^-, I^-, SCN^- with Ag^+	Yellow → pink	Titration medium must be neutral to weakly alkaline
2. Eosin	Br^-, I^-, SCN^- with Ag^+	Pink → reddish violet	Best in acetic acid solution
3. Rose Bengal	I^- in presence of Cl^- with Ag^+	Red → purple	Best if $(NH_4)_2CO_3$ added
4. Rhodamine	Ag^+ with Br^-	Orange → reddish violet	Best in dil. HNO_3

d) Turbidity formation

In this procedure the appearance of turbidity in the titration medium is utilized to detect the end point. If after end point the addition of one solution to the other yields formation of turbidity, then there is no need to use of any indicator in this titration and titration will continue till there is no turbidity in the medium. A very simple example of this type of reaction is the titration of sodium chloride with silver nitrate or vice-versa.

15.4.8 Complex Formation Reactions

A simple example of this type of reaction is the titration of alkali cyanide with silver nitrate. When $AgNO_3$ is added to the cyanide solution a white precipitate is formed and on stirring it dissolves owing to the formation of stable soluble complex of cyanide.

$$AgNO_3 + 2KCN \rightleftharpoons \underset{\text{(Stable complex)}}{K[Ag(CN)_2]} + KNO_3$$

If the titration proceeds further addition of $AgNO_3$ produces the insoluble complex of silver argentocyanide and the end point of the reaction is indicated by the formation of turbidity or permanent precipitation.

$$Ag(CN)_2^- + Ag^+ \rightleftharpoons Ag[Ag(CN)_2]$$

This reaction is applicable in quantitative purpose, but only problem is the identification of correct end point. Because cyanide precipitated by the local excess concentration of silver ion somewhat prior to the equivalence point, is very slowly redissolved and makes the titration time consuming. To avoid it iodide ion is used as indicator in presence of ammonia solution which dissolves silver argentocyanide.

$$Ag[Ag(CN)_2] + 4NH_3 = 2[Ag(NH_3)_2]^+ + 2CN^-$$

$$[Ag(NH_3)_2]^+ + I^- \rightleftharpoons AgI + 2NH_3$$

The formation of silver iodide precipitation indicates the correct end point. During titration the presence of excess cyanide ion prevents the formation of silver iodide precipitation until the equivalence point is reached.

$$AgI + 2CN^- \rightleftharpoons Ag(CN)_2^- + I^-$$

This method can also be adopted for indirect determination of several metals e.g., nickel, cobalt and zinc which form stable complexes with cyanide ion.

The vast majority of complexation titrations are carried out using multidentate ligands such as EDTA or similar substances as the complexone.

15.4.8.1 Titration curve

Like acid-base titration, in the EDTA titration, if pM (-log $[M^{n+}]$) is plotted against the volume of EDTA solution added, a point of inflexion occurs at the equivalence point. In some instances this sudden increase may exceed 10 pM units. Like acid-base titration, in the EDTA titration also a metal ion sensitive indicator (metal-ion indicator) is employed to detect changes in pM.

15.4.8.2 Metal Ion Indicators

The success of an EDTA titration depends upon the precise determination of the end point. The requisites of a metal ion indicator for use in the visual detection of end point include:

- The metal-indicator complex must possess sufficient stability, otherwise, because of dissociation; a sharp colour change is not obtained. The stability must be less than that of metal-EDTA complex to ensure complete removal of metal ions from metal indicator complex at the end point.
- The colour contrast between the free indicator and metal-indicator complex should be readily observed.
- The indicator must be sensitive to metal ions (i.e. to pM).
- The above requirements must be fulfilled within the pH range at which the titration is performed.

Table 15.4: Some metal ion indicators

Indicator	pH range	Metal-indicator complex colour	Free indicator colour
1. Solochrome black (Eriochrome black T), sodium 1-(1-hydroxy-2-naphthylazo)-6-nitro -2-naphthol-4-sulphonate	7-11	Red (Ca & Mg)	blue
2. Murexide, ammonium salt of purpuric acid	10-11	Red (Ca), yellow (Ni), orange (Cu)	Blue -violet
3. Patton and Reeder's indicator (HHSNNA), 2-hydroxy-1-(2-hydroxy-4-sulpho-1-naphthylazo) -3–naphthoic acid	12-14	Red (Ca)	Pure blue
4. Solochrome dark blue(calcon), sodium 1-(2-hydroxy-1-naphthylazo)-2-naphthol-4-sulphonate	12-14	Pink (Ca)	Pure blue
5. Calmagite, 1-(1-hydroxy-4-methyl-2-phenylazo)- 2-naphthol-4-sulphonic acid	10	Red (Mg)	Clear blue

15.4.9 Oxidation –Reduction Reaction

When a metal is immersed in a solution containing its own salt, say Zn in $ZnSO_4$, Cu in $CuSO_4$, a potential is established between metal and the solution and the metal will function as an electrode. The potential, E for an electrode reaction M^{n+} + ne M is given by the equation:

$$E = E^{o} + \frac{R.T}{nF} \ln a_{M}^{n+} \tag{15.13}$$

Where, R is the gas constant, T is the absolute temperature, F is Faraday constant, n is the valence of the ion, a_M^{n+} is the activity of the ion in the solution, E^0 is the constant dependent on the metal. The equation (15.13) can be simplified by putting the values of R (8.31 joules per degree per mole), F (96,500 coulombs) and converting natural logarithm to base 10 by multiplying with 2.3026 it becomes:

$$E = E^0 + \frac{0.0001983T}{n} \log a_M^{n+}$$

At temperature 25°C (T = 298K)

$$E = E^0 + \frac{0.0591}{n} \log a_M^{n+} \ (14.14)$$

If in the equation (8.14) $a_M^{n+} = 1$, then $E = E^0$, E^0 is called standard potential of the metal.

15.4.9.1 Oxidation –Reduction Cell

As discussed earlier, reduction is accompanied by gain of electron and oxidation by loss of electron. In a system containing both an oxidant and its reduction product, there will be equilibrium between them and electrons. And if an inert electrode is immersed in this type of system, for example, platinum electrode is dipped into a solution containing ferric and ferrous ions, a potential will be established indicative to the position of equilibrium. For example, if the system acts as an oxidizing agent, then Fe^{3+} ($Fe^{3+} + e \rightarrow Fe^{2+}$) takes electron from platinum leaving platinum positively charged; If the system acts as a reducing agent then Fe^{2+} ($Fe^{2+} - e \rightarrow Fe^{3+}$) gives up electron to the metal which will then be negatively charged. The magnitude of the potential is actually the measure of oxidising or reducing properties of the system. The strength of an oxidizing agent can be obtained by measuring the potential developed between platinum and the solution relative to standard of reference. For a redox system, the standard of reference is the condition in which the ratio of the activity of oxidant to that of reductant is unity. In Fe^{3+}- Fe^{2+} electrode, the redox cell would be

$$Pt, H_2 \left| H^+ (a = 1) \right\| \begin{matrix} Fe^{3+} (a = 1) \\ Fe^{2+} (a = 1) \end{matrix} \left| Pt \right.$$

The potential thus measured is termed as standard (reduction) potential.

Table 15.5: Standard (reduction) potential, E^0 at 25°C

Half reaction	E^0 (volts)
$S_2O_8^{2-} + 2e \rightleftharpoons 2SO_4^{2-}$	+2.01
$MnO_4^- + 4H + 3e \rightleftharpoons MnO_2 + 2H_2O$	+1.69
$MnO_4^- + 8H + 5e \rightleftharpoons Mn^{2+} + 4H_2O$	+1.52
$Cl_2 + 2e \rightleftharpoons 2Cl^-$	+1.36
$Cr_2O_7^{2-} + 14H + 6e \rightleftharpoons 2Cr^{3+} + 7H_2O$	+1.33
$IO_3^- + 6H + 5e \rightleftharpoons \frac{1}{2}I_2 + 3H_2O$	+1.20
$Hg^{2+} + 2e \rightleftharpoons 2Hg$	+0.79
$Fe^{3+} + e \rightleftharpoons Fe^{2+}$	+0.77
$I_2 + 2e \rightleftharpoons 2I^-$	+0.54
$[Fe(CN)_6]^{3+} + e \rightleftharpoons [Fe(CN)_6]^{4-}$	+0.36
$Cu^{2+} + e \rightleftharpoons Cu^+$	+0.15
$Sn^{4+} + 2e \rightleftharpoons Sn^{2+}$	+0.15
$2H^+ + e \rightleftharpoons H_2$	0.0
$Cr^{3+} + e \rightleftharpoons Cr^{2+}$	-0.41
$Fe(OH)_3 + e \rightleftharpoons Fe(OH)_2 + OH^-$	-0.56
$Zn(OH)_4^{2+} + 2e \rightleftharpoons Zn + 4OH^-$	-1.22

The standard potential enables one to tell which one can oxidise the other at unit activity. Higher the positive value of standard potential higher is its oxidizing power and higher the absolute value of standard potential with negative sign higher is its reducing power. Thus, $Cr_2O_7^{2-}$ ion can oxidize I^-, Fe^{2+}, $[Fe(CN)_6]^{4-}$ but not Cl^- or SO_4^{2-}. Another typical example, both MnO_4^- - Mn^{2+} system and $Cr_2O_7^{2-}$ - Cr^{3+} system can oxidize Fe^{3+} - Fe^{2+} system, but only MnO_4^- can oxidize Cl^- ion. That is why $Cr_2O_7^{2-}$ but not MnO_4^- can be used for the titration of Fe^{2+} in HCl medium.

15.4.9.2 Calculation of Standard (reduction) Potential

A reversible oxidation – reduction reaction can be written in this form

Oxidant + ne $\rightleftharpoons$ Reductant

$$(Fe^{3+} + e \rightleftharpoons Fe^{2+})$$

The electrode potential established due to immersion of inert electrode in a solution containing oxidant and reductant will be:

$$E_T = E^0 + \frac{R.T}{nF} In \frac{a_{Ox}}{a_{Red}} \tag{15.15}$$

E_T = electrode potential at temperature T relative to standard hydrogen electrode ($E^0 = 0$).

Activity term can be substituted by concentration without introducing too much error. Now, putting the value of R, T, F and changing natural logarithm to base 10, the equation (15.15) can be written as follows:

$$E_{25°C} = E^0 + \frac{0.0591}{n} \log \frac{c_{Ox}}{c_{Red}}$$

At unit concentration of both oxidant and reductant $E_{25°C} = E^0$

15.4.9.3 Oxidation- Reduction Curve

Let us consider the titration of 100 ml 0.1 **N** ferrous iron with 0.1 **N** ceric cerium in presence of dilute sulphuric acid.

$$Ce^{4+} + Fe^{2+} \rightleftharpoons Ce^{3+} + Fe^{3+} \tag{15.16}$$

Here, we have two systems or two half reactions, (i) Fe^{3+} - Fe^{2+} ion electrode and (ii) Ce^{4+} - Ce^{3+} ion electrode.

In Fe^{3+} - Fe^{2+} system at 25°C:

$$E_1 = E_1^0 + \frac{0.0591}{1}\log\frac{[Fe3+]}{[Fe2+]}$$

$$= +0.75 + 0.0591 \log\frac{[Fe3+]}{[Fe2+]}$$

In Ce^{4+} - Ce^{3+} system at 25°C:

$$E_2 = E_1^0 + \frac{0.0591}{1}\log\frac{[Ce4+]}{[Ce3+]}$$

$$= +1.45 + 0.0591 \log\frac{[Ce4+]}{[Ce3+]}$$

Equilibrium constant of the reaction (14.16) is given by:

$$K = \frac{([Ce3+]x[Fe3+])}{([Ce4+]x[Fe2+])}$$

At equilibrium the two potentials will be same

$E_1 = E_2$

$$\text{or, } +0.75 + 0.0591 \log\frac{[Fe3+]}{[Fe2+]} = +1.45 + 0.0591 \log\frac{[Ce4+]}{[Ce3+]}$$

$$\text{or, } 0.0591 \log\frac{([Ce3+]x[Fe3+])}{([Ce4+]x[Fe2+])} = 0.70$$

$$\text{or, } 0.0591 \text{ x } \log K = 0.70$$

$$\log K = 11.84$$

$$K = 7 \text{ x } 10^{11}$$

The large value of the equilibrium constant signifies that the reaction will proceed from left to right till almost completion. Alternately, addition of ceric solution up to equivalence point its only effect will be to oxidize the Fe^{2+} and consequently change the ratio of $[Fe^{3+}] / [Fe^{2+}]$.

When 10 ml of the oxidizing agent (Ce^{4+}) is added, out of 100 ml ferrous solution 10 ml solution will be oxidized to ferric iron and rest amount will remain as ferrous state. Thus, the ratio of $[Fe^{3+}] / [Fe^{2+}]$ will be 10 / 90 (approx.) and

$E_1 = 0.75 + 0.0591 \log (10 / 90) = 0.69$ v

With 50 ml ceric solution, $E_1 = E_1^0 = 0.75$ v

With 90 ml ceric solution, $E_1 = 0.75 + 0.0591 \log (90 / 10) = 0.81$ v

With 99.9 ml ceric solution, $E_1 = 0.75 + 0.0591 \log (99.9 / 0.1) = 0.93$ v

At equivalence point the electrode potential is given by

$(E_1^0 + E_2^0) / 2 = (0.75 + 1.45) / 2 = 1.10$ v

[aOXI + b RedII $\rightleftharpoons$ bOXII + a RedI, the potential at equivalence is given by $E_e = (bE_1^0 + aE_2^0) / (a+b)$, where E_1^0 refers to $Ox_I - Red_I$ and E_2^0 refers to $Ox_{II} - Red_{II}$ system]

After attaining equivalence point, the subsequent addition of ceric solution will merely increase the ratio of $[Ce^{4+}] / [Ce^{3+}]$.

With 100.1 ml ceric solution, $E_2 = 1.45 + 0.0591 \log (0.1 / 100) = 1.27$ v ($[Ce^{3+}]$ will remain unaltered).

With 110 ml ceric solution, $E_2 = 1.45 + 0.0591 \log (10 / 100) = 1.39$ v

With 190 ml ceric solution, $E_2 = 1.45 + 0.0591 \log (90 / 100) = 1.45$ v

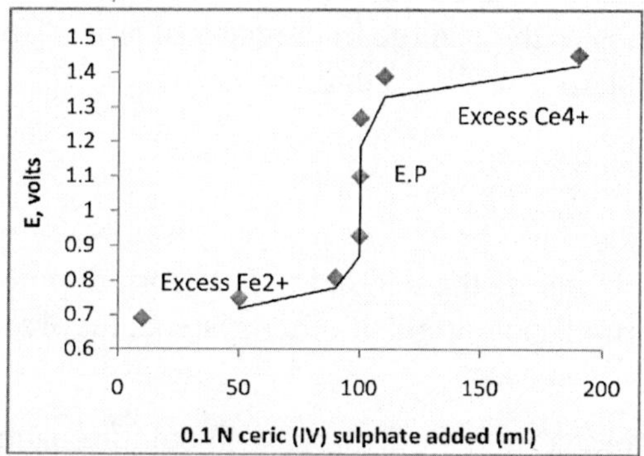

Fig. 15.5: Titration curve of Fe (II) with with 0.1 N ceric (IV) sulphate

It is evident that sharp change of potential at the proximity of equivalence point is dependent upon the standard potential of oxidation – reduction systems and independent of concentration unless these are extremely small.

15.4.9.4 Formal Potential

Standard potentials, E^0 are evaluated giving full regard to activities of ions. But in practice activities of the species in question are much smaller than the concentration of the solution in presence of other electrolyte and use of concentration term may lead to erroneous conclusion. For this reason formal

potential term is used in quantitative analysis to supplement standard potential. Formal potential is the potential that is observed experimentally in the solution containing formal concentration of substance i.e., formal weight substance dissolved per litre. Some examples of formal potential ($E^{0/}$) for Fe^{2+} - Fe^{3+}system, $E^0 = +0.77$ v, $E^{0/} = +0.73$ v in 1 M $HClO_4$, + 0.70 v in 1M HCl, + 0.68 v in H_2SO_4 and + 0.61 v in 0.5 M H_3PO_4 +1 M H_2SO_4.

15.4.9.5 Redox Indicators

Like acid – base indicators (pH indicators) in acid – base titration redox indicators are used to detect the correct end point in oxidation – reduction titration. The ideal redox indicator, in addition to the property of exhibiting sharp detectable colour change in the neighbourhood of equivalence point, must have the oxidation potential between the two formal potentials of the solutions of titrate and titrant.

Redox indicators exhibit different colours in the oxidized and reduced form: $In_{Ox} + ne \rightleftharpoons In_{Red}$. According to Nernst equation:

$$E = E^0_{In} + \frac{R.T}{nF} \ln \frac{[InOx]}{[InRed]}$$

If the colour intensity of the two forms are comparable, the colour change interval corresponds to the change in the ratio of $[In_{Ox}]/[In_{Red}]$ from 10 to1/10 which again corresponds to the potentials;

$$E = E^0_{In} \pm \frac{0.0591}{n} \text{ volts at } 25°C$$

For sharp colour change at the end point, E^0_{In} (formal potential) should differ from standard (formal) potentials of other systems involved in the reaction by at least 0.15v.

A list of selected redox indicators along with colour change in two forms and standard reduction potential at unit hydrogen ion activity are given in table 15.6.

Table 15.6: Some redox indicators

Indicator	Colour change		Reduction potential (E^0) in volt at pH =0
	Oxidized form	Reduced form	
Nitro-ferroin	Pale blue	Red	1.25
Ferroin	Pale blue	Red	1.06
5,6-dimethyl ferroin	Pale blue	Red	0.97
Diphenylbenzidine	Violet	Colourless	0.76
Diphenylamine	Violet	Colourless	0.76
Starch I_3^-, KI	Blue	colourless	0.53
Methylene blue	Blue	colourless	0.52

Problems, Questions and Hints

P1. Find the pH of a solution of which $[H^+] = 4 \times 10^{-5}$

Solution: $pH = -\log_{10}[H^+] = -\log_{10}[4 \times 10^{-5}] = -\log_{10}4 + 5\log_{10}10 = -0.602 + 5 = 4.398$

P2. Calculate the pH of N/200 solution of HCl as well as of NaOH at 25°C (Assume complete dissociation of the acid and the alkali).

Solution: a) In N/200 HCl, $[H^+] = 0.005$gram equivalent/litre

$pH = -\log_{10}[0.005] = -\log_{10}(5 \times 10^{-3}) = -\log_{10}5 + 3\log_{10}10 = -0.7 + 3 = 2.3$

b) In N/200 NaOH, $[OH^-] = 0.005$gram equivalent/litre

$[H^+] = K_w / [OH^-] = 10^{-14} / 5 \times 10^{-3} = 2 \times 10^{-12}$

$pH = -\log_{10}(2 \times 10^{-12}) = -\log 2 + 12\log_{10}10 = -0.3 + 12 = 11.7$

P3. Find the $[H^+]$ corresponding to pH 5.643.

Solution: $pH = -\log_{10}[H^+] = 5.643$

or, $\log[H^+] = -5.643 = \bar{6}.357 = \log(2.28 \times 10^{-6})$ [taking antilog of .357]

$[H^+] = 2.28 \times 10^{-6}$

P4. Calculate the pH of a 0.01M solution of acetic acid (Assume the degree of dissociation is 12.5%).

Solution: $[H^+] = (12.5 \times 0.01) / 100 = 1.25 \times 10^{-3}$

$pH = -\log_{10}(1.25 \times 10^{-3}) = -\log_{10}1.25 + 3\log_{10}10 = -0.097 + 3 = 2.903$

Q.1 Whether pH value less than 0 or greater than 14 is possible?

Hint: If in the solution $[H^+]$ or $[OH^-]$ is more than 1 g ion per litre then the pH value less than 0 and greater than 14, respectively, is possible. But in both the cases the concentration of the acid or alkali is very very high. pH value less than 0 and greater than 14 is of no use because the terminology pH is only applicable for dilute solution.

Q.2 Whether the pH of (i) a solution containing 10^{-8} g ion per litre H^+ ion and (ii) a solution of 10^{-8} **N** (or 10^{-8} **M**) HCl will be the same?

Hint: No. pH of 1st solution will be $-\log 10^{-8} = 8$ i.e. alkaline, but the pH of 2nd solution will be less than 7 i.e. acidic, because total $[H^+]$ in the solution will be $[H^+]$ coming from HCl i.e. 10^{-8} g ion per litre plus $[H^+]$ coming from dissociation of water, say X g ion per litre.

As, $[H^+] \times [OH^-] = 10^{-14}$ at 25°C, then $(X + 10^{-8}) \times X = 10^{-14}$

Hence, X = 9.51×10^{-8} g ion per litre

pH of the solution will be = $-\log(10^{-8} + 9.51 \times 10^{-8}) = 6.978$

Q.3 What are criteria of a substance to be considered as primary standard?

Hint: See section 8.4.5

Q.4 Why $K_2Cr_2O_7$ is used for titration of Fe^{2+} in HCl medium but not $KMnO_4$?

Hint: MnO_4^--Mn^{2+} system having higher standard (reduction) potential (+1.52v) than Cl_2-$2Cl^-$ (+1.36v) and Fe^{3+}-Fe^{2+} (+0.77v) can oxidize both of them and introduce error in the titration, but $Cr_2O_7^{2-}$-Cr^{3+} (+1.33v) can oxidize only Fe^{2+} but not Cl^-.

Q.5 What do you mean by formal potential?

Hint: See section 8.4.9.4

Q.6 What is the criterion of selection of a redox indicator in any redox reaction?

Hint: The formal potential of indicator ($E^{0/}_{in}$) should differ by at least 0.15v from the standard (formal) potentials of two redox systems involved in the titration. That indicator is the best, if the formal potential value of indicator is around the average of standard (formal) potentials of the two redox systems.

Q.7 What are the criteria of a metal ion indicator?

Hint: See section 15.4.8.2

Q.8 Prepare a 250 ml solution of (a) 0.25 M, (b) 0.25 **N** solution of Na_2CO_3.

Hint: (a) The gram molecular weight of Na_2CO_3 is (2 x 23 + 12 + 3 x 16) = 106 g

Now, 1000 ml 1 **M** Na_2CO_3 solution contains 106 g Na_2CO_3

So, 250 ml 0.25 **M** ,, ,, ,, ,, ,, $\frac{106 \times 250 \times 0.25}{1000} = 6.625g\ Na_2CO_3$

(b) The gram equivalent weight of Na_2CO_3 is $\frac{106}{2} = 53$ g

Now, 1000 ml 1 **N** Na_2CO_3 solution contains 53 g Na_2CO_3

So, 250 ml 0.25 **N** ,, ,, ,, ,, ,, $\frac{53 \times 250 \times 0.25}{1000} = 3.3125g\ Na_2CO_3$

Q.9 Prepare a 500 ml 1000 ppm K solution with KCl.

Hint: 1000 ppm K solution means 1000 mg K will remain per litre of solution

The gram molecular weight of KCl is (39.1 + 35.457) = 74.55 g

1 litre 1000 ppm K solution contains 1000 mg K $\equiv \frac{74.55\times1000}{39.1} = 1906.7$mg Kcl

KCl

Thus, to get 1 litre 1000 ppm K solution dissolve 1906.7 mg or 1.9067g KCl in water and make up volume to 1 litre.

Thus, to prepare 500 ml 1000 ppm K solution with KCl, dissolves $\frac{1.9067}{2} = 0.9533$ g KCl in water and make up volume to 500 ml litre.

Q.10 Prepare a 100 ml of 20 ppm Na solution from 1000 ppm Na solution.

Hint: Here, V_1= 100 ml, S_1 = 20 ppm, S_2 = 1000 ppm, V_2 = ?

$$= V_2 = \frac{V_1 \times S_1}{S_2} = \frac{100\times20}{1000} = 2 \text{ ml}$$

So, to get 20 ppm Na solution of 100 ml, dilute 2 ml 1000 ppm Na solution to 100 ml with water.

Q.11 What is the pH of the solution made up by mixing 50 ml of solution A having pH 1 with 50 ml of solution B having pH 2?

Hint: 1000 ml of solution A contains 10^{-1} g mol of H^+. Its 50 ml contains

$\frac{1\times50}{10\times1000} = \frac{1}{200}$ g mol of H^+.

Similarly, 50 ml of solution B contains $= \frac{1}{2000}$ g mol of H^+.

Therefore, 100 ml of final solution contains $\frac{1}{200} + \frac{1}{2000} = \frac{11}{2000}$ g mol of H^+.

H^+ concentration of the final solution is $\frac{11}{200}$ g mol L^{-1}.

Therefore, the pH of the final solution is $-\log \frac{11}{200}$ = -log11 +log200 = -1.04 + 2.30 = 1.26

16

Instrument and Its Principle of Functioning

16.1 Balance and Weighing

16.1.1 Principle of Functioning of a Balance

It is one of the important tools used in the laboratory and one should have thorough knowledge about its assembly, use and care, otherwise it may be the initial step for introducing error in any physical or chemical analysis. Every balance has its maximum load limit and sensitivity depending upon its construction.

Essentially balance may be regarded as a rigid beam, BC having a central fulcrum, O and two arms of equal length, d_1, d_2 (Fig.16.1). The both ends of beam have prism edges upon which the balance pans are suspended by means of suitable suspension.

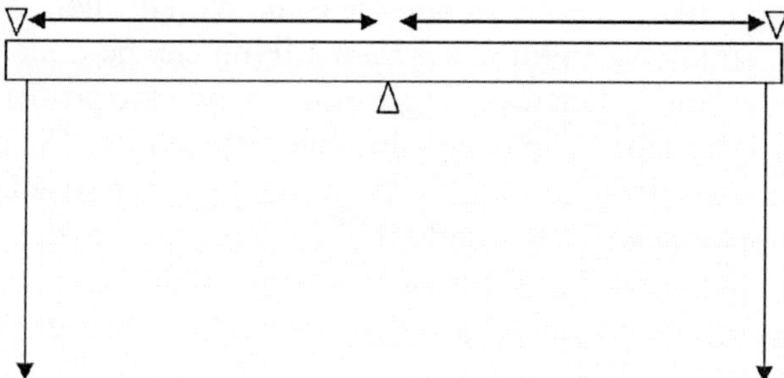

Fig.16.1: Principle of a balance

According to the principle of lever the forces, F_1 and F_2 acting on the two prisms edges at equilibrium is:

$$F_1 \text{ x } d_1 = F_2 \text{ x } d_2 \quad (16.1)$$

or, $F_1 = F_2$ (as two arms have equal length)

Again, if M_1 and M_2 are the masses placed at the left and right pan, respectively and g, the acceleration due to gravity, then $F_1 = M_1 g$ and $F_2 = M_2 g$. At equilibrium:

$$F_1 / F_2 = M_1 / M_2$$

Though actually we compare the masses placed on two balance pans, but customarily we use the term "weight" synonymous with mass at a given place. However, in strictly speaking, the analytical balance determines the mass, not weight.

16.1.2 Description of an Analytical Balance

The beam of the balance is light in weight and rigid. When the balance is in use the beam is mounted on a knife edged agate at the centre. At the two ends of the beam there are two terminal knife-edged agates equally distant from the central agate. From the terminal agate balance pans are suspended with the help of suitable suspension. At the centre of the beam a pointer is attached which can move over the scale fixed at the bottom of the pillar. The pointer is used to indicate the deflection of the beam from the equilibrium position.

Fig. 16.2: Analytical Balance

By moving rider, platinum wire along the beam with the help of rider hook and rider carriage smaller weighing less than 1.0 mg can be done. The balance is leveled with the help of two leveling screws. To indicate proper leveling plumb line suspended from the column or spirit level is provided. For equilibrium two adjustment screws attached to the sides of the beam are used. The balance is provided with pan arrest at the bottom of each pan to prevent or minimize the injury at the agate and the disturbance of adjustment during operation. The balance is operated by moving a screw usually attached at the centre of the base.

Now-a-days most of the modern laboratories are equipped with electronic balance, which provides lot of conveniences such as, selecting and removing weights, smooth release of balance beam and pan support, returning the beam to rest are eliminated. With an electronic balance, operation of single on-off control permits the operator to read the weight of an object on the balance pan immediately from digital display or in the form of printed record. Taring facility permits the operator to balance out the weight of the container, thus on addition of material to the container the weight of the material only is obtained.

Electronic balances operate by applying an electromagnetic restoring force to the support to which the balance pan is attached, so that when an object is added to the pan, the resultant displacement of the support is cancelled out. The magnitude of the restoring force is governed by the amount of current flowing in the coils of the electromagnetic compensation system and this is proportional to the weight placed on the balance pan.

16.1.3 Quality of a Good Analytical Balance

a) The balance must be accurate and give same result in successive weighing – this can be achieved if the arms are of equal length, the beam is rigid and three prisms lie in the same plane and parallel to one another.

b) The balance must be stable i.e. the beam must return to the horizontal position after swinging – it depends upon proper adjustment of centre of gravity.

c) the balance must be sensitive, i.e. 0.1 mg should be readily detectable with average load-sensitivity of balance which increases with the length of balance arm and lightness of beam. [Sensitivity of a balance is the number of scale divisions that the rest point (or equilibrium point) is displaced by an excess weight of 1 mg].

d) The period of oscillation should be short – universally accepted analytical balance (short arm) should have the oscillating period of 5-10 seconds.

16.1.4 Use and Care of Balance

1. The balance should be placed upon the foundation which is free from vibration, preferably upon the concrete or stone slab. The balance should be kept away from fumes of acid.
2. The balance should be properly leveled with the help of leveling screws.
3. The balance beam and pan support should be raised to protect the agate prisms against any unnecessary load impact when the balance not in use.
4. To release the balance, the arresting screw should be moved very slowly to minimize the swing period and to avoid any disturbance in balance setting.
5. Objects to be weighed must be brought to the room temperature if it has been heated before weighing.
6. Substances must be weighed in a suitable container or on the nonabsorbent paper placing over the balance pan.

7. Both weights and substances must be placed at the centre of the pan.
8. During weighing any removal or addition of weights or substances should be done always by raising the balance pan support to avoid any injury in the prisms and disturbance of balance setting.
9. Any substance which is corrosive to the balance pan should not be placed directly on it.
10. The balance must not be overloaded.
11. Weights should only be handled with the forceps provided with the weight box not by hand directly.
12. Nothing would be left on the pan after the completion of weighing. The balance should be kept clean and dirt free with the help of camel hair brush.

16.1.5 Sources of Error in Weighing

1. Incorrect balance setting.
2. Inaccuracy of the weights.
3. Change of condition of the substance during and between weighing - this is happened occasionally during weighing of substance which has the tendency to absorb moisture from the atmosphere.
4. Effect of the buoyancy of the air upon the object and weights.

16.2 pH meter

It has been discussed in chapter 8 that when a metal is immersed in a solution of its own ion, a potential is developed which is termed as electrode potential. In order to determine the electrode potential it is necessary to have another electrode of known potential. The two electrodes then are combined to form a voltaic cell, the E.M.F of which can be directly measured. The E.M.F of the cell will be the arithmetic sum or difference (depending upon the sign) of the two electrode potentials.

An electrode potential varies with the concentration of the ion. Thus, when the two electrodes of same metal are dipped into the solution of its ion of different concentration, a cell may form which is known as concentration cell. The E.M.F of this cell will be the arithmetic difference of electrode potentials, if the liquid-liquid junction potential can be eliminated by inserting a salt bridge.

16.2.1 Normal / Standard Hydrogen Electrode

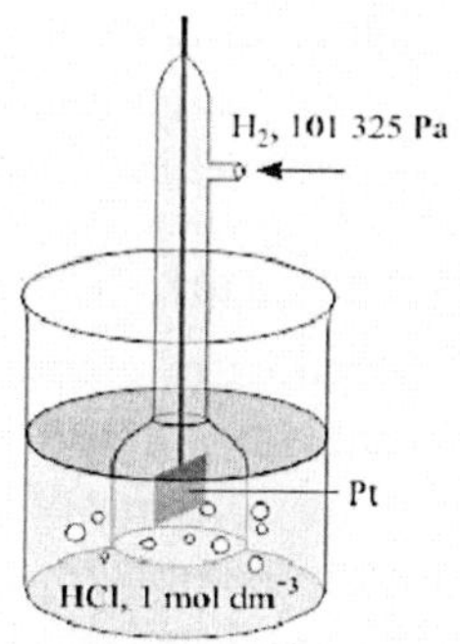

Fig.16.3: Hydrogen electrode

The electrode potentials are quoted with reference to the standard hydrogen electrode, and hence this is regarded as the **primary reference electrode**. This electrode consists of a platinum foil coated electrolytically with platinum black dipped in a solution of hydrochloric acid of unit hydrogen ion activity (This corresponds to 1.18 **M** HCl at 25°C). Hydrogen gas at a pressure of 1 atmosphere passed over the platinum foil through the pipe.

The platinum black has the unique property of adsorbing hydrogen gas and permits the change from gaseous state to ionic state and reverse without any hindrance. Therefore, it behaves like a hydrogen electrode. Conventionally, the potential of hydrogen electrode immersed in a solution of unit hydrogen activity in equilibrium with hydrogen gas at one atmospheric pressure at all temperature is equal to zero.

Standard Hydrogen Electrode: Pt, H_2 (1 atm) | H^+ (a = 1); E^0_{H2}= zero

16.2.2 Electrodes in pH meter

A. Reference Electrode

In routine work, the use standard hydrogen electrode has some practical difficulties. However, some special cases where pH is above 12 require its application as no other electrode can be used satisfactorily. The standard hydrogen electrode can be substituted with other reference electrode, the secondary standard half cell, the potential of which is known accurately relative to the standard hydrogen electrode. The most widely used reference electrode is the **'calomel electrode'**, consists of mercury and calomel (Hg_2Cl_2) covered with potassium chloride solution of definite concentration (0.1**M**, 1**M** or saturated). The electrode reaction is:

$$Hg_2Cl_2 \text{ (s)} + 2e \rightleftharpoons 2Hg \text{ (liq)} + 2Cl^-$$

The potential (including the liquid junction potential) of the saturated (prepared with saturated KCl) calomel electrode is determined with reference to standard hydrogen electrode.

Calomel (sat) Electrode Potential: Hg | Hg_2Cl_2, KCl (sat.) || H^+ (a=1), H_2 | Pt; E_{25}= 0.244 v

This potential cancels out when the pH meter is standardized with standard pH buffer.

Another important reference electrode is the Ag-AgCl electrode, consists of Ag wire coated eletrolytically with a thin layer of silver chloride dipped in a potassium chloride solution of known concentration.

B. Indicator Electrode

The indicator electrode of a cell is one whose potential is dependent upon the activity (also the concentration) of a particular ionic species whose concentration is to be determined. When hydrogen ions are involved, a hydrogen electrode can obviously be used as indicator electrode, but its function can also be performed by other electrode, foremost among which is the **glass electrode.** It is the most widely used hydrogen ion responsive indicator electrode. It is made up of special type glass (low melting point, soft, relatively high electrical conductivity and high hygroscopicity). For example, composition of Corning 015 glass is 72% SiO_2, 6% CaO and 22% Na_2O. In this half cell within the glass bulb calomel electrode (or Ag-AgCl electrode) remains dipped into dilute HCl (a_{H1}^+). The electrode potentials developed by this electrode are membrane potential [This potential develops between inside dilute HCl solution (a_{H1}^+) and outside unknown solution (a_{H2}^+) across the glass membrane], plus the potential of Ag-AgCl-HCl reaction inside the electrode.

$$AgCl + e^- \rightleftharpoons Ag + Cl^-$$

Since the activities of Ag and AgCl are unity and Cl^- activity is fixed by the constant HCl concentration, the Ag-AgCl potential ($E°_{25} = 0.222$ v) is constant. This potential is also cancelled out when the electrode is standardised against a standard pH buffer solution.

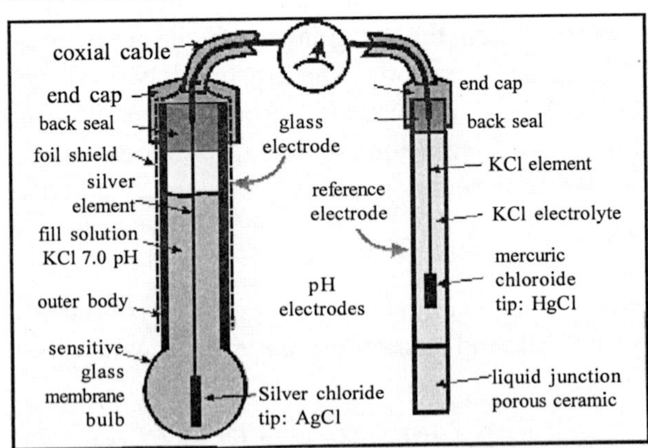

Fig. 16.4: Electrodes in pH meter

$$\underbrace{Ag, AgCl, KCl\ (sat.)\ |\ H^+\ (a_1)\ |\ Glass}_{\text{Glass electrode}}\ \underbrace{|\ H^+\ (a_2)}_{\text{Test sol}^n}\|\ \underbrace{KCl\ (sat.),\ Hg_2Cl_2,\ Hg}_{\text{Reference electrode}}$$

The glass electrode has several advantages over the others:

a) It does not expel CO_2 from the system during estimation.

b) It is adaptable to the solution of any consistency.

c) Unlike hydrogen electrode and quinhydrone electrode, it is not affected by oxidizing or reducing agents.

d) It does not require H_2 gas.

16.2.3 Measurement of pH or H^+ ion Concentration

It has been shown in chapter 15 that electrode potential of a metal in solution of its ion at 25°C is given by the expression:

$$E_{25^\circ} = E_0 + \frac{0.0591}{n} \log a_{M+} \tag{16.2}$$

The expression for a hydrogen electrode at 25°C is:

$$E_{25^\circ} = E_H{}^0 + 0.0591 \; \log a_H^+$$

Where, $E_H{}^0$ is the standard potential of hydrogen electrode, conventionally equal to zero. Thus, the above expression transforms into

$$E_{25^\circ} = 0.0591 \; \log a_{H+} \text{ (at 25°C)}$$

For determination of pH or H^+ ion concentration using hydrogen electrode, the platinized platinum electrode is dipped into the solution whose pH is to be determined and a steady stream of moist hydrogen at 1 atmosphere is passed through the solution over the electrode. If $a_H{}^+$ is the activity of hydrogen ions in the given solution then its electrode potential would be,

$$E_{HI} = E^0{}_{HI} + \frac{R.T}{F} \; \ln a_H{}^+$$

This half-cell (half cell$_1$) is now coupled with reference electrode (half cell$_2$), say a saturated calomel electrode (E_{Cal} = 0.244 volts) through a KCl salt bridge.

$$\underbrace{\text{Pt, } H_2 \text{ (1 atm)} \mid H^+ \text{ (a = x)}}_{\text{Half cell}_1} \parallel \underbrace{Hg_2Cl_2\text{, Hg}}_{\text{Half cell}_2}$$

In the above scheme, the double vertical line connotes a liquid junction, whose potential (liquid junction potential) is considered to be eliminated by a salt bridge and single vertical lines represents a metal electrode boundary at which potential difference is taken into account.

If E is the observed e.m.f of the cell, then,

$$E_{observed} = E_{calomel} - E_{H1 electrode}$$

$$= E_{calomel} - 0.0591 \log a_{H}^{+}$$

$$\text{or}, pH = \frac{(E_{observed} - E_{calomel})}{0.0591} \qquad (16.3)$$

However, the most commonly used electrode for pH determination now-a-days is the glass electrode. A membrane potential will develop across glass membrane separating two solutions and the magnitude of this potential depends chiefly on the pH of the solutions. In one of the solutions kept constant and the pH of the other solution varied, then the electrode potential follows the relation

$$E_G = E^0{}_G + \frac{R.T}{F} \ln a_H{}^{+}$$

That means the electrode functions in the same way as a hydrogen electrode does. $E^0{}_G$ includes a small asymmetry potential which exists across the glass memebrane due to internal strain. When this electrode is coupled with a reference electrode, the e.m.f of the cell will be

$$E = E_{cal} - E_G$$

In practice, the assembly of glass electrode is first calibrated with a solution of known pH (say, pH_1), and say, its e.m.f is E_1. This solution is then substituted with the unknown solution pH (say, pH_2) and its e.m.f is E_2. Then,

$$\Delta E = E_1 - E_2 = 2.303 \frac{R.T}{F} (pH_1 - pH_2) = 0.0591 (pH_1 - pH_2)$$

The pH of the solution can thus be measured. It is thus immaterial what reference electrode is being employed provided the same is used in both the measurements.

16.2.4 Types of pH meter

a) Potentiometric or Slide wire type

The glass electrode is incorporated in an ordinary potentiometric circuit and the off balance current is measured by milliammeter through detection of the balance point.

b) Direct Reading type

The E.M.F of glass electrode cell is impressed across a high resistance and D.C. current flowing through the resistance is measured by milliammeter. The deflection directly indicates the pH. Slide wire type is characterized by high precision, easy servicing and easy detection of fault in the electric circuit, while the direct reading type is characterized by greater rapidity of reading but with less precision.

16.2.5 Principle of Functioning of pH meter

The line CD represents a uniform resistance wire of a potentiometer. A standard cell (E= 1.0183 v) is attached to CD wire by P and P/ terminals and R is a variable resistance (Fig.16.5). Now, the variable resistance, R is so adjusted that the potential difference between D and C does not allow any current to flow through the galvanometer i.e. no deflection in the galvanometer needle. This is only possible when the potential drop between P and P/ is balanced by potential of the standard cell, E and ultimately current from the working cell flows through the main circuit i.e. DP/ PC.

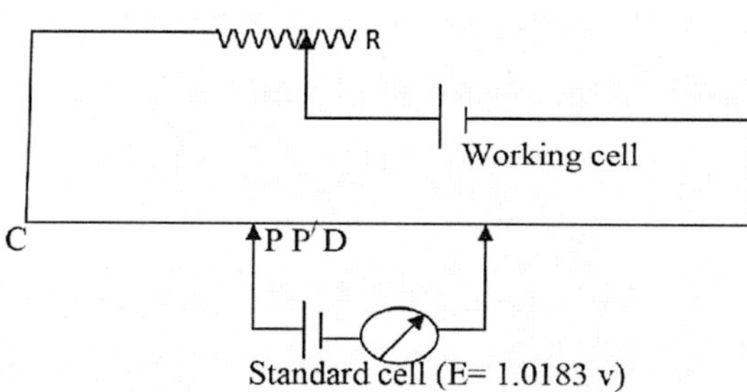

Fig.16.5: Principle of the pH meter with its electric circuit

If the space between P and P/ is divided off into a scale of 1018.3 equal units, then each unit will correspond to one millivolt. Since, the potential of glass electrode is directly proportional to the pH, the space between P and P/ can be calibrated directly into pH scale. It has been found that at 25°C temperature 59.16 millivolts (0.0591v) correspond to one pH unit.

When the potential of an unknown cell is to be measured, the unknown cell is substituted for the standard cell, E and by setting of R by moving P/ along CD keeping P constant till the galvanometer shows no deflection. At the balance point the spacing between P and P/ represents either (i) potential of the unknown cell, (ii) pH as scale described.

16.2.6 Care of Glass Electrode

The glass electrode is fragile, expensive and subject to excessively rapid deterioration if proper care is not taken. Following precautions would help to increase the longevity of the glass electrode.

(i) The electrode should not be kept dipped in the test solution for longer period particularly if it is highly alkaline, pH more than 9.

(ii) After use, the electrode should be washed off thoroughly with strong stream of distilled water from wash bottle; if necessary by mild rubbing with tissue paper to remove the film of $CaCO_3$. Electrode should be stored dipping the bulb in distilled water and drying out of electrode must be avoided.

(iii) Failure of electrode is indicated when after standardization it gives slow response to larger pH changes.

16.2.7 Operation of pH meter

- Switch on the pH meter.
- Adjust the temperature setting at the laboratory condition.
- Keep it on for at least 30 minutes for warm up.
- Standardize the instrument by alternately immersing the electrodes in three buffer solutions (pH 4.0, 7.0, 9.2 or other buffer solution of pH near the test solution) for few times.
- Each time flush off the electrodes with distilled water.
- Place the electrodes in test solution or suspension and turn pH dial
- Record the reading when the galvanometer balance reaches.
- Repeat the test until stability is assured.
- Turn off pH meter and wash off the electrodes thoroughly with distilled water.

16.3 Conductivity Bridge

By definition electrical conductance or conductivity is the reciprocal of the resistance.

$$C = \frac{1}{R}$$

Where, R is the resistance in ohms and C is the conductivity of a solution expressed in mhos.

Likewise, specific conductance of a solution is the reciprocal of the specific resistance and can be defined as the conductance between electrodes of 1 sq. cm cross section placed 1cm apart. Specific conductance is generally expressed in mhos/cm or millimhos/cm (1 mhos/cm = 1000 millimhos/cm).

16.3.1 Principle of Conductivity Bridge

The well-known Wheatstone's meter bridge method is directly not suitable for measuring specific conductance, because flow of current will be accompanied by electrolysis of the solute and eventually accumulation of the product of electrolysis on the electrodes will change the effective conductance. To avoid this problem alternating current (A.C) is employed so that polarizing effect of the products of electrolysis due to current flow in one direction will be neutralized by the effect of current flow in other direction.

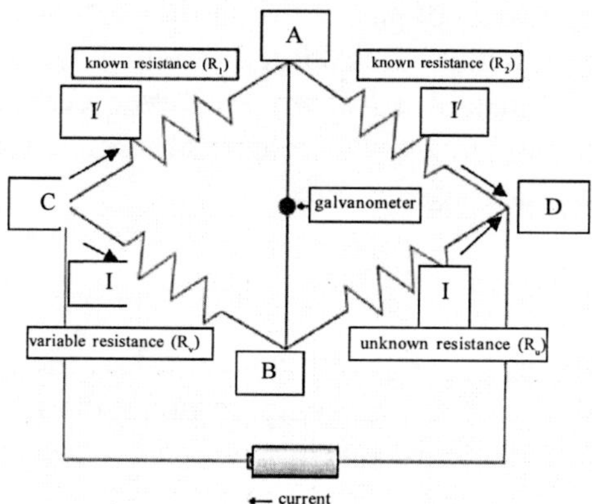

Fig.16.6: Schematic diagram of Wheatstone's bridge

In a conductivity bridge there are two fixed resistance R_1 and R_2 and a variable resistance, R_V are connected in a branched circuit with conductance cell having resistance cell, R_U. An AC potential is applied at C and D.

Usually 1000 cycle per second source is used but some bridges operate on a 60 cycle per second power source. The variable resistance, R_V is so adjusted that no current flows through AB indicated by no sound in the earphone or shift in electric eye. At this R_V, the potential of A and B will be the same and the voltage drop between A and D, $I'R_2$ must be equal to voltage drop between B and D, IR_U

$I'R_2 = IR_U$

Similarly, $I'R_1 = IR_V$

$$\text{Then, } \frac{IR_U}{IR_V} = \frac{I'R_2}{I'R_1}$$

$$\text{Then, } \frac{IR_U}{IR_V} = \frac{I'R_2}{I'R_1}$$

Since, R_1 and R_2 are constant; dial on R_V can be calibrated to read $1 / R_U$, the conductance of the test solution.

16.3.2 Determination of Cell Constant of Conductance Cell

In conductance cell, two electrodes of platinum foil are placed in a cylindrical vessel. The electrodes are coated with platinum black deposited on it by electrolysis of platinum chloride. Connections are made by platinum wires fused through glass tubes. Each conductance cell has a definite cell constant and this cell constant is known. However, cell constant of a conductance cell can be determined from the electrical conductance of a standard KCl solution using the following formula:

$$K = \frac{L}{C} \quad (16.4)$$

Where, K is the cell constant, L is the specific conductance of standard KCl solution at a given temperature, mmhos/cm and C is the observed conductance of the standard solution with the given conductance cell, mmhos.

Thus, specific conductance = Observed conductance × Cell constant

[Specific conductance of 0.02 M KCl at 18°C and 25°C are 2.39 and 2.768 mmhos/ cm, respectively and of 0.01 M KCl at 25°C is 1.413 mmhos/cm]

16.3.3 Determination of Conductance of Solution

The conductance cell is generally stored immersed in distilled water.

- Before use rinse the conductance cell thoroughly with distilled water and if the volume permits, wash at least twice with the test solution. If the test solution is insufficient, then wash with distilled water and dry with acetone to avoid the dilution of the test solution.
- Adjust cell constant if known by turning the cell constant knob.
- If cell constant is not known, then determine it using standard KCl solution.
- Dip the cell in test solution and record the reading as conductance when bridge is balanced.
- Calculate specific conductance of test solution by multiplying conductance with the cell constant.
- Make the temperature correction of conductance before giving the final result. Usually electrical conductance increases by 2 per cent per degree centigrade temperature increase.

16.4 Colorimeter and Absorption Spectrophotometer

Light absorption is very common phenomenon as we see different coloured substance in the earth. White colour changes to coloured light when some wavelengths are absorbed by the substance and some are reflected or passing through the substance. The colour of a substance is the characteristics of the wavelength not absorbed by it.

Colorimeter is concerned with the determination of the concentration by the measurement of relative absorption of visible light with respect to a known concentration of the substance. Absorption spectrophotometry concerns with the determination of the concentration by measurement of relative absorption of radiation of visible, ultraviolet or infrared ranges of electromagnetic radiation. Quantitative estimation of colour intensity began with the comparison of standard and test solution in the tube of similar shape and size with the help of naked eyes. Later to avoid the personal error in the estimation photo electric cell was introduced. Each type of photo cell has its specific band of wavelength of maximum sensitivity. Wavelength of specific band is obtained by passing white light through filter (plates of coloured glass or gelatin) or prism.

The advantage of colorimetric and spetrophotometric method is that with the help of simple, less expensive instrument one can measure very minute quantities of substance. Light consists of electromagnetic waves of different wavelengths. White light covers the entire visible spectrum between 400 to 760 mμ or nm (1 mμ or nm = 10-6 mm) which human can detect.

Table 16.1: Approximate wave lengths of coloured light

Colour of the light	Wavelength, mμ
Ultra violet	< 400
Violet	400 – 450
Blue	450 – 500
Green	500 – 570
Yellow	570 – 590
Orange	590 – 620
Red	620 – 760
Infrared	>760

16.4.1 Principles of Colorimetry and Absorption Spectrophotometry

When light falls upon a homogeneous medium a portion of the incident light is absorbed by the medium, a small portion is reflected from the medium and rest is transmitted through the medium.

$$I_0 = I_a + I_r + I_t \qquad (16.5)$$

Where, I_0 is the intensity of incident light, I_a, I_r and I_t are that of absorbed, reflected and transmitted, respectively. At air-glass interface the value of I_r is very less (usually < 4 %) and we can discard the fraction and the equation (16.5) can be rewritten as:

$$I_0 = I_a + I_t$$

Lambert (1760) and Beer (1852) investigated the relation between I_0 and I_t. Lambert's and Beer's laws are the basis of any colorimetric and spectrophotometric analysis.

Lambert's Law

This law states that when monochromatic light passes through a transparent medium the rate of decrease of light intensity with the thickness of the medium is proportional to the intensity of light. It can be expressed as:

$-dI / dt \propto I$

or, $-dI / dt = k I$

I is the intensity of incident light and t is the thickness of the medium, k is the proportionality factor.

$-dI / I = k\, dt$

$-\ln I + A = kt$ (on integration) (16.6)

A is a constant, when $t = 0$, then $I = I_0$ and equation (16.6) transforms into:

$\ln I_0 = A$ (16.7)

Substituting the value of A of the equation (16.7) in equation (16.6), we get

$\ln I_0 - \ln I_t = kt$

$$or, \ln \frac{I_0}{I_t} = kt$$

or, $I_t = I_0 e^{-kt}$ (16.8)

$= I_0 10^{-0.4343kt}$ [converting Natural logarithm to Briggsian logarithm]

or, $I_t = I_0 10^{-Kt}$ [$K = 0.4343k$]

Where, I_0 is the intensity of incident light, I_t is the intensity of light transmitted through the medium of thickness, t; K is termed as extinction coefficient. Extinction coefficient may be defined as the reciprocal of thickness requires to reduce the ratio of I_t / I_0 to 1/10.

$I_t / I_0 = 10^{-Kt}$

or, $10^{-1} = 10^{-Kt}$

$$or, K = \frac{1}{t}$$

The ratio of I_t / I_0 is termed as transmittance or transmission **(T)** and I_0 / I_t is the opacity. Thus, transmittance may be defined as the ratio of the intensity of transmitted light to that of incident light. Logarithm of opacity is the **optical density (D) or absorbance (A).**

$$A = \log \frac{I_o}{I_t} = \log \frac{1}{T} = -\log T \tag{16.9}$$

Beer's Law

Lambert in his law described the relation between the intensity of the transmitted light and thickness of absorbing medium, while Beer (1852) established the relation between the concentrations of the absorbing medium i.e. coloured substance with the intensities of the transmitted light. He also found the similar relationship between the concentration and the transmission as was observed by Lambert between the thicknesses of the medium and transmission [Equation (16.6)] i.e. the intensity of transmitted light decreases exponentially as the concentration of then absorbing substance increases arithmetically. This may be expressed as:

$$I_t = I_0 e^{-kc} = I_0 10^{-0.4343kc} = I_0 10^{-Kc} \tag{16.10}$$

Where, c is the concentration. k' and K' are the constants.

Combining the Equation (16.5) and (16.6) we get

$$I_{t=I_{0.}e^{-\varepsilon ct}}$$

$$\text{or}, \log \frac{I_0}{I_t} = \varepsilon ct$$

This is the fundamental law of colorimetric or spectrophotometric analysis known as **Lambert-Beer Law.** The value of ε depends upon the method of expression of concentration. If c is expressed as gram moles pet litre and t in centimeter then ε is the molecular extinction coefficient.

15.4.2 Deviation from Beer's Law

Usually, Beer's law holds good over a wide range of concentration of coloured ion or coloured non electrolyte dissolved in solution, if the structure of ion does not change with concentration. But deviation is found when the coloured substance ionizes, dissociates or associates in solution, as the nature of the species in solution will vary with concentration. Thus, it is advisable to test whether the substance follows Beer's law or not by plotting log (1/T) or the optical density (D) against concentration. If the substance follows Beer's law then the straight line must pass through the origin.

The solution which does not follow Beer's law; it is advisable best to prepare a standard curve using a series of standard solution. The dilution or the concentration of the test solution must lie within the range of standard solutions used.

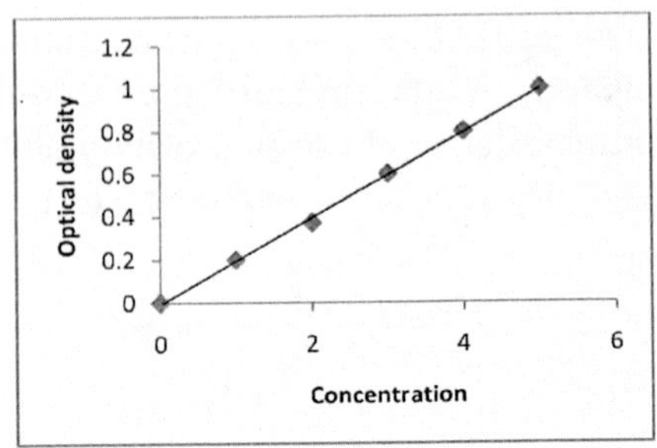

Fig. 16.7: Optical density against concentration

16.4.3 Colorimeter

The essential components of a colorimeter consist of (i) light source, (ii) a suitable light filter to obtain monochromatic light of desirable wavelength, (iii) a glass cell for holding the solution, (iv) a photo electric cell to receive the radiation of light transmitted through the solution and to convert the light energy to electrical energy and (v) a measuring device i.e. galvanometer to read the output of photo electric cell.

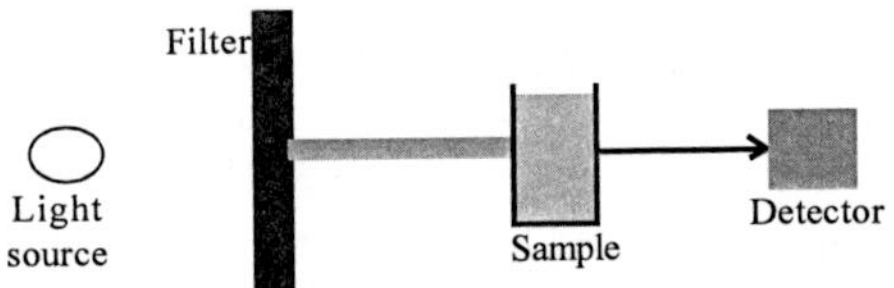

Fig.16.8: Diagram of a typical colorimeter

The components of spectrophotometer are the same with the colorimeter only difference is that instead of light filter as in colorimeter, spectrophotometer has a monochromator used for getting a narrow band of radiant energy i.e. monochromatic ight.

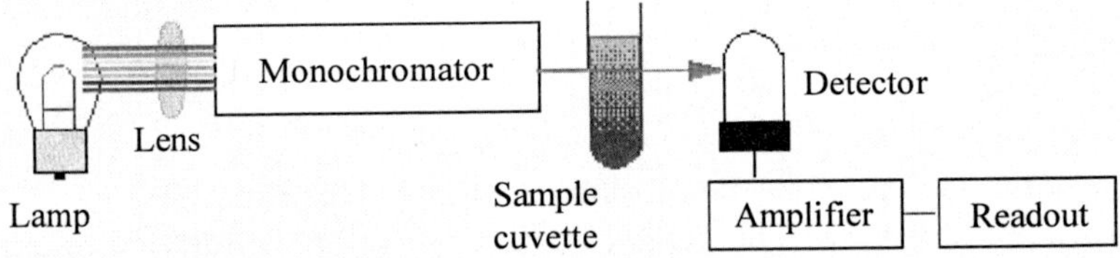

Fig. 16.9: Diagram of a typical spectrophotometer

16.4.4 Operation of Colorimeter / Spectrophotometer

- Allow the instrument to get warmed for at least 20-30 minutes.
- Rotating wavelength dial select the required wavelength.
- Closing the light shutter keep the dark current adjustment at zero.
- Place the blank solution in glass cell or sample holder and adjust the percent transmission to 100 by turning the blank set control knob. [Blank solution is prepared by adding all the reagents in similar proportion as in the test solution or sample except the ion or element concerned].

- Replace the blank with standard solutions prepared similarly as the blank plus adding known amount of ion concerned.
- Read percent transmission directly from the galvanometer and prepare a standard curve by plotting percent transmission in abscissa and concentration in ordinate in a semi logarithmic graph paper.
- Now replace standard solution with the test solution. From the galvanometer reading determine the concentration of the test solution.

16.4.5 Criteria for Satisfactory Colorimetric Determination

- Colour should be highly specific for the element to be determined.
- With changing concentration the colour intensity must change.
- The colour must be stable for at least time sufficient for its determination.
- The solution must be free from any turbidity.
- The colour should be developed rapidly under specific experimental condition.

16.5 Flame Spectroscopy

If a solution containing metallic salt (or some other metallic compound) is aspirated into a flame (acetylene burning in air), a vapour is formed containing atoms of that metal. Some of these gaseous metal atoms may rise to an energy level (**excited state**) which is sufficiently high to permit the emission of radiation characteristics of that metal e.g., the characteristic yellow colour imparted to the flames by compounds of sodium. This is the basis of **flame emission spectroscopy (FES)**, often referred to as flame photometry. However, a much larger number of the gaseous metal atoms will normally remain in an unexcited or in other words in **ground state**. These ground state atoms are capable of absorbing radiant energy of their own specific resonance wavelength, which in general is the wavelength of the radiation that the concerned atoms would emit if excited from the ground state. Thus, if light of the resonance wavelength is passed through a flame containing atoms in question, then a fraction of the incident light will be absorbed by the atoms and the extent of absorption will be proportional to the number of atoms present in ground state in the flame. This is basis of **atomic absorption spectroscopy (AAS)**. The steps by which the metal atom present in the flame get excited by absorbing either thermal energy from the flame or by absorbing radiant energy of their own specific wave length may be summarized as follows:

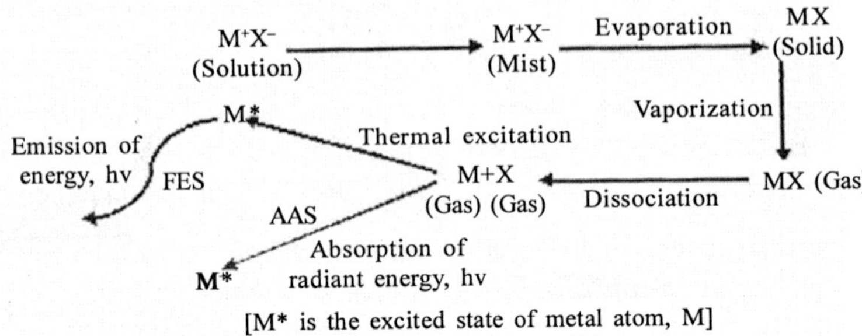

Fig. 16.10: Steps behind excitation of atoms in the flame

Consider the simplified energy level diagram shown below (Fig.16.11) where E_0 represents the ground state in which the electrons of a given atom are at their lowest energy level and E_1, E_2, E_3, etc. represent the higher and excited energy levels. Transition between two quantized energy levels, say from $E_0 \rightarrow E_1$ corresponds to absorption of radiant energy and the amount of energy absorbed is given by Bohr's equation

$$\Delta E = E_1 - E_0 = h\upsilon = hc/\lambda \tag{6.11}$$

Where; c = velocity of light, h = Planck's constant, υ = frequency (No. of wave or cycle pass through a point in a second) and λ = wavelength of radiation absorbed. Eventually, the transition from $E_1 \rightarrow E_0$ corresponds to the emission of radiation of energy, ΔE of frequency, í. Since an atom of a given element gives rise to a definite, characteristic line of spectrum, it follows that there are different excitation states associated with different elements. Consequence is that the emission spectra not only involves the transition from excited state to the ground state e.g. $E_3 \rightarrow E_0$, $E_2 \rightarrow E_0$, but also the transition from one excited state to another exicted state before going to ground state ($E_3 \rightarrow E_2$, $E_3 \rightarrow E_1$).

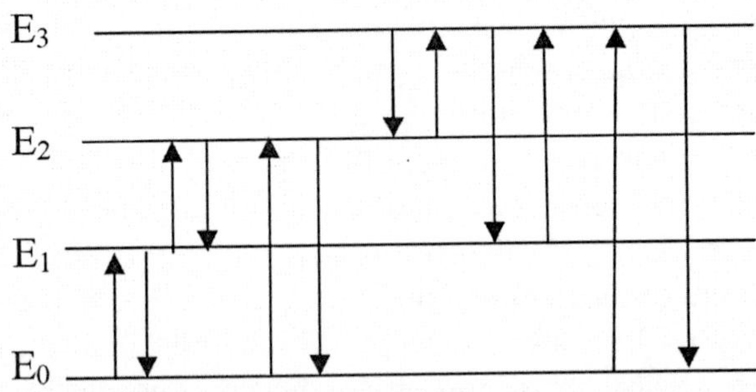

Fig. 16.11: Transition between energy levels

Thus it indicates that emission spectra of a given element is quite complex. Theoretically, it is always possible for absorption of radiation by already excited states to occur, e.g., $E_1 \rightarrow E_2$, $E_2 \rightarrow E_3$, etc. But in practice amount of excited atoms in relation to amount of atoms in ground state is very small. Thus, absorption spectrum of a given element is usually only associated with the transition from ground state to higher energy state and is much simpler in characteristics than emission spectrum.

The relationship between ground state and excited state atom population is given by the Boltzmann equation:

$$\frac{N_1}{N_0} = \left(\frac{g_1}{g_0}\right) e^{\frac{-\Delta E}{kT}} \qquad (16.12)$$

Where, N_1 = No. of atoms in the excited state, N_0 = No. of atoms in ground state, g_1 / g_0 = ratio of statistical weights for excited and ground states, "E = energy of excitation = hõ, k = Boltzmann constant and T = absolute temperature.

It is evident from the equation that (N_1/N_0) is dependent on both the excitation energy, "E and the temperature (T). An increase in temperature and decrease in excitation energy will result in a higher value for the ratio.

Atomic absorption spectroscopy is less prone to inter-element interferences than in flame emission spectroscopy. Further, due to high proportion of atoms in ground state to excited state in the flame it would appear that atomic absorption should be more sensitive than flame emission spectroscopy. The wavelength of the resonance line is critical factor as the resonance lines of low energy values are more sensitive than that of higher energy value for flame emission spectroscopy. Again, energy value is inversely proportional to the wavelength of the emission lines. Thus, sodium having an emission line wavelength of 589.0 nm can be successfully determined by flame emission spectroscopy but not zinc having emission line wave length 213.9 nm.

In case of atomic absorption spectroscopy, the absorbance, A is given by:

$A = \log \frac{I_o}{I_t} = K.L.N_o$ Where, I_0 is the intensity of incident light, I_t is the intensity of transmitted light, K is constant related to absorption coefficient, L is the path length and N_0 is the concentration of atom in the flame (No. of atoms per litre).

In case of flame emission spectroscopy, the detector response, E is expressed by

$$E = K.\alpha.C$$

Where, K is related to number of factors including atomization efficiency, α is the efficiency of atomic excitation and C is the concentration of the test solution.

Table 16.2: Flame temperature with various fuels

Fuel gas	Temperature (K)	
	In air	In nitrous oxide
Acetylene	2400	3200
Hydrogen	2300	2900
Propane	2200	3000

16.5.1 Flame Photometry

16.5.1.1 Flame Photometer

When a molecule or atom of a substance gets excited by absorbing energy from flame it will emit radiant energy upon coming back to its ground or stable state in the form of emission spectrum. The spectral characteristics may be different.

- **Continuous spectrum** without any individual lines emitted by incandescent solid.
- **Band spectrum** composed of individual bands closely packed lines emitted by excited molecules / atoms.
- **Line spectrum** consists of widely spaced distinct lines emitted by atoms.

Flame photometer is actually an instrument which measures the intensity of emitted radiation characteristic of the element concerned. The emitted radiation may be composed of several wavelengths and intensity of the wavelength corresponding to the most probable transition will be the highest. The characteristic line for each element is already established. Each excited atom releases one quantum (packet) of radiation upon returning to its ground state; thus total intensity of radiation will simply be the multiple of quantum of radiation or multiplication of quantum of radiation by the number of atom present in the flame i.e. the concentration of the test solution.

There are three discrete component systems in flame photometer.

- **Sample excitation system or Nebulizer burner system** – Nebulizer produces mist of the solution while flame helps the conversion of solution to excited gaseous atoms.
- **Radiation dispersion system or Monochromation system**– Filter or prism interposed in between flame and detector separates out desirable radiation from the other undesirable radiations.

- **Radiation measurement system or Photometric system**– It measures the intensity of the emited radiation.

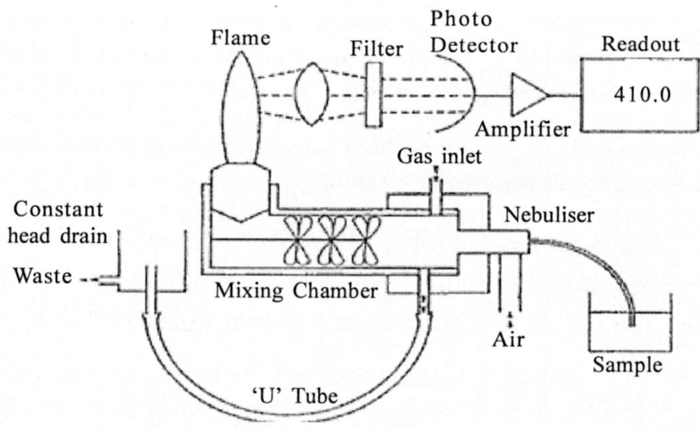

Fig. 16.12: Layout of flame photometer

Compressed air is passed into the nebulizer and the solution is thus drawn because of suction created at the nebulizer. Fuel (acetylene, propane or liquefied petroleum gas) is supplied to the burner at a given pressure mixed with the mist of solution in the way and ultimately the mixture is burnt at the burner. Radiation thus produced at the flame passes through the lens and then through iris diaphragm and finally through the filter specific for the element or prism. The filter interposed between burner and photo detector transmits only the strong lines of radiation of the element. The photo detector converts the incident radiation into the direct current which can be measured in the galvanometer.

Table.16.3: Wave length and minimum detection limit of some elements by flame photometer

Element	Wave length (mì)	Minimum detection limit (mg kg^{-1})
Calcium	422.7	0.2
Copper	327.4	10
	324.8	1.0
Barium	553.6	1.0
Lithium	670.8	0.02
Magnesium	285.2	1.0
Potassium	766.5	0.05
	404.4	2.0
Sodium	589.3	0.001
Iron	386.0	2.0
	372.0	1.0

16.5.1.2 Operation of Flame Photometer

- Run compressor and adjust the pressure (usually 0.5 kg/sq. m) according to the specification mentioned by the manufacturer.
- Fill the sample holder/ beaker with distilled water and be confirmed about the suction of solution by draining of water through drain pipe.
- Keep all the controlling knobs at their minimum position (extreme left) and turn on galvanometer switch.
- Turn on the gas supply and lit the gas at the burner.
- Adjust fuel supply by turning the fuel controlling knob so that ten separate blue coloured cones (one for each burner hole) of flame in two rows.
- Adjust galvanometer reading to zero with blank by turning zero set knob.
- With the suitable concentration of standard solution adjust galvanometer reading to 100 by turning full scale set knob.
- Repeat earlier two steps for at least twice to be confirmed about setting of galvanometer.
- Prepare a standard curve with the solutions of known concentration less than the concentration of full scale adjustment. Plot the standard curve putting galvanometer reading in abscissa against concentration of the solution in ordinate.
- Place test solution in the sample holder/beaker and note the galvanometer reading. Determine concentration of the solution by comparing the reading in the standard curve.
- In between two test solution suctions place sample holder / beaker filled with distilled water.
- After estimation first cut off fuel supply but run the compressor for few minutes to wash the nebulizer, mixing chamber and the burner with distilled water. Then turn off the compressor.

16.5.1.3 Advantage

Large number of sample can be analyzed within a very short period of time, thus, it is most suitable for routine soil analysis. However, for few numbers of samples conventional chemical method is faster.

16.5.1.4 Accuracy

Accuracy in flame photometric analysis depends largely upon the following factors-

- Absence of extraneous cations other than the element concerned.
- No formation of complex of high boiling point or high melting point with the anion present in the flame.
- Precision of filter which should only allow the characteristic spectrum to pass through.

Though most of the manufacturers claim nearly 1-3% accuracy in the determination of K, Na, Ca, Mg, and Li, but in practice less than 5% accuracy in the determination of Ca, Mg, and Li is rare.

16.5.2 Atomic Absorption Spectrophotometry

16.5.2.1 Atomic Absorption Spectrophotometer

In addition to three essential components as were found in flame photometer i.e. **(i) Nebulizer-Burner System, (ii) Monochromatic System** and **(iii) Detector and Read Out System**, atomic absorption spectrophotometer (AAS) also contains another important component i.e. **Resonance Line Source.**

- **Nebulizer-Burner System:** Like flame photometer the purpose of nebulizer is to produce mist of the solution drawn up through a capillary tube by the venturi action of a jet of air blown across the top of capillary. Nebulizer is capable for adjustment of solution uptake rate from 1-5 ml/min. Burner section is somehow modified from that of flame photometer. Here, burner is a long horizontal tube with narrow slit along its length. Burner head is usually made up of titanium to avoid overestimation of iron and copper in the acidic test solution due to their contamination from stainless steel head burner. The flame path of burner using air as an oxidant is about 10-12 cm in length, but using nitrous oxide as an oxidant it is reduced to 5 cm due to higher burning velocity. Height of the burner head is adjustable.
- **Resonance Line Source:** For AAS, a resonance line source which is actually a hollow cathode lamp is essential. Hollow cathode lamp is made up of a cathode of the same element as that being studied in the flame. The cathode is in the form of cylinder and electrodes are enclosed in a borosilicate or quartz envelope containing an inert gas (neon or argon) at a pressure of approximately 5 torr. The application of high potential across the electrodes causes a discharge which results ionization of inert gas.

The ions of the inert gas are accelerated to the cathode, and on collision, cathode gets excited and emits radiation.

- **Monochromatic System:** In AAS, the function of monochromator is to isolate the resonance line from all other non-absorbing lines emitted by the radiation source. Diffraction gratings rather than prisms are used in most commercial instrument to have more uniform dispersion of lines and ultimately a higher resolution over a longer range of wavelength.

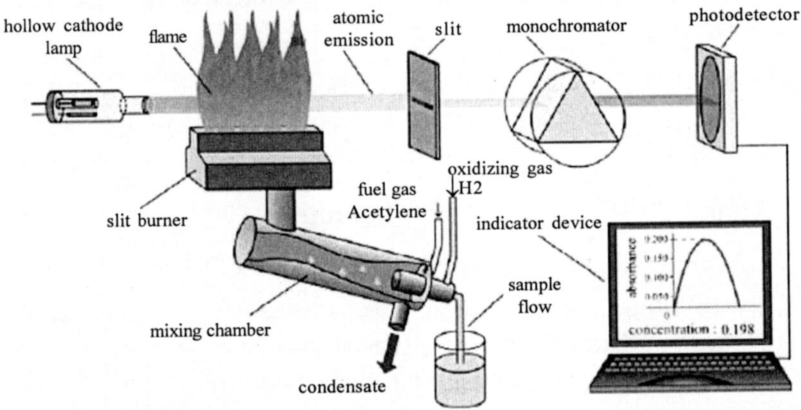

Fig. 16.13: Layout of Atomic Absorption Spectrophotometer

- **Detector and Read out System:** In view of the getting improved spectral sensitivity a photomultiplier is employed. The radiations received by the photomultiplier not only concern about the radiation of absorbed resonance line but also emitted from the flame. The emission may be from excited atoms and/ or molecule of the element concerned. Hence, the detector received absorption signal intensity, I_A plus the intensity of the emitted radiation from the excited atom, S. But our concern here is to determine the concentration of the test solution through measuring the intensity of absorbed resonance line by the ground state atoms. Thus, it is important to separate out the emitted radiation generated from the excited atoms in the flame from the non- absorbed resonance line. This can be achieved either by a mechanical chopper device or electronically by using alternating current appropriate to the particular frequency of the resonance line (signal arising from flame is direct current in character). The ultimate output is then fed to a read out system consisting of a meter or chart recorder or digital display.

16.5.2.2 Operation of Atomic Absorption Spectrophotometer

- Switch on the exhaust fan of the room to expel the inflammable gas present (if any) in the room.
- Attach the cathode lamp in the lamp holder and adjust the burner slit height to bring the radiation line and burner at the same height.
- Turn on the instrument, adjust lamp current requisite for the concerned lamp and allow it to get warmed for at least 15 to 20 minutes.
- Open fuel gas (acetylene) and air supply line and adjust their flow rate as specified for the instrument.
- Check whether nebulizer is capable of drawing sample (the draining of condensed solution through drain pipe).
- Lit the flame and adjust nebulizer to get requisite suction rate (1–5 ml/min) as evidenced by bluish flame with distilled (double) water.
- Calibrate the instrument with standard solutions of the element concerned and record the reading in concentration mode.
- Feed the test solution and record the concentration of the solution.
- After estimation first cut off the fuel supply line and drain off the gas in the pipeline by on and off the burner for 2–3 times.
- Continue to run the air supply for few minutes which facilitates the washing of the sample line with distilled water.
- Bring down the lamp current to zero and off the instrument.
- Remove the cathode lamp after few minutes.

16.5.2.3 Interferences

Several factors may affect the measurement of concentration of any solution through spectroscopy. These factors are broadly classified as (i) spectral interference and (ii) chemical interference.

Spectral Interference: In AAS spectral interference may arise mainly from overlapping of frequencies of the resonance line with the lines emitted by other element in the flame. Since the line width of an absorption line is about 0.005 nm, only a few cases of spectral overlapping between resonance lines of hollow cathode lamp and absorption lines of other metal atoms in flame are reported.

Table 16.4: Example of some typical spectral interference

Resonance source	Wave length, ë (nm)	Analyte	Wave length, ë (nm)
Aluminium	308.216	Vanadium	308.211
Antimony	231.147	Nickel	231.095
Copper	324.754	Europium	324.755
Gallium	403.307	Manganese	403.307
Iron	271.903	Platinum	271.904
Mercury	253.652	Cobalt	253.649

However, these interferences are only observed with relatively minor resonance line or with higher concentration of interfering element. Mercury estimation may be interfered by cobalt at concentration higher than 200 mg L^{-1}. Interference may also arise from emission band spectra produced by molecule or molecular fragments (-OH or -CN radical) present in the flame in many cases.

Chemical Interference: The production of ground state gaseous atoms is the basis of flame spectroscopy which may be hindered by two types of chemical interferences.

- **Stable compound formation:** It leads to incomplete dissociation of the substance in the flame e.g. estimation of calcium in presence of sulphate or phosphate. However, this interference may be overcome by (i) increasing the flame temperature e.g. by using nitrous oxide as an oxidant in place of air (ii) using releasing agent which forms a stable compound with the anionic part of the metal solution and thus increases the concentration of the free metal atoms e.g. strontium forms a stable complex with phosphate from calcium phosphate and estimation of calcium then be possible in an acetylene-air flame without any interference from phosphate (iii) removal of interfering elements through performing a simple solvent extraction.
- **Ionization of the atoms in the flame:** $M \rightarrow M^+ + e^-$ will reduce the extent of absorption of the resonance line by the ground state gaseous atoms. Possibility of ionization can be minimized by: (i) performing atomization at the lowest possible temperature, or (ii) using ionization suppressant having low ionization potential than the concerned cation e.g. potassium at higher concentration can act as an ionization suppressant for calcium, barium and strontium.

Other Effects

a) **Matrix effect:** Variation in the physical properties of the solution prepared from different solvents is behind this interference. To avoid this effect same solvent should be used for each solution (both in standard solution and the test solution) and solution should not differ too widely in their bulk composition also.

b) **Molecular absorption:** In acetylene – air flame a high concentration of sodium chloride will absorb radiation of wavelength around 213.9 nm, which is the wave length of the major zinc resonance line. Thus, sodium chloride is the interfering molecule in zinc determination which can be avoided by: (i) using different resonance line or (ii) high temperature flame to dissociate the interfering molecule.

c) **Background absorption:** Interference is due to presence of gaseous molecule or particulate materials in the flame particularly from organic solvent. In atomic absorption spectrophotometer background correction methods are always used.

Safety Measures both for FES and AAS

The following recommendations are the summary of code of practices recommended by Scientific Apparatus Makers' Association (SAMA) of the USA:

1. Ensure that the laboratory is well ventilated provided with adequate exhaust system.
2. Gas cylinders must be well marked and kept securely in an adequately ventilated room well away from heat.
3. When equipment is turned off, close the fuel gas cylinder tightly and bleed the gas line to the atmosphere through exhaust.
4. Time to time check for leaks in the joint and seal applying soap solution.
5. Special precaution for acetylene cylinder

a) Never run acetylene gas at a pressure higher than 15 psi, otherwise can explode spontaneously.

b) Avoid the use of copper tubing.

c) Avoid contact between acetylene gas with silver, mercury or chlorine.

d) Never run acetylene cylinder after pressure has dropped to 50 psi. At lower pressure the gas will be contaminated with acetone.

6. Care must be taken when using volatile flammable organic solvent for aspiration into the flame.
7. Never try to view the flame or hollow cathode lamp directly.
8. Never leave the flame unattended.

16.6 Inductively Coupled Plasma –Atomic Emission Spectroscopy (ICP-AES)

Inductively coupled plasma (ICP) is a very high temperture (7000-8000K) excitation source that efficiently desolvates, vaporizes, excites, and ionizes atoms. Molecular interferences are greatly reduced with this excitation source but are not eliminated completely. ICP sources are used to excite atoms for atomic-emission spectroscopy and to ionize atoms for mass spectrometry. Inductively coupled plasma atomic emission spectroscopy (ICP-AES) is used for simultaneous multiple element analysis of soil and biological materials.

Principle

Plasma refers to a cloud of hot highly ionized gas composed of ions, electrons and neutral particles. Typically, in plasma over 1 percent of the total atoms in a gas are in ionised state. The ICP-AES is based on the observation of atomic emission spectra while samples in the form of aerosol (thermally generated vapour or powder) are injected into an ICP atomization and excitation source. Plasma is electrically conducted, thus can be referred as electrical flame, as no combustion takes place. In ICP-AES the gas, usually argon (Ar), is ionized by the influence of a storng electrical field by radio frequency.

The ICP source comprises three concentric silica quartz tubes open at the top. The inner channel is for introduction of sample into the plasma. The outer channel conducts Ar gas at the rate of about 15 to 17 L/min to the plasma to sustain the plasma and to isolate the quartz tube from high temperature. The middle channel also auxillarily conduct the Ar gas at the rate of about 1L/min. The ICP is produced by passing initially ionized Argon gas through a quartz torch located inside a Cu coil connected to a radio frequency (RF) generator. The RF generator provides 1.5 to 3 kW forward power at a frequency of 27.1 MHz. The high frequency currents flowing in the coil generate oscillating magnetic fields whose lines of force are axially oriented inside the quartz tube and follow elliptical closed paths outside the coil.

Electrons and ions passing through the oscillating electromagnetic field flow at high acceleration rates in closed annular paths varying in their direction and strength, resulting in electron acceleration in each half cycle. Collisions between accelerated electrons and ions, and ensuing unionized Ar gas cause further ionization. The collisions cause ohmic heating and give thermal temperatures ranging from 6000 to 10000 K.

The sample is injected into the plasma by using Ar carrier gas at the rate of about 1 litre per minute. The ICP generated has unique physical properties that make it an excellent source for elemental vaporization/ atomization/ionization/

excitation. The aerosol droplets containing the analyte are dissolvated, the analyte salts/oxides are vaporized and analyte is atomized at the high temperature region of the plasma near the Cu coil.

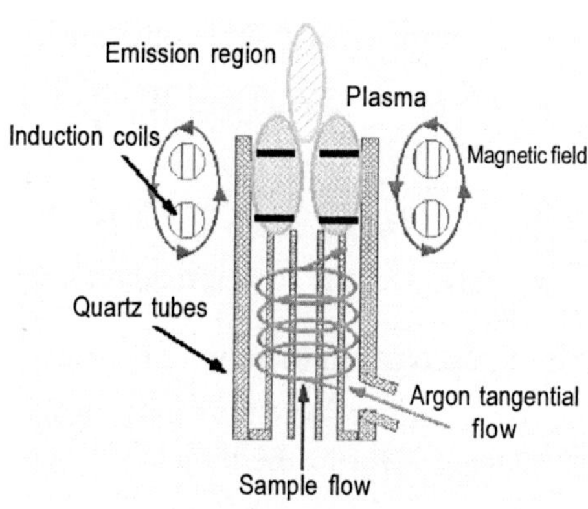

Fig.16.14: Schematic cross-section of an ICP

Excited atoms of elements in the sample emit light of characteristic wavelengths with intensity directly proportional to the element concentration. The light is focused on the entrance slit of the spectrometer to illuminate the diffraction grating. The diffracting grating separates light into its components wavelengths of lines (spectrum). The spectral line of an analyte passes through the aperature of an exit slit and strikes a photomultiplier tubes that produce signals directly proportional to the intensity of the spectral line.

The signal is fed to the readout system, which displays intensities, concentration, or both. Readout system is computer controlled. The computer stores the intensities of standards and uses these data to calculate the concentration of unknown solution.

Procedure

(A) Preparation of Sample

(1) Preparation of Soil Extract for Available Nutrients

Universal/multielement soil extractants can be used for the extraction of soil samples (Soltanpour *et al.*, 1998). After extraction, ICP-AES is used to simultaneously analyze these extracts for P, K, Zn, Fe, Cu, and Mn. AB-DTPA solution should be acidified to get rid of the carbonate-bicarbonate matrix in order to prevent clogging of the capillary tip.

(2) Preparation of Soil Sample for Total Elemental Analysis

- Weigh 1.000 g of the 100 mesh soil into a 100 ml Teflon beaker.
- Add 10 ml of HNO_3 and 10 ml of $HClO_4$. Cover with a Teflon watch cover.
- Heat at 200°C for 1 hour. Remove cover and continue heating to reduce the volume to 2 to 3 ml.
- Cool the sample; add 5 ml of $HClO_4$ and 10 ml HF. Cover with watch cover and heat overnight at 200°C.

- Remove cover and continue heating to reduce the volume to 2 to 3 ml.
- Cool the digest; add 10 ml of 50% HCl, cover, and heat at 100°C for 30 minutes.
- Remove from hot plate, allow it to cool.
- Transfer the solution quantitatively into 50 ml volumetric flask.

(B) Preparation of Stock Standard Solution

It is recommended that ready to use standard solution of highest purity be purchased. But in the developing countries, these ready to use standard may not be available. Prepare primary standard solution as mentioned in Appendix XII.

Secondary standards should be prepared in such a way that standard solutions match the sample solution in concentration of acids. For AB-DTPA extracts, standards should be made in AB-DTPA solution that has been neutralized with concentrated HNO_3. For acid digest, standard solution should be made to contain 5% (v/v) $HClO_4$ and 10% (v/v) HCl.

(C) Analysis

The general operation principles are similar to that for AAS. Since actual operation of different ICP-AES differs with model, make and computer software it is advisable to consult the operation manual supplied with the instrument.

Points to be considered during analysis

- In calibrating the ICP spectrometer, one should consider the concentration range to be used, the interelement interference correction, and the stability of the standards.
- Mixture of chemicals that cause precipitation should be avoided. McQuaker *et al.* (1979) devised a calibration scheme for 30 elements that satisfies the needs of analysis of soil, water, plant, and particulate matter analysis.
- In soil analysis, one set of secondary standards is required for each extracting solution and another set for the total soil digest.
- In making a multielement standard solution, one should avoid making solution containing high concentration of affecting and low concentration of affected elements.

Questions and Hints

Q.1 What we actually measure by analytical balance – mass or weight?

Hint: Mass; though we customarily use the term weight. However, at a given place these two terms are synonymous.

Q.2 What are the characteristics of a good analytical balance?

Hint: *See section 16.1.3 Page no 189*

Q.3 Why calibration of pH meter is needed with buffer solutions?

Hint: An asymmetry potential develops across the glass membrane of the glass electrode due to differing condition of strain in the inner and outer surfaces even when solution of the same H^+ ion activity are on the both sides of the glass electrode. To nullify this asymmetry potential calibration of pH meter with buffer solutions is needed

Q.4 Why KCl is used as salt bridge?

Hint: At the junction between calomel electrode and the solution / suspension a potential termed as liquid junction potential may develop due to variation in the mobility of cations and anions present in salt bridge to carry electric current. The cation and anion of KCl salt have nearly equal mobility (same transport number) in solution. Thus, use of KCl (or NH_4NO_3) helps to keep liquid junction potential low (nearly zero) and constant as far as practicable.

Q.5 State Beer's law.

Hint: *See section 16.4.1 page no 199*

Q 6. What are reasons behind the deviation of Beer's law?

Hint: Sometimes, the coloured species may ionize, dissociate or associate in the solution, thus the nature of the species changes with concentration causing deviation from Beer's law. Beer's law holds good till the structure of the coloured ion or non-electrolyte remains unchanged with the variation of concentration.

Q 7. What is role of light filter or monochromator in colorimeter or absorption spectrophotometer?

Hint: Light filter or monochromator discards undesirable wavelength and allows the narrow band of desirable wavelength to pass through.

Q 8. What type of current is passed in conductivity bridge and why?

Hint: Alternating current (AC). To prevent the accumulation of the products of hydrolysis of soluble salts on the electrodes which results reduction of effective conductance of solution.

Q 9. What are the differences between fundamental principles of emission spectroscopy and absorption spectroscopy?

Hint: In emission spectroscopy the concentration of the solution is measured in relation to the quantum of energy emitted by the excited atoms or molecules (getting energy from the flame) of the substance during their coming down to ground state. But, in case of AAS the concentration of the solution is measured in relation to the quantum of energy absorbed by the atoms or molecules to get excited from their ground state absorbing radiation energy emitted by the hollow cathode lamp of the element concern.

Q 10. Why Na concentration can be estimated in the flame photometer, but not Zn?

Hint: Resonance line of low energy value is more sensitive than that of higher energy value; again the wavelength of the resonance line is inversely proportional to the energy value. The wavelength of the resonance line of Na is 589.0 nm, while that for Zn is 213.9 nm.

Q 11. Explain how does emission spectra originate?

Hint: When atoms, ions or their groupings are subjected to some form of energy excitation, as in a flame or in an electric arc, an emission spectrum originates. Energy is first absorbed by electron shifts to positions more distant from the nucleus. As they regain or partially regain their stable or ground state, the previously absorbed energy is re-emitted in the form of emission spectra.

Q 12. What are the advantages of ICP-AES over AAS method?

Hint: (a) ICP is superior source in accuracy, precision, detection limit, freedom from interference, and dynamic range. (b) One can analyze a solution for many elements in 1 minute. Therefore, large volume of data can be generated very fast. (c) Elements like Al, P, S and B which are either poorly measured at low concentration or at all not measured with AAS are successfully measured with high precision by ICP-AES.

17
Soil Reaction

17.1 Determination of Soil pH

The soil pH value is the measure of hydrogen or hydroxyl ion activity in the soil solution and it indicates whether the soil is acidic, neutral or alkaline in reaction. The pH is the most important property of soil as it controls the availability of plant nutrients, microbial activity and physical condition of soil. Since crop growth is affected both under very low (strongly acidic) and very high (strongly alkaline) soil pH, reclamation of these soils are necessary. Soil pH in fact indicates the extent of active acidity (activity of hydrogen ion in soil solution) of soil. Depending upon the purpose of measurement and soil condition, soil pH is measured in several soil-water or soil-salt solution ratios. The saturated paste is used for identifying specific soil problems like degree of acidity or alkalinity. For making fertilizer recommendation, 1: 2 soil-water suspension is generally adopted. To mask the variability in salt content of the soil, the pH is measured in soil-$CaCl_2$ suspension (1:2 soil-0.1 M $CaCl_2$ solution). The soil pH value is independent of dilution over a wide range of soil-salt solution ratio, thus more reproducible than those obtained with soil-water suspension.

By determining soil pH in distilled water, in 0.01 M $CaCl_2$ and 1 M KCl lime potential and net charge of the soil colloid can be ascertained from the following equations:

$\Delta pH_1 = pH_{CaCl2} - ½pCa$

Where, ΔpH_1 is the lime potential

$\Delta pH_2 = pH_w - pH_{KCl}$

ΔpH_2 with positive value means colloidal surface is net negatively charged while negative value indicates that the colloidal surface is net positively charged. Thus, zero point charge (ZPC) is attained when ΔpH_2 equals to zero.

Equipments and Materials

pH meter with glass and calomel reference electrode, 100 and 500 ml beaker, glass rod.

Reagents

- Standard buffer solutions: Certified buffer tablets are available for different pH values. Buffer tablets of pH 4.0, 7.0 and 9.2 are mostly used for pH meter calibration. Dissolve one tablet in specified volume (usually 100 ml) of water and preserve in refrigerator.
- 0.01 **M** $CaCl_2$ solution: Dissolve 1.47 g $CaCl_2.2H_2O$ in water and make up the volume to 1 litre with distilled water.
- 1 **M** KCl solution: Dissolve 74.5 g KCl in water and make up the volume to 1 litre with distilled water.

Procedure

1) Sample Preparation

a) Saturated Soil Paste

- Take 100 g processed air dry soil in 500 ml beaker.
- Prepare a soil paste with gradual addition of distilled water to the soil while working with spatula. At saturation the soil surface will glisten and flow slightly on tilting the beaker.
- Allow it to stand with a cover above the beaker for about an hour.
- Observe whether any free water is present on the soil surface or the paste is very stiff. The perfect paste will neither show the presence of free water nor lose its glistening characteristic on standing.
- If the paste is too wet or too stiff, then add some dry soil or water, respectively and follow the steps as discussed above.

b) Soil – Water Suspension

- Take 20 g processed air dry soil in 100 ml beaker and add 40 ml (for 1: 2 soil - water ratio) or 50 ml (for 1: 2.5 soil - water ratio) of distilled water.
- Stir the soil with glass rod occasionally for about 30 minutes.

a) Soil – $CaCl_2$ or -KCl Suspension

- Take 20 g processed air dry soil in 100 ml beaker and add 40 ml (for 1: 2 soil - water ratio) 0.01 **M** $CaCl_2$ or 1 **M** KCl solution.

- Allow the soil to absorb the solution and then stir thoroughly with glass rod.
- Intermittently stir the soil the suspension for the next 30 minutes.

2) pH Determination

- Switch on the pH meter and allow it to warm up for half an hour.
- Set the temperature of the instrument at room temperature by turning temperature set knob.
- Calibrate the pH meter with three buffer solutions, one in acidic range (usually pH 4.0), one in alkaline range (usually pH 9.2) and other in neutral (pH 7.0) by using calibration knob.
- During each transition wash the electrodes thoroughly with distilled water and soak the adhering water with tissue paper.
- Repeat the calibration step 2 – 3 times till the instrument is completely calibrated at both the pH buffers.
- Carefully insert or dip the electrodes in the paste or suspension and measure the pH by pushing the pH measuring knob.
- Always withdraw electrodes from the solution when the instrument is in stand by position.
- Wash the electrodes thoroughly with distilled water.
- During rest keep the electrodes immersing in distilled water.

Rating

Soil Reaction Class	pH Range
Extremely acidic	< 4.5
Very strongly acidic	4.5 – 5.0
Strongly acidic	5.0 – 5.5
Moderately acidic	5.5 – 6.0
Mildly acidic	6.0 – 6.5
Neutral	6.5 - 7.5
Mildly alkaline	7.5 - 8.0
Strongly alkaline	8.0 – 9.0
Very strongly alkaline	> 9.0

17.2 Determination of Lime Potential of Soil

Soil pH values are affected by number of factors *viz*. dilution effect (soil: water ratio), suspension effect and the salt concentration. If, instead of single ion activity (H^+) which varies with soil suspension ratios, we consider ionic activity ratio, the values are found to be relatively constant and can be considered as the characteristic for a soil. Determination of soil pH with 0.01 **M** $CaCl_2$ gives a stable reading over a wide range of soil: suspension ratio and salt content. Lime potential is an index of reserve acidity and buffering capacity of soils. Comparing the lime potential of two soils, lesser the value of lime potential greater will be the buffering capacity of soil.

If the soil exchange surface is saturated with H^+ and Ca^{2+} ions, then according to Schofield and Taylor (1955) at equilibrium

$$a_{H+} / \sqrt{a_{Ca2+}} = K \text{ (constant)}$$

$$\text{or, } -\log K = -\log a_{H+} + ½\log a_{Ca2+}$$

$$= pH - ½pCa \quad (17.1)$$

The negative log K is called the lime potential.

Principle

Soil pH is measured with 0.01 **M** $CaCl_2$ at 1: 2 soil: suspension ratio. Lime potential then can be calculated using equation (10.1) considering the activity of Ca^{2+} in the suspension is equal to the activity of Ca^{2+} at pure 0.01 **M** $CaCl_2$. As the activity of Ca^{2+} at pure 0.01 **M** $CaCl_2$ solution is 5.19 x 10^{-3} ML^{-1}, the value of ½pCa at 25°C is 1.14.

Reagents

- 0.01 **M** $CaCl_2$ solution: Dissolve 1.47 g $CaCl_2.2H_2O$ in water and make up the volume to 1 litre with distilled water.
- Standard buffer solutions of pH 4.0 and 7.0.

Equipments and Materials

pH meter with glass and calomel reference electrode, 100 ml beaker, glass rod, balance.

Procedure

- Take 20 g processed air dry soil in 100 ml beaker and add 40 ml 0.01 **M** $CaCl_2$ solution.
- Shake for half an hour and measure the pH of the suspension.

Calculation

$$\text{Lime Potential} = pH_{CaCl2} - ½pCa$$

$$= pH_{CaCl2} - 1.14$$

17.3 Forms of Soil Acidity

17.3.1 Exchange Acidity (Sokonov, 1939)

The soil pH actually measures the concentration of hydrogen ion (proton) present in soil solution i.e. **active acidity**, but does not take into account the amount of exchangeable acidic cations, H^+ and Al^{3+}, present on the soil surface. Again, exchangeable Al coming into the soil solution phase liberates protons on hydrolysis.

$$Al^{3+} + 3H_2O \rightarrow Al(OH)_3 + 3H^+$$

Exchangeable H^+ and Al^{3+} also termed as electrostatically bound H^+ and Al^{3+}, respectively which together constitute the exchange acidity.

Principle

The conventional method of determining exchange acidity as proposed by Yuan (1959) is to leach the soil with a solution of neutral salt (1 **N** KCl) to displace exchangeable H^+ and Al^{3+} and the amount of displaced H^+ and Al^{3+} are then determined by titration with standard alkali in presence of a suitable indicator. To estimate the amount of exchangeable Al^{3+}, NaF or KF is added to the solution. In the titrated solution aluminium present in the form of $Al(OH)_3$ which on reaction with NaF/KF forms complex of sodium or potassium fluro-aluminate and releases equivalent amount of alkali. The released alkali is then titrated with standard acid.

$$Al(OH)_3 + 6NaF \rightarrow Na_3AlF_6 + 3NaOH$$

Equipments and Materials

Conical flasks of 100 and 250 ml capacity, 50 ml measuring cylinder, 100 ml volumetric flask, burette, Whatman No. 40 filter paper.

Reagents

- 1**N** KCl: Dissolve 74.56g AR grade KCl in distilled water and dilute it to the volume of 1 litre.
- 1**N** NaF: Dissolve 42g of NaF in about 900 ml of distilled water and add few drops of phenolphthalein. If the colour of the solution is not pink then add dilute NaOH (0.1**N**) drop by drop until it turns pink. Discharge colour by adding 0.1 **N** HCl in drops. Dilute it to 1 litre.
- Standard 0.1 **N** NaOH solution (approx.): Dissolve 4 g NaOH in water and dilute it to 1 litre. Standardize it with standard 0.1 **N** oxalic acid in presence of phenolphthalein as an indicator.

- Standard 0.1 **N** HCl solution (approx.): Dilute 8 ml concentrated HCl with distilled water and make the volume up to 1 litre. Standardize it against standard NaOH in presence of methyl red indicator.
- 1% phenolphthalein indicator solution: Dissolve 1 g of indicator in 95% 100 ml ethanol.

Procedure

- Take 10g air dry soil in 100 ml conical flask.
- Add 50 ml of 1 **N** KCl solution, mix thoroughly and shake the content on a mechanical shaker for 30 minutes.
- Filter through Whatman No. 40 filter paper into a 100 ml volumetric flask.
- Leach the soil on the filter paper four times with small volume (about 10 ml in each time) of KCl solution and make the volume upto the mark with KCl solution.
- Transfer the leachate to a 250 ml conical flask.
- Add 4 to 5 drops of phenolphthalein indicator and titrate with standard NaOH solution to the first permanent (at least 30 minutes persistance) pink colour. This will give the estimate of exchange acidity (EB-H^++ EB-Al^{3+}).
- Preserve the titrated solution for exchangeable Al estimation.
- Run a blank without soil in similar way with 100 ml 1 **N** KCl solution and titrate against standard NaOH solution.
- To determine exchangeable Al add 10 ml 1 **N** NaF solution to the titrated solution.
- Titrate it with standard HCl solution till the pink colour disappears.
- Leave it for at least 10 minutes and add few more drops of HCl to get a lasting end point colour. [Release of last hydroxyl group from Al $(OH)_3$ takes some time].

Calculation

Say,

Weight of the soil taken = W g

Volume of standard NaOH required for sample = S ml

Volume of standard NaOH required for blank = B ml

Strength of standard NaOH = A

Volume of standard HCl = V ml

Strength of standard HCl = C

$$\text{Exchange acidity (EB} - H^{+} + \text{EB} - \text{Al}^{3+}3^{+}\text{), meq / 100 g soil} = (\text{S-B}) \times \text{A} \times \frac{100}{\text{W}}$$

$$\text{Exchangeable Al}^{3+}\text{(EB-Al}^{3+}\text{), meq/100 g soil} = \text{V} \times \text{C} \times \frac{100}{\text{W}}$$

Exchangeable H^+ (EB-H^+), meq /100 g soil = Exchange acidity – Exchangeable Al^{3+}

17.3.2 Extractable Al or Extractable Acidity

Extractable Al includes Al present in the form of exchangeable Al on the soil surface plus soluble aluminium hydroxide and some hydroxy Al monomer or polymers remain as strongly adsorbed on the colloidal complex or trapped in the interlayer position of the expanding silicate clay as non-exchangeable form. This form of acidity not only measures the extractable acidity but also indicates the weathering status of the soil.

Principle

1 **N** ammonium acetate (NH_4OAc) of pH 4.8 extractable Al is treated with a colour developing reagent, aluminon in presence of thioglycollic acid and the intensity of the colour is measured colorimetrically at 535 nm wavelength.

Equipments and Materials

Hot water bath, colorimeter, Buchner funnel, 100 ml beaker, 50 ml measuring cylinder, 25 ml volumetric flask, graduated pipette, Whatman No. 40 filter paper.

Reagents

- 1 **N** NH_4OAc (pH 4.8) solution: Add 58 ml of glacial acetic acid in 1 litre volumetric flask containing about 400 ml of distilled water. To it add 70 ml of concentrated ammonia solution slowly and make up the volume to 1 litre. Adjust the pH of the solution at 4.8 with 1 **N** acetic acid.
- Aluminon solution: Dissolve 0.75g aluminon (aurine tricarboxylic acid) and 15 g gum acacia in a 500 ml beaker with distilled water. Mix thoroughly and add 342 ml of concentrated HCl. Filter it through Buchner funnel and dilute the content to 1 litre.
- Thioglycollic acid solution: Dilute 1 ml of pure thioglycollic acid to 100 ml with distilled water.

- Standard Al solution: Dissolve exactly 0.8094 g $AlCl_3 . 6H_2O$ in water and dilute to 1 litre which gives 100 ppm Al solution. Dilute ten times to make it 10 ppm.

Procedure

- Take 10 g air dry processed soil in 100 ml beaker and add 50 ml of 1**N** NH_4OAc (pH 4.8) solution.
- Mix thoroughly and allow it to stand for 2 hours.
- Filter the suspension through Whatman No. 40 filter paper fitted on a Buchner funnel under suction.
- Leach the soil four times with small volume (10 ml in each time) of NH_4OAc solution and make up volume to the mark with NH_4OAc.
- Take 5 ml aliquot in a 25 ml volumetric flask and add 5 ml aluminon solution.
- Dilute it to about 20 ml with distilled water and add 0.2 ml thioglycollic acid solution.
- Mix thoroughly and make up the volume to the mark with distilled water.
- Place the volumetric flask on hot water bath (100°C) for about 10-15 minutes until brilliant red colour appears.
- Cool the flask at room temperature and measure the colour intensity in colorimeter at 535 nm wavelength.
- Prepare a standard curve with 0, 1, 2, 4, 8 and 10 ml of 10 ppm standard Al stock solution following the same procedure as for test solution.

Calculation

Say,

Weight of the soil = W g

Volume of the leachate made up after washings with1**N** $NH_4OAc = V_A$ ml

Volume of the aliquot taken = V_B ml

Final volume of the coloured solution = V_C ml

Concentration of Al in the test solution obtained from standard curve = S ppm

$$\text{Then, Extractable Al, ppm} = S \times \frac{V_C}{V_B} \times \frac{V_A}{W} = Z$$

$$\text{Extractable Al, meq/100 g of soil} = \frac{Z}{9} \times \frac{100}{1000} = \frac{Z}{9 \times 10}$$

[As the equivalent weight of Al = 9]

17.3.3 Non-exchangeable Al

This fraction of soil Al retained by the soil colloids as non-exchangeable form against extraction by neutral normal NH_4OAc

Non-exchangeable Al = Extractable Al – Exchange acidity

17.3.4 Total Acidity (Kappen, 1934)

Total soil acidity includes the sum of acidic cations present in soil solution as well as retained by the adsorptive complex (clay and organic matter) which contributes to the acidity on hydrolysis with increase in soil pH.

$$\boxed{\text{Soil}}\begin{matrix} H \\ \\ Al \end{matrix} + 4CH_3COONa + 3H_2O \rightleftharpoons \boxed{\text{Soil}}\begin{matrix} Na \\ Na \\ Na \\ Na \end{matrix} + 4CH_3COOH + Al(OH)_3$$

The liberated acid is then titrated with standard alkali using suitable indicator.

Equipments and Materials

Balance, 250 ml conical flask, 100 ml measuring cylinder, Whatman No. 42 filter paper.

Reagents

- 1 **N** sodium acetate (NaOAc) solution (pH 8.2): Dissolve 82 g NaOAc in about 900 ml distilled water and adjust the pH at 8.2 by adding 1 **N** NaOH solution. Make up volume to 1 litre with distilled water.
- 0.1 **N** NaOH solution: Dissolve 4 g NaOH and make up volume to 1 litre with distilled water and standardize against oxalic acid as in section 17.3.1.
- 1% phenolphthalein indicator solution: Dissolve 1g of indicator in 95% 100 ml ethanol.

Procedure

- Take 40 g air dry soil in a 250 ml conical flask and add 100 ml 1**N** NaOAc solution to it.
- Shake the flask for an hour and filter through Whatman No. 42 filter paper.

- Titrate the filtrate with standard 0.1 **N** NaOH solution after adding 2-3 drops of phenolphthalein indicator solution till a permanent pink colour appears.

Calculation

Say,

Weight of soil taken = W g

Volume of standard NaOH solution required for titration = V ml

Strength (N) of NaOH = S

$$\text{Total acidity, meq/100g of soil} = \frac{(V \times S \times 100)}{W}$$

Hydrolytic Acidity = Total acidity – Exchange acidity

17.3.5 Total Potential Acidity

Total potential acidity refers to all soil acidity components including exchangeable H^+ and Al^{3+}, partially neutralized hydroxyl polymer of Al and weakly acidic carboxylic (-COOH) and phenolic (-OH) groups present in soil organic matter at pH 7.0.

Principle

The soil is leached with a solution of $BaCl_2$ plus triethanol amine buffered at pH 8.2. Barium displaces H^+ and Al^{3+} from the soil surface. Soluble Al in turn on hydrolysis releases proton to the solution. These protons and proton liberated on dissociation of carboxylic and phenolic groups are neutralized by triethanol amine. The leachate is then titrated against a standard acid using suitable indicator. The titer value will indicate the amount of unconsumed triethanol amine.

Equipments and Materials

Mechanical shaker, conical flasks of 25 and 500 ml capacity, 250 ml volumetric flask, 100 ml measuring cylinder, Whatman No. 42 filter paper, burette.

Reagents

- 1**N** Barium Chloride ($BaCl_2$) solution: Dissolve 208 g $BaCl_2$ in water and make up the volume to 1 litre with distilled water.
- Buffer extractant of 0.5 N $BaCl_2$ plus triethanol amine (pH 8.2): Dilute 25 ml of triethanol amine to 250 ml with water and adjust pH to 8.2 with 1**N** HCl. Dilute the solution to 500 ml with distilled water and add 500 ml of 1**N** $BaCl_2$ solution.

- Standard 0.2 **N** HCl (approx.): Dilute about 17 ml of concentrated HCl to 1 litre with distilled water and standardize against standard sodium carbonate using bromo cresol green – methyl red indicator.
- Bromo cresol green-methyl red indicator: Dissolve 0.5 g bromo cresol green and 0.1 g methyl red in 100 ml 95% ethanol and adjust pH to 4.5.

Procedure

- Take 10 g soil in a 250 ml conical flask.
- Add 100 ml buffer extractant and shake for half an hour and leave it for overnight.
- Filter it through Whatman No. 42 filter paper fitted on a Buchner funnel.
- Wash the conical flask with additional small volume of extractant.
- Transfer the leachate in 250 ml volumetric flask and make up the volume with the extractant.
- Pour the leachate in a 500 ml conical flask, add 4-5 drops of indicator and titrate with standard HCl until pink colour appears.
- Run a blank in the same way but without soil.

Calculation

Let,

Weight of soil taken = W g

Volume of standard HCl required for blank titration = B ml

Volume of standard HCl required for sample titration = S ml

Volume of HCl consumed for neutralization of total potential acidity = (B –S) ml

Strength (N) of standard HCl = Z

Then, total potential acidity, meq/100 g soil = $(B - S) \times Z \times \frac{100}{W}$

17.3.6 pH Dependent Acidity

pH dependent acidity is that fraction of total potential acidity which is due to dissociation of carboxylic (-COOH) and phenolic (-OH) groups present in soil organic matter and hydroxyl group (OH) exposed at the broken edge of soil mineral.

pH dependent acidity = Total potential acidity – Exchange acidity

17.4 Lime Requirement

Agriculturally neutral soil means the soil having pH between 6.5 and 7.5. In this soil pH range the availability of most of the plant nutrients are satisfactory. Though there are some crops which prefer soil pH less than 6.5 e.g. tea or more than 7.5 e.g. sugar beet. Satisfactory plant growth in acid soil can only be achieved by reclamation of acid soil (pH < 6.0) with lime through improving physical, chemical and biological environment of the soil. Among the several methods of determining lime requirement of acid soil, soil-lime incubation method and soil – buffer equilibration method are most commonly used. Soil – buffer equilibration method is very rapid and suitable for soils having appreciable amount of Al, but unsuitable for low lime requirement soil.

17.4.1 Determination of Lime Requirement by Incubation Method

Principle

This method involves equilibration of acid soil with different amount of lime and finally measurement of pH of the equilibrated soil solutions. The pH values are plotted against the amount of lime added in the graph paper. The amount of lime required to raise the pH of the soil to the desired level can be calculated from the graph and lime of requirement of soil then can be expressed in terms of tons of agricultural lime per hectare of land.

Equipments and Materials

pH meter with glass and calomel electrodes, mechanical shaker, balance, 100 ml volumetric jar with lid, 2 litre jar, glass rod.

Reagents

- Saturated calcium hydroxide [$Ca(OH)_2$] solution: Take sufficient amount of $Ca(OH)_2$ in 2 litre jar and add about 1.5 litre of distilled water. Shake thoroughly on a mechanical shaker and leave it for 2 days with occasional shaking. Siphoned out the clear supernatant liquid and determine the equivalent $CaCO_3$ content from the Ca concentration in the solution estimated by versenate titration method as described in section 18.2.4.

- Standard buffer solutions of pH 4.0 and 7.0.

Procedure

- Take 10 g air dry processed acid soil (pH < 6.0) in a series of 100 ml volumetric jars and moisten the soil with distilled water.

- Add graded volume of (0, 10, 15, 20, 25, 30, 35 ml) saturated solution of $Ca(OH)_2$ to the soil.

- Make the volume of each jar to 50 ml mark with distilled water.
- Shake well with the glass rod and cover with the lid.
- Keep for 3 days with at least 15 minutes regular shaking with glass rod.
- Finally shake for 15 minutes and allow settling down the soil particles for half an hour.
- Measure pH of the supernatant liquid after calibrating the pH meter with standard buffer solutions. Plot the pH of the equilibrium solution against the amount of $CaCO_3$ supplied through graded volume of saturated Ca $(OH)_2$ solutions (Fig. 17.1).

Calculation

a) Ca concentration determination by versenate method

Say,

Volume of aliquot of supernatant liquid saturated with Ca $(OH)_2$ taken = V_A ml

Volume of EDTA solution required for titration = V_B ml

Strength (N) of EDTA solution = M

$$\text{Ca content in the solution (mg / ml)} = \frac{(V_B \times M \times 20.04)}{V_A} = Y$$

$CaCO_3$ equivalent (mg/ml) = Y x 2.5 = A

b) Lime requirement

Equivalent $CaCO_3$ (mg/100g soil) supplied through graded volume of Ca $(OH)_2$ = (A x V_d x 100) / W

Where, V_d is the volume of saturated $Ca(OH)_2$ solution added to the soil, ml

W is the weight of soil, g.

Plot the pH of the equilibrium solution against the amount of $CaCO_3$ added.

Amount of pure $CaCO_3$ (mg/100g soil) required to raise the pH to a desired level = B (from Fig.17.1)

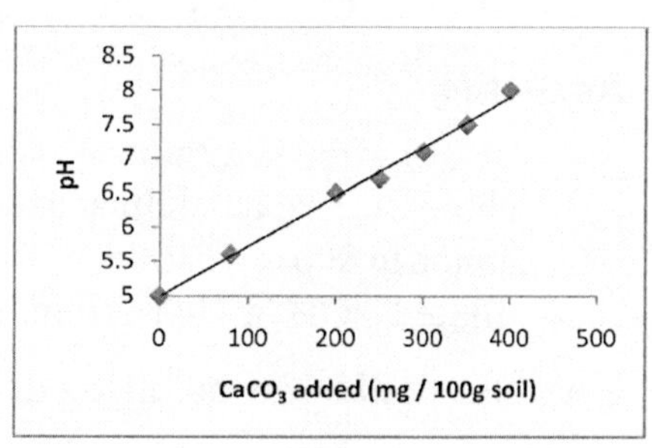

Fig. 17.1

Lime requirement in terms of pure $CaCO_3$ (kg /ha) = $\frac{B}{10^6} \times \frac{1000}{100} \times 2.2 \times 10^6$

= B x 22 = Z

$$\text{Agril.lime required (t / ha)} = \frac{Z}{10^3} \div \text{purity} \div \text{CaCO3 equivalent of Agril.lime}$$

$$[CaCO_3 \text{ equivalent of any Agril.lime (\%)} = \frac{50 \times 100}{\textit{Equivalent weight of that agril. lime}}$$

$$\text{CaCO3 equivalent of CaO} = \frac{50 \times 100}{28} = 179\%$$

If agril lime is CaO with 90% purity,then CaO requirement (t/ha) =

$$\frac{Z}{10^3} \div \frac{90}{100} \div \frac{179}{100} = \frac{Z}{9 \times 179}$$

17.4.2 Determination of Lime Requirement by SMP Buffer Method

Principle

Shoemaker *et al.* (1961) original single buffer method i.e. SMP buffer method involves equilibrating soil with SMP buffer solution (pH 7.5) and recording the changes of pH values of the suspension. The lime requirement to a desired level of pH is then determined from the table 17.1 against the soil pH value.

Equipments and Materials

pH meter with glass and calomel electrodes, balance, 100 ml beaker, 50 ml measuring cylinder, glass rod.

Reagents

- SMP buffer solution: Dissolve 1.8 g of p-nitrophenol, 3 g potassium chromate, 2 g calcium acetate, 53.1 g $CaCl_2.2H_2O$ and 2.5 ml triethanol amine in 800 ml distilled water. Adjust the pH of the buffer at 7.5 by using dilute NaOH or HCl and make up the volume to 1 litre.
- Standard buffer solutions of pH 4.0 and 7.0.

Procedure

- Take 10 g soil in a 50 ml beaker and add 20 ml of SMP buffer solution.
- Stir continuously for 10 minutes.

- After calibrating the pH meter with standard buffer solutions measure the pH of the equilibrated soil solution.
- Determine lime requirement on the basis of the soil pH from the ready reckoner table given below.

Table 17.1: Lime requirement based on soil-buffer suspension pH

pH of soil-buffer suspension	Lime required (t / ha) to raise the pH to the desired level pH 6.0	pH 6.4	pH 6.8
6.7	2.43	2.92	3.40
6.6	3.40	4.13	4.62
6.5	4.37	5.35	6.08
6.4	5.59	6.56	7.53
6.3	6.56	7.78	8.99
6.2	7.53	8.99	10.21
6.1	8.51	10.21	11.66
6.0	9.48	11.42	13.12
5.9	10.69	12.64	14.58
5.8	11.66	13.85	15.80
5.7	12.64	15.07	17.01
5.6	13.61	16.28	18.71
5.5	14.58	17.50	20.17
5.4	15.80	18.71	21.63
5.3	16.77	19.93	22.84
5.2	17.98	20.90	24.30
5.1	18.95	22.11	25.76
5.0	19.93	23.33	27.22
4.9	20.90	24.54	28.67
4.8	22.11	25.76	30.13

Calculation

Say,

Amount of pure $CaCO_3$ required (t/ha) to raise soil pH at desired level = A (from the table 17.1)

Agril. lime required (t/ha) = A ÷ purity ÷ $CaCO_3$ equivalent of Agril.lime

17.5 Determination of Gypsum Requirement of Soil (Schoonover, 1952)

Presence of large amount of sodium as high as ≥ 15% on the exchange complex results in high soil pH (>8.5) e.g., sodic (alkali) and saline-sodic soil. Presence of huge amount of sodium as well as high soil pH causes nutritional imbalance, poor soil physical condition like, poor structure, very slow infiltration, poor hydraulic conductivity and poor biological environment also. For reclamation of these soils gypsum ($CaSO_4.2H_2O$), pyrites (FeS_2) and sulphur, etc are used. However, gypsum application on sodic soil is very popular.

Principle

A known amount of soil is equilibrated with a known excess amount of saturated gypsum solution and unutilized amount of gypsum is then estimated by versenate titration. The difference between the amount of Ca added and Ca left in the soil solution indicates the amount of Ca consumed for exchange of sodium, from which gypsum requirement can be calculated.

$$\boxed{\text{Soil}}\text{Na} + CaSO_4.2H_2O \rightarrow \boxed{\text{Soil}}\ \text{Ca} + Na_2SO_4 + 2H_2O\ \text{Na (Gypsum)}$$

Equipment and Materials

Mechanical shaker, 250 ml conical flask, 100 ml porcelain dish, 100 ml measuring cylinder, Whatman No. 1 filter paper, single mark 5 ml pipette, burette.

Reagents

- Saturated gypsum solution: Add 5 g gypsum ($CaSO_4.2H_2O$) to 1 litre of distilled water, shake vigorously for 15 minutes on mechanical shaker and filter through Whatman No. 1 filter paper.
- Ammonium chloride-ammonium hydroxide buffer solution: Dissolve 67.5 g of ammonium chloride (NH_4Cl) in 570 ml of ammonium hydroxide (NH_4OH) and make up volume to 1 litre with distilled water.
- Standard EDTA (0.01 **N**) solution: Dissolve 2 g disodium salt of ethylene diamine tetra acetic acid (EDTA) and 0.05 g $MgCl_2.6H_2O$ in water and dilute to 1 litre. Standardize it against standard calcium solution.
- Standard calcium chloride (0.01 **N**) solution: Dissolve exactly 0.5 g AR grade $CaCO_3$ in 10 ml 0.5 **N** HCl. Make up the volume to 1 litre with distilled water.
- Eriochrome black T (EBT) indicator: Dissolve 0.5 g of EBT dye and 4.5 g hydroxyl amine hydrochloride in 100 ml of 95% ethanol.
- Potassium (or sodium) cyanide solution (1% w/v): Dissolve 1 g reagent KCN (or NaCN) and dilute to 100 ml with distilled water.
- Potassium ferro cyanide (4%) solution: Dissolve 4 g reagent [$K_4Fe(CN)_6$] and dilute to 100 ml with distilled water.

Procedure

- Take 5 g air dry processed soil in a 250 ml conical flask and add 100 ml of saturated gypsum solution.
- Shake for 5 minutes on the mechanical shaker and filter through Whatman No. 1 filter paper.

- Pipette out 5 ml aliquot of the filtrate into a 100 ml porcelain dish.
- Add 1 ml of NH_4Cl-NH_4OH buffer solution, 10 drops each of KCN (or NaCN), $K_4Fe(CN)_6$ and 2-3 drops of EBT indicator.
- Titrate with standard EDTA solution until colour changes from wine red to sky blue.
- Run a blank in a similar way with 5 ml of saturated gypsum solution separately in place of soil filtrate aliquot.

Calculation

a) Standardization of EDTA

Say,

Volume of Ca solution taken = V_A ml

Strength (N) of Ca solution = S_A

Volume of EDTA required for titration of Ca solution = V_B ml

So, strength (N) of EDTA solution = $\frac{V_A \times S_A}{V_B} = Z$

b) Gypsum requirement calculation

Say,

Weight of soil taken = W g

Volume of saturated gypsum solution added to the soil = V_C ml

Volume of aliquot of the filtrate taken for titration = V_F ml

Volume of EDTA solution required for titration of filtrate aliquot = V_T ml

Volume of EDTA required for titration of blank = V_D ml

Caused for exchange of Na, meq/100 g soil = $(V_D - V_T) \times Z \times \frac{V_C}{V_F} \times \frac{100}{W} = A$

[As 1 ml 1 **N** EDTA ≡ 1 meq of gypsum ≡ 1 meq of Ca]

So, amount of gypsum required per kg of soil, kg = $\frac{A \times 86}{10^6} \times \frac{1000}{100} = \frac{A \times 86}{10^5} = B$

[Equivalent weight of gypsum = 86 or 1 meq = 86 mg gypsum]

$$\text{Amount of gypsum required per hectare of soil,tons} = \frac{B}{10^3} \times 2.2 \times 10^6$$

[Considering weight of 1 hectare furrow slice (15 cm) soil = 2.2 x 10^6 kg]

[*Note*: Practically it is observed that gypsum addition of about S! of the gypsum requirement value obtained in this method is satisfactory in most of the cases.]

Questions and Hints

Q.1 What do you mean by suspension effect in soil pH determination?

Hint: If the soil suspension is allowed to settle down and pH is measured placing the electrode pair in supernatant liquid and in sediment, then the difference in the soil pH reading is known as suspension effect. The pH value of soil sediment is always lower than the pH of the supernatant liquid.

Q.2 What happens in the soil pH values with changing soil-water ratio?

Hint: In general, irrespective of soil type (acidic or alkaline), with dilution i.e. increasing soil: water ratio, the pH value increases and reverse is also true. On dilution, H^+ concentration decreases but at the same time activity coefficient of H^+ increases. However, the overall effect is decrease in the activity of H^+ with dilution. The effect of dilution is more pronounced in alkaline soil due to hydrolysis of exchangeable Na ions with the formation of NaOH.

Q.3 How salt content affects the pH values?

Hint: Increase in salt content of the soil decreases the pH value. In neutral soil, increase in salt content increases the apparent strength of acidic groups; while in alkaline soil, the pH value is lowered due to reduction of hydrolysis of exchangeable cation like Na.

Q.4 Why the soil pH value remains unaffected over a wide range of soil: solution (0.01 M $CaCl_2$) ratio?

Hint: This is related to Ratio Law. As Ca is the most abundant basic metal cation in most soils, addition of $CaCl_2$ solution does not usually change the proportion of exchangeable cations (a_H+/$\sqrt{a_{Ca}}$2+).

Q.5 What are factors governing the lime requirement of soil?

Hint: (a) The desired change of pH, (b) the buffering capacity, (c) the chemical nature of the liming material and (d) the fineness of the liming material.

Q.6 What is the role of thioglycollic acid in the determination of extractable Al?

Hint: To eliminate the interference of iron by masking.

Q.7 Why the pH of the extractant is maintained at 8.2 in total potential acidity determination?

Hint: This pH value corresponds to the pH of complete neutralization of hydroxy Al compounds present in the soil.

Q.8 Why $MgCl_2$ is added in the standard EDTA solution?

Hint: Stability of Ca-EBT indicator complex is less. That's why detection of sharp end point may be difficult. Use of Mg which forms relatively stable complex with indicator helps to get sharp colour change at the end point (*also see hint of Q.8 of chapter 18 page no 256*).

Q.9 Why $CaCl_2$ can not be used for preparation of standard Ca solution?

Hint: Because of its hygroscopicity, accurate weighing of the material is quite difficult.

18

Electrical Conductivity and Water Soluble Salt

18.1 Determination of Electrical Conductivity of Soil Solution

The conductivity or more precisely the specific conductance of a soil at a given soil-water ratio is the reciprocal of the specific resistance offered by the solution at 25°C between the electrodes of 1 sq cm cross section kept 1 cm apart. Thus, the electrical conductivity of any solution depends on the amount of soluble salt present in it. Knowledge on the extent of salinity is important for management of saline soil including the choice of the crop suitable for that soil. Soluble salt content in saturation extract is more reliable index of salinity level than in soil: water ratio of 1 : 2 or 1 : 2.5, because it simulates the extent of salinity experienced by crop in field moisture condition.

Salinity class of soil is based on the electrical conductivity (EC) of saturation extract at 25°C. However, EC of water extract from 1:2 soil-water suspension can be transformed into the EC of saturation extract following the formula (except for soil containing gypsum).

$$EC_{sat} = \mathrm{EC}_{1:2} \times \frac{200}{Saturation\ percentage} \qquad (18.1)$$

Where, saturation percentage $= \dfrac{Moisture\ content\ at\ saturation}{Oven\ dry\ weight\ of\ soil} \times 100$

Principle

A simple Wheatstone bridge principle (except the type of current) is used for the measurement of EC of the soil solution, as EC of any solution is directly proportional to the concentration of soluble salt. Electrical conductivity is expressed as millimhos/cm (mmhos/cm) or decisiemens/meter (dS/m).

Equipments and Materials

Conductivity meter with conductance cell, mechanical shaker, Buchner funnel, 100 ml conical flask, Whatman No. 42 filter paper.

Reagents

- Standard potassium chloride solution, 0.01 **N:** Dissolve 0.7456 g of AR grade potassium chloride (KCl) and make up volume to 1 litre with distilled water. The EC_{25} of 0.01 **N** KCl solution is 1.413 dS/m.

Procedure

a) Preparation of soil-water extract

(i) Saturation extract

- Prepare a saturated soil paste as described in section 17.1.
- Spread the saturated paste on a Whatman No. 42 filter paper fitted on a Buchner funnel.
- Apply suction by vacuum pump until air starts to pass through the filter paper and collect the extract.

(ii) 1:2 or 1:2.5 soil-water extract

- Take 20g air dry soil in a 100 ml conical flask and add 40 ml (for 1:2) or 50 ml (for 1:2.5) distilled water.
- Shake on a mechanical shaker for 30 minutes and allow settling down for another 30 minutes.
- Filter the supernatant liquid through Whatman No. 42 filter paper into a beaker.

b) Determination of Electrical Conductivity

- Switch on the conductivity meter and warm up it for 30 minutes.
- Adjust temperature at 25°C (if provided with the instrument) and cell constant (if known).
- If cell constant value is not known then calibrate the instrument using 0.01 **N** KCl solution with its standard specific conductance value at room temperature (Appendix V).
- Thoroughly wash the cell with distilled water and then rinse with test solution.

- Measure the conductance of the solution directly from digital display and calculate specific conductance by multiplying electrical conductance with the cell constant value.
- Make necessary temperature correction at 25°C (if not provided with the instrument) by multiplying electrical conductivity with appropriate correction factor at room temperature (Table 18.1)

Calculation

Electrical conductivity of the extract, mmhos/cm or dS/m = Conductance of the extract (mmhos) × cell constant (cm^{-1}) temperature correction factor (tf) (if correction facility is not provided with the instrument).

Table 18.1: Temperature correction factor (tf) for correcting conductivity to 25°C

°C	tf	°C	tf	°C	tf	°C	tf	°C	tf
10.0	1.411	20.0	1.112	23.6	1.029	27.2	0.956	30.8	0.894
11.0	1.375	20.2	1.107	23.8	1.025	27.4	0.953	31.0	0.890
12.0	1.341	20.4	1.102	24.0	1.020	27.6	0.950	31.2	0.887
13.0	1.309	20.6	1.097	24.2	1.016	27.8	0.947	31.4	0.884
14.0	1.277	20.8	1.092	24.4	1.012	28.0	0.943	31.6	0.880
15.0	1.247	21.0	1.087	24.6	1.008	28.2.	0.940	31.8	0.877
16.0	1.218	21.2	1.082	24.8	1.004	28.4	0.936	32.0	0.873
17.0	1.189	21.4	1.078	25.0	1.000	28.6	0.932	32.2	0.870
18.0	1.163	21.6	1.073	25.2	0.996	28.8	0.929	32.4	0.867
18.2	1.157	21.8	1.068	25.4	0.992	29.0	0.925	32.6	0.864
18.4	1.152	22.0	1.064	25.6	0.988	29.2	0.921	32.8	0.861
18.6	1.147	22.2	1.060	25.8	0.983	29.4	0.918	33.0	0.858
18.8	1.42	22.4	1.055	26.0	0.979	29.6	0.914	34.0	0.843
19.0	1.136	22.6	1.051	26.2	0.975	29.8	0.911	35.0	0.829
19.2	1.131	22.8	1.047	26.4	0.971	30.0	0.907	36.0	0.815
19.4	1.127	23.0	1.043	26.6	0.967	30.2	0.904	37.0	0.801
19.6	1.122	23.2	1.038	26.8	0.964	30.4	0.901	38.0	0.788
19.8	1.117	23.4	1.034	27.0	0.960	30.6	0.897	39.0	0.775

Rating

Salinity class	$EC_{sat\ at\ 25}$ (dS/m)
Non saline (Normal)	0 – 2
Very slightly saline	2 – 4
Moderately saline	4 – 8
Strongly saline	8 – 16
Very strongly saline	> 16

Crop sensitivity to salt content

$EC_{sat\ at\ 25}$ (dS/m)	Crop suitability
0 - 2	Suitable for most of the crops
2 - 4	Unsuitable for very high salt sensitive crops e.g. citrus, potato, tomato
4 – 8	Yield of sesame, soybean will be affected
8 – 16	Only salt tolerant crops like barley, sorghum, corn should be grown
> 16	Only high salt tolerant crops like cotton, sunflower, sugar beet can be grown

Some soil parameter related to $EC_{sat\ at\ 25}$

- Total Dissolved Solid (TDS), mg/L $\approx 640\ EC_{sat\ at\ 25}$ (dS/m)
- Total Cation or Anion Concentration, mmol/L $\approx 10\ EC_{sat\ at\ 25}$ (dS/m)
- Osmotic Pressure at 25°C, bars $\approx 0.36\ EC_{sat\ at\ 25}$ (dS/m)

18.2 Water Soluble Salts

After estimating electrical conductivity if it is found that the EC is more than 2 dS/m for saturation extract or 1 dS/m for 1:2 soil-water extract, then estimation of individual ions is necessary. Usually, the dominant cations present in soil are Ca^{2+}, Mg^{2+}, K^+ and Na^+, though cations like Al^{3+}, Fe^{3+} and Mn^{2+} may be prevalent in acid soils. The primary anions in soil are SO_4^{2-}, Cl^-, NO_3^-, CO_3^{2-} and HCO_3^-. The water soluble salts in solution are generally determined either in the saturated extract or 1:2 soil-water extract.

Extraction of Soluble Salts

As discussed in the section 18.1.

18.2.1 Determination of Carbonate and Bicarbonate in Soil Solution

Principle

Carbonate and bicarbonate can be determined by titrating the solution with standard acid in presence of phenolphthalein and methyl orange indicator, respectively. Carbonate neutralization takes place in two stage neutralization reactions; in first stage carbonate is converted to bicarbonate as indicated by phenolphthalein indicator and in second stage total bicarbonate (originally present in the solution plus bicarbonate coming from half neutralization of carbonate) undergoes complete neutralization by acid as indicated by methyl orange indicator.

1st stage reaction: $Na_2CO_3 + H_2SO_4 \xrightarrow[\text{Phenolphthalein}]{\text{Pink - Colourless}} NaHCO_3 + NaHSO_4$

2nd stage reaction: $NaHCO_3 + H_2SO_4 \xrightarrow[\text{Methyl orange}]{\text{Yellow - Rose red}} Na_2SO_4 + 2CO_2 + 2H_2O$

It is evident from the above reactions that amount of H_2SO_4 required in both stages of neutralization of carbonate is equal.

Equipments and Materials

100 ml porcelain dish, 10 pipette, burette.

Reagents

- Standard sulphuric acid (H_2SO_4), 0.01 **N** (approx.): Dilute 0.3 ml of concentrated H_2SO_4 to 1 litre with distilled water and standardize with standard alkali (Na_2CO_3) in presence of indicator methyl orange. Colour changes from rose red to yellow.
- Phenolphthalein indicator, 0.25%: Dissolve 0.25 g indicator in 100 ml of 60% ethanol.
- Methyl orange, 0.5%: Dissolve 0.5 g indicator in 100 ml of 95% ethanol.

Procedure

- Pipette out 10 ml of extract in a porcelain dish and add 5 drops of phenolphthalein indicator.
- Titrate with standard sulphuric acid until just disappearance of pink colour. This end point indicates the half neutralization of carbonate ($CO_3^{2-} \rightarrow HCO_3$). Record the burette reading.
- Add 4 drops of methyl orange indicator to that colourless solution.
- Titrate with standard acid until yellow colour changes to rose red which indicates complete neutralization of total bicarbonate.
- Record the burette reading.

Calculation

Say,

Weight of soil taken = W g

Volume of water added for extraction of salt = V_W ml

Volume of aliquot of extract taken for titration = V_A ml

Volume of H_2SO_4 consumed for 1st titration = V_1 ml

Volume of H_2SO_4 consumed for 2nd titration = V_2 ml

Strength (N) of sulphuric acid = S

$$\text{Carbonate concentration in the extract,meq/ L} = \frac{2V_1 \times S}{V_A} \times 1000 = Z$$

$$\text{Carbonate concentration in the extract, g / L} = \frac{Z}{1000} \times 30$$

(As equivalent weight of carbonate = 30)

$$\text{Carbonate concentration,meq/100g of soil} = 2V_1 \times S \times \frac{V_W}{V_A} \times \frac{100}{W} = M$$

$$\text{Carbonate concentration in soil (\%)} = \frac{M}{1000} \times 30$$

$$\text{Bicarbonate conc.in the extract,meq/ L} = \frac{(V_2 - V_1) \times S}{V_A} \times 1000 = Y$$

$$\text{Bicarbonate concentration in the extract,g / L} = \frac{Y}{1000} \times 61$$

(As equivalent weight of bicarbonate = 61)

$$\text{Bicarbonate concentration,meq/100g of soil} = (V_2 - V_1) \times S \times \frac{V_W}{V_A} \times \frac{100}{W} = N$$

Rating

Soluble salts(%) in soil of 20 cm depth		Remarks
CO_3^{2-}	HCO_3^-	
Nil	< 0.06	Non-salinized
< 0.005	-	Weakly salinized
0.005 – 0.010	-	Average salinized
0.011 – 0.030	-	Strongly salinized

18.2.2 Determination of Chloride in Soil Solution

Principle

Chloride in the solution can be determined by titration with silver nitrate ($AgNO_3$) using potassium chromate (K_2CrO_4) as an adsorption indicator. Silver nitrate reacting with chloride ion forms silver chloride (AgCl) precipitate. At the end of neutralization when no chloride ion remains in the solution chromate ions combine with silver ions to form brick red coloured precipitate of silver chromate.

$$NaCl + AgNO_3 \rightarrow AgCl \downarrow + NaNO_3$$

$$K_2CrO_4 + AgNO_3 \rightarrow Ag_2CrO_4 + 2KNO_3$$

(Brick red)

Equipments and Materials

As in section 18.2.1

Reagents

- Standard silver nitrate solution, 0.02 **N**: Dissolve 3.4 g silver nitrate in distilled water and make up volume to 1 litre. Standardize against standard NaCl (0.02 **N**) solution. Store $AgNO_3$ solution in amber colour bottle away from light.
- Standard sodium chloride solution, 0.02 **N**: Dissolve exactly 1.169 g pure dry NaCl and volume make to 1 litre with distilled water.
- Potassium chromate solution, 5%: Dissolve 5 g pure K_2CrO_4 in 50 ml distilled water and add saturated solution of $AgNO_3$ drop by drop until slightly red permanent precipitate is formed. Filter it and dilute to 100 ml.
- Standard sulphuric acid (H_2SO_4), 0.01 **N** (approx.): As in section 18.2.1.

Procedure

- Either pipette out fresh 10 ml of extract in a 100 ml porcelain dish and add equal volume of standard sulphuric acid (0.01 **N**) as required for complete neutralization of carbonate and bicarbonate or use the same extract after neutralization of carbonate and bicarbonate.
- Add a little sodium bicarbonate to neutralize excess acid added and 5-6 drops of potassium chromate solution.
- Titrate the solution with standard $AgNO_3$ until the first permanent brick red precipitate forms.

Calculation

Say,

Weight of soil taken = W g

Volume of water added for extraction of salt = V_W ml

Volume of aliquot of extract taken for titration = V_A ml

Volume of $AgNO_3$ required for titration = V_T ml

Strength (N) of $AgNO_3$ = S

$$\text{Chloride concentration in the extract,meq / L} = \frac{V_T \times S}{V_A} \times 1000 = Z$$

$$\text{Chloride concentration in the extract,g / L} = \frac{Z}{1000} \times 35.5$$

[As equivalent weight of chloride = 35.5]

$$\text{Chloride concentration, meq/100g of soil} = V_T \times S \times \frac{V_W}{V_A} \times \frac{100}{W} = Y$$

$$\text{Chloride concentration in soil (\%)} = \frac{Y}{1000} \times 35.5$$

Rating

Chloride concentration (%) in soil of 20 cm depth	Remarks
< 0. 020	Non-salinized
0.020 – 0.050	Weakly salinized
0.051- 0.120	Average salinized
0.121 - 0.200	Strongly salinized

18.2.3 Determination of Sulphate in Soil Solution

Principle

Sulphate in solution can be determined gravimetrically by precipitating sulphate as barium sulphate in acidic medium. The precipitate is dried, ignited and weighed to calculate the amount of sulphate present.

$Na_2SO_4 + BaCl_2 \rightarrow BaSO_4 + 2NaCl$

Equipments and Materials

Muffle furnace, oven, silica crucible, water bath, desiccator, balance, 50 ml measuring cylinder, 10 ml graduated pipette, Whatman No. 42 filter paper,

Reagents

- Barium chloride solution, 5%: Dissolve 5g AR grade barium chloride ($BaCl_2$) and make the volume to 100 ml with distilled water.
- Concentrated hydrochloric acid (HCl)

Procedure

- Transfer 50 ml clear soil extract in a 100 ml beaker and add 5 ml concentrated HCl and boil.

- Add $BaCl_2$ solution drop by drop and stir constantly till total amount of sulphate is precipitated. Add an excess amount of precipitating agent ($BaCl_2$ solution).
- Keep the content in the beaker simmering for half an hour over a hot water bath or low temperature hot plate to allow for complete precipitation.
- Filter through Whatman No. 42 filter paper and wash with hot water to remove all the free chloride ions (confirm by detection of no precipitate in the leachate with $AgNO_3$).
- Transfer the precipitate in a silica crucible previously ignited to redness, cooled in desiccator and weighed.
- Dry the crucible and precipitate in an oven at 100-110°C temperature until it turns into white (if the precipitate is slightly discoloured, add 1 or 2 drops of dilute H_2SO_4 and evaporate). Transfer the crucible along with its content in muffle furnace and ignite to 550°C for the period of 15 minutes until constant weight is attained.
- Cool in desiccator and weigh the crucible with its content.

Calculation

Say,

Weight of soil taken = W g

Volume of water added for extraction of salt = V_W ml

Volume of aliquot of extract taken = V_A ml

Weight of empty crucible = X g

Weight of crucible plus $BaSO_4$ ppt = Y g

Weight of dry $BaSO_4$ precipitate = (Y - X) = A g

$$\text{Therefore, weight of sulphate in the aliquot, g} = A \times \frac{96}{223.4} = Z$$

[As 96 g sulphate present in 223.4 g $BaSO_4$ (Mol. Wt)]

$$\text{Sulphate content in the extract, meq / L} = \frac{Z \times 1000}{48} \times \frac{1000}{V_A}$$

[As equivalent weight of sulphate = 48]

$$\text{or, Sulphate content in the extract, g / L} = \frac{Z}{V_A} \times 1000$$

$$\text{Sulphate content,meq / 100g of soil} = \frac{Z \times 1000}{48} \times \frac{V_W}{V_A} \times \frac{100}{W}$$

18.2.4 Determination of Calcium and Magnesium in Soil Solution

Principle

The method is based on the formation of soluble complex of cations including calcium and magnesium with versene (disodium salt of ethylene diamine tetra acetic acid). Disodium salt of ethylene diamine tetra acetic acid (EDTA) can be expressed as Na_2H_2Y which reacts with metals in 1:1 molar ratio. It is evident from the above reaction that the dissociation (backward reaction) of $(MY)^{(n-4)+}$ complex will be governed by the H^+ concentration. Higher H^+ concentration (lower pH) will favour dissociation i.e., lowers the stability of the metal-EDTA complex.

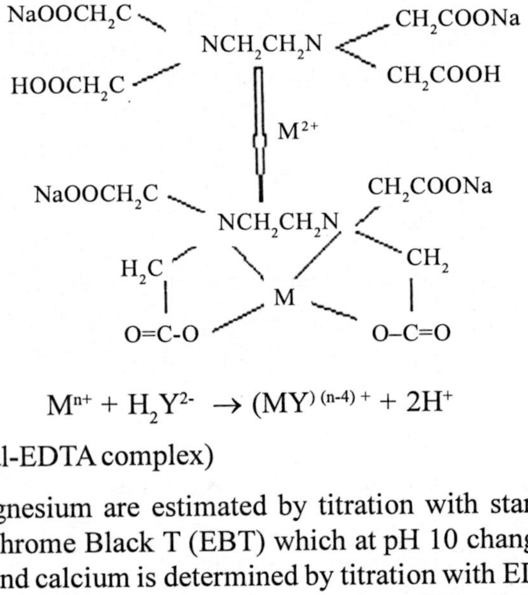

$NaOOCH_2C$ — CH_2COONa
NCH_2CH_2N
$HOOCH_2C$ — CH_2COOH

$\Downarrow$ M^{2+}

$NaOOCH_2C$ — CH_2COONa
NCH_2CH_2N
H_2C — CH_2
M
O=C-O — O–C=O

$$M^{n+} + H_2Y^{2-} \rightarrow (MY)^{(n-4)+} + 2H^+$$

(Metal ion) (Metal-EDTA complex)

Calcium and magnesium are estimated by titration with standard EDTA in presence of Eriochrome Black T (EBT) which at pH 10 changes colour from wine red to blue and calcium is determined by titration with EDTA at pH 12 in presence of calcon indicator which changes colour from pink to blue at end point. At this pH, magnesium interference is avoided by precipitating it quantitatively as $Mg(OH)_2$.

The success of an EDTA titration depends upon the correct detection of end point by use of suitable metal-ion indicator.

M-In + EDTA $\rightarrow$ M-EDTA + In

(Metal-ion indicator) (Free indicator)

The above reaction suggests that the success of EDTA titration will be achieved if M-EDTA complex becomes more stable than M-In, so that at the end of titration M from the M-In complex gets free and forms complex with EDTA and free indicator (In) will indicate the end point colour. On the basis of stability, the complexes formed during Ca plus Mg titration can be arranged in the following sequence:

Ca-EDTA complex > Mg-EDTA complex > Mg-In complex > Ca-In complex

Thus, during titration of Ca plus Mg with EDTA using EBT as indicator, EDTA first reacts with free Ca^{2+}, then free Mg^{2+} and finally with Mg being dissociated from Mg-In complex (As stability of Ca-In complex is less, Ca will be easily dissociated from this complex and forms complex with EDTA prior to release of Mg from Mg-In complex). Since Mg-EBT complex is wine red in colour and free indicator is blue between pH 7 - 11, the colour of the solution changes from wine red to blue at the end point.

$$\underset{\text{(Wine red)}}{\text{Mg-EBT}} + H_2Y^{2-} = MgY^{2-} + \underset{\text{(Blue)}}{\text{H-EBT}^{2-}} + H^+$$

During estimation of Ca and Mg number of other cations may present in the system which may interfere in the determination of Ca and Mg. To eliminate the interference of other cations different masking agents are used. Hydroxylamine in acidic medium reduces higher valent cations like Cu^{2+} and Fe^{3+} to their lower valence state, triethanolamine can mask Ti^{4+}, Al^{3+} and Fe^{3+}. Cyanide ion can act as an effective masking agent with the formation of stable cyanide complex with cations like Cd^{2+}, Zn^{2+}, Hg^{2+}, Cu^{2+}, Co^{2+}, Ni^{2+}, Ag^{+}, and Fe^{2+}.

Equipments and Materials

100 ml porcelain dish, 10 ml pipette, burette.

Reagents

- Ammonium chloride – ammonium hydroxide buffer solution: Dissolve 67.5 g of ammonium chloride (NH_4Cl) in 570 ml of concentrated ammonia solution and make up volume to 1 litre with distilled water.
- Standard EDTA (0.01 **N**) solution: Dissolve 2 g disodium salt of ethylene diamine tetra acetic acid (EDTA) in water and dilute to 1 litre. Standardize it against standard calcium solution.

- Standard calcium chloride (0.01 **N**) solution: Dissolve exactly 0.5 g AR grade $CaCO_3$ in 10 ml 0.5 **N** HCl. Make up the volume to 1 litre with distilled water.
- Hydroxylamine hydrochloride ($NH_2OH.HCl$), 5% (w/v): Dissolve 5 g hydroxylamine hydrochloride in 100 ml with distilled water.
- Triethanolamine
- Eriochrome black T (EBT) or solochrome black indicator: Dissolve 0.2 g of EBT dye [1-(1-hydroxy-2-naphthylazo)-6-nitro-2-naphthol-4-sulphonate] in 50 ml methanol.
- Potassium (or sodium) cyanide solution (1 % w/v): Dissolve 1 g reagent KCN (or NaCN) and dilute to 100 ml with distilled water.
- Potassium ferro cyanide solution (4%): Dissolve 4 g reagent [$K_4Fe(CN)_6$] and dilute to 100 ml with distilled water.
- Calcon or solochrome dark blue indicator: Dissolve 0.2 g calcon [1- (2-hydroxy-1-naphthylazo) -2 naphthol-4-sulphonate] in 50 ml methanol/ ethanol.
- Sodium hydroxide solution, 10%: Dissolve 10g NaOH in distilled water and make up volume to 100 ml.

Procedure

a) Titration for Ca plus Mg

- Pipette out 10 ml (or suitable aliquot containing not more than 0.1 meq of Ca + Mg) of the extract in a porcelain dish and dilute to about 25 ml with distilled water.
- Add 10 ml NH_4Cl-NH_4OH buffer and 10 drops each of KCN (or NaCN), hydroxylamine hydrochloride, $K_4Fe(CN)_6$ and triethanolamine. Warm the solution for 3 minutes.
- After cooling add 3-4 drops of EBT indicator and titrate with standard EDTA solution until the colour changes from wine red to blue.
- Run a blank (replacing the soil extract with water) with 25 ml distilled water.

[*Note:* During versene titration of Ca plus Mg in a solution containing no or very trace amount of Mg using EBT as indicator, Mg should be added with standard EDTA solution to obtain a sharp end point (Section 10.5). Prepare separate standard EDTA solution adding small amount of $MgCl_2$ (0.05g $MgCl_2.6H_2O$ per litre of 0.01**N** EDTA solution)].

b) Titration for Ca

- Pipette out 10 ml (or suitable aliquot containing not more than 0.1 me of Ca) of the extract in a porcelain dish and dilute to about 25 ml with distilled water.
- Add 10 drops each of KCN (or NaCN), hydroxylamine hydrochloride and triethanolamine.
- Add 1 ml or more 10% NaOH solution to raise the pH of the solution to 12 (check by pH meter).
- Add 5 drops of calcon indicator and titrate with standard EDTA solution until the colour changes from pink to blue.
- Run a blank replacing the soil extract with 25 ml distilled water in similar way.

Calculation

Say,

Weight of the soil taken = W g

Volume of water added for extraction of salt = V_W ml

Volume aliquot taken for Ca plus Mg estimation = V_1 ml

Volume aliquot taken for Ca estimation = V_2 ml

Volume of EDTA consumed for Ca plus Mg estimation= V_X ml

Volume of EDTA consumed for Ca estimation = V_Y ml

Volume of EDTA consumed for blank (Ca plus Mg titration) = V_A ml

Volume of EDTA consumed for blank (Ca titration) = V_B ml

Strength (N) of EDTA = S

Hence,concentration of Ca plus Mg in the extract, meq/L = $(V_X - V_A) \times S \times \frac{1000}{V_1} = A$

Concentration of Ca plus Mg,meq/100 g of soil =$(V_X - V_A) \times S \times \frac{V_W}{V_1} \times \frac{100}{W} = B$

Concentration of Ca in the extract, meq/L = $(V_Y - V_B) \times S \times \frac{1000}{V_2} = C$

Concentration of Ca,meq/100 g of soil = $(V_Y - V_B) \times S \times \frac{V_W}{V_2} \times \frac{100}{W} = D$

Concentration of Mg in the extract, meq/L = A - C

Concentration of Mg, meq/100 g of soil = B - D

18.2.5 Determination of Sodium in Soil Solution

Principle

Sodium can be determined flame photometrically as it is easily excited in the flame with a characteristic strong radiation of 589.6 nm wavelength of yellow colour. Other less intensity wavelengths are discarded by Na filter. The intensity of the emission of characteristic wavelength of Na is measured photelectrically.

Equipments and Materials

Flame photometer with Na filter, volumetric flask of 1 litre and 100 ml capacity.

Reagents

- Standard NaCl solution: Dissolve exactly 2.542 g of AR grade NaCl in distilled water and make up volume to 1 litre with distilled water. This gives 1000 ppm Na solution. From 1000 ppm stock solution prepare 1, 2, 4, 6, 8, and 10 ppm Na solution by proper dilution.

Procedure

- As per the direction in the operation manual adjust gas supply and air pressure.
- Set Na filter and adjust zero with distilled water and 100 with maximum concentration (say, 10 ppm of Na).
- Record flame photometer readings for other standard Na solutions.
- Feed the soil extract in the flame photometer and note the reading.
- Construct a standard curve by plotting concentration (ppm) along X-axis against flame photometer reading along Y- axis (Fig. 18.1).
- Draw an average line passing through the origin.

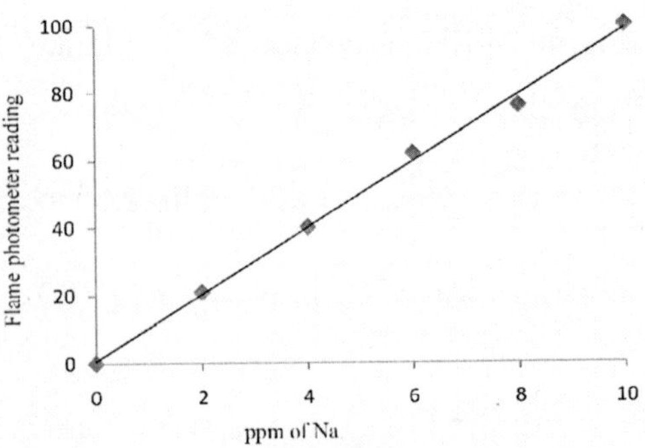

Fig. 18.1: Standard curve for Na

Calculation

Say,

Weight of the soil taken = W g

Volume of water added for extraction of salts $=V_W$ ml

Concentration of Na in the soil extract obtained from the standard curve against the corresponding flame photometer reading = A ppm.

Aliquot dilution (if required during estimation) factor = D

$$\text{Concentration of Na in the extract, meq/L} = \frac{A \times D}{23}$$

[As equivalent weight of Na = 23]

$$\text{Concentration of Na, meq/100 g of soil} = \frac{A \times D}{23} \times \frac{V_W}{1000} \times \frac{100}{W}$$

18.2.6 Determination of Potassium in Soil Solution

Principle

Like sodium, potassium can be determined by flame photometer as its atoms are readily excited in the flame producing reddish light which is primarily due to the emission of 766.5 nm wavelength, characteristic radiation of potassium. Red filter only allows passing through 766.5 nm wavelength blocking other low energy radiation of different wavelengths. The intensity of the emitted radiation which is proportional to the concentration of potassium in the flame is measured.

Equipments and Materials

Flame photometer with K filter, volumetric flask of 1 litre and 100 ml capacity

Reagents

- Standard KCl solution: Dissolve exactly 1.907 g of AR grade KCl and make up volume to 1 litre with distilled water. This gives 1000 ppm K solution.
- From 1000 ppm stock solution prepare 2, 5, 10, 15, and 20 ppm K solution by proper dilution.

Procedure

- As described in the determination of Na, except adjust 100 flame photometer reading with 20 ppm of K.

Calculation

Say,

Weight of the soil taken = W g

Volume of water added for extraction of salts =V_w ml

Concentration of K in the soil extract obtained from the standard curve against the corresponding flame photometer reading = A ppm.

Aliquot dilution (if required during estimation) factor = D

$$\text{Concentration of K in the extract,meq /L} = \frac{A \times D}{39}$$

[As equivalent weight of K = 39]

$$\text{Concentration of K,meq /100 g of soil} = \frac{A \times D}{39} \times \frac{V_W}{1000} \times \frac{100}{W}$$

Questions and Hints

Q.1 What are the principal ions that are likely to be present as soluble salt in soils?

Hint: Water soluble salts occurring in soils usually consist principally of four cations Na^+, K^+, Ca^{2+} and Mg^{2+}, linked mainly to Cl^- and SO_4^{2-}, sometimes to NO_3^- and CO_3^{2-}, and to a limited extent to HCO_3^-.

Q.2 What are the ions mainly responsible for soil salinity?

Hint: Soil salinity problem mainly arises from Na^+, Cl^- and SO_4^{2-}, and seldom from Ca^{2+}, Mg^{2+} and CO_3^{2-}.

Q.3 Why the specific conductance of the soil saturation extract is a more reliable measurement of soil salinity than that of an extract made with a constant soil: water ratio?

Hint: Soil: water ratio influences the amount and composition of salt extracted. Higher water to soil ratio extracts more salt than is present in soil at field moisture condition. Since the wilting percentage is lesser for sandy soil than for finer textured soil, a given absolute amount of salt per unit weight of sandy soil creates a greater salt concentration in the solution at wilting point than for finer textured soil.owing to the fundamental relation of the saturation moisture percentage to soil moisture constants. The saturation extract is equipotential soil moisture content for all soils. Thus, specific electrical conductance of the saturation extract, which is linearly related to osmotic pressure as well as salt concentration in the solution, can be interpretated directly in terms of plant growth by means of salinity scale.

Q.4 What are factors for conversion of specific conductance (say, L dS/m) to (a) meq of salt per litre, (b) ppm of salt in solution, (c) per cent salt in solution, (d) per cent salt in soil and (e) osmotic pressure of solution?

Hint: (a) Meq of salt per litre =12.5 L

(b) ppm of salt in solution = 640 L

(c) % salts in solution = 0.064 L

(d) % salt in soil $= 0.064\ \text{L} \times \dfrac{\%\ \textit{water in soil at extraction}}{100}$ *and*

(e) osmotic pressure of solution, bar = 0.36 L.

Q.5 Why during calculation of carbonate concentration, the volume of H_2SO_4 consumed in 1st titration is multiplied by 2?

Hint: Carbonate neutralization is a two step reaction: in 1st step carbonate is converted to bicarbonate and in the 2nd step bicarbonate is neutralized by acid. In both steps the amount of H_2SO_4 consumed is equal. That is why during calculation of carbonate concentration; the volume of H_2SO_4 consumed in 1st titration is multiplied by 2.

Q.6 Why two pH indicators are used for estimation of carbonate and bicarbonate?

Hint: The choice of pH indicator depends on the pH at equivalence point. The pH at equivalence point for conversion of carbonate to bicarbonate is 8.3 and for conversion of bicarbonate to carbonic acid is 3.7. That's why two different pH indicators are used for estimation of carbonate and bicarbonate. Phenolphthalein changes colour between pH 8.3 and 10.0, while methyl orange changes colour between 2.8 and 4.8.

Q.7 Define the role of triethanolamine, hydroxylamine and cyanide during estimation of Ca and Mg by versene titration.

Hint: In addition to Ca and Mg soil solution contains other cations which may interfere in the estimation of Ca and Mg. To eliminate this interference different masking agents are used. Hydroxylamine in acidic medium reduces higher valent cations like Cu^{2+} and Fe^{3+} to their lower valence state, triethanolamine can mask Ti^{4+}, Al^{3+} and Fe^{3+}. Cyanide ion can act as an effective masking agent with the formation of stable cyanide complex with cations like Cd^{2+}, Zn^{2+}, Hg^{2+}, Cu^{2+}, Co^{2+}, Ni^{2+}, Ag^{+}, and Fe^{2+}.

Q.8 Why small amount of Mg should be added with EDTA during estimation of Ca plus Mg in the soil extract of Mg deficient soil?

Hint: Ca-In complex is very unstable. To get a sharp colour change in Mg deficient soil extract small amount of Mg is added during preparation of standard EDTA as during titration of Ca plus Mg with EDTA in presence of EBT indicator, the actual change of colour takes place due to conversion of Mg-In complex (wine red) to free indicator (blue).

Q.9 Why during (a) estimation of Ca plus Mg ammonium chloride-ammonium hydroxide buffer and during (b) estimation of Ca, NaOH/ KOH are added?

Hint: (a) The stability of metal –EDTA complex is pH dependent. Lowering of solution pH favours dissociation of the complex or increases the instability. Again, the success of an EDTA titration depends upon the correct detection of end point by use of suitable metal-ion indicator. EBT changes colour from wine red to blue at pH 10. That is reason of using the buffer solution.

(b) During estimation of Ca, NaOH/KOH is used to raise the pH of the medium to 12. As because at this pH Mg interference is avoided by precipitation of Mg as Mg $(OH)_2$ and calcon changes colour from pink to blue at this pH.

Q.10 Establish relationship between ppm and meq/100g.

Hint: X ppm $= \dfrac{X\ mg}{kg} = \dfrac{\frac{X\ mg}{10}}{\frac{kg}{10}} = \dfrac{\frac{X\ mg}{10}}{100g}$

Again, $\dfrac{mg}{Equivalent\ Wt.} = meq$

So, $\dfrac{\frac{X\ mg}{10}}{100g} = \dfrac{\frac{X\ mg}{10 \times Equivalent\ Wt.}}{100g} = \dfrac{Y\ meq}{100g}$ [if, Y meq $= \dfrac{X\ mg}{10 \times Equivalnet\ Wt.}$]

Thus, $\dfrac{ppm}{10 \times Equivalent\ Wt.} = \dfrac{meq}{100g}$

19

Soil Organic Matter

Carbon may be present in soil in four different forms:

a) Carbonates: Mainly as $CaCO_3$ and $MgCO_3$, active CO_3^{2-} and HCO_3^- ions and gaseous CO_2.

b) Highly condensed elemental organic form: Charcoal, graphite.

c) Completely decomposed resistant organic residues: Humus.

d) Undecomposed and partially decomposed organic residues.

The latter three fractions constitute the total organic carbon of which last two are chemically active form.

Total organic matter content in soil can be determined (a) directly from loss in weight by (i) oxidation with H_2O_2 or (ii) ignition and (b) indirectly by estimating organic carbon content and multiplying with a factor. The conventional organic carbon to organic matter factor which is known as **Van Bemmelen factor** is 1.724 based on the assumption that soil organic matter contains 58% carbon. The above factor may vary not only among the soils but also within a profile along the depth and may be as high as 2.5 for subsoil. Direct methods have lots of limitation such as incomplete oxidation of organic matter by H_2O_2 or overestimation in ignition method because of influence of salts, H_2O and OH on weight loss. Among the indirect methods dry combustion provides the most accurate value as it includes elemental carbon such as charcoal, graphite and thus considered as reference for the other methods. But it is a time consuming and cost involving method.

19.1 Direct Method of Soil Organic Matter Determination

19.1.1 Determination of Total Organic Matter by Weight Loss on Ignition

Principle

If soil sample can be heated at 350-400°C temperature it is found that organic matter is destroyed without any loss of structural water and OH of inorganic soil components including $CaCO_3$. However, this method is far from satisfaction particularly for soils containing amorphous materials.

Equipments and Materials

Crucible, muffle furnace, desiccator, balance.

Procedure

- Take equal amount (about 2 g) of air dry processed (<0.2 mm) soil in two previously weighed separate crucibles.
- Dry one crucible in oven at 110°C for 6 to 8 hours, cool in desiccator and weigh to get the oven dry weight of sample.
- Place another crucible in muffle furnace and heated at 400°C for 7 to 8 hours, cool in desiccator and weigh it.

Calculation

Say,

Weight of empty crucible No. 1 = W_1 g

Weight of empty crucible No. 2 = W_2 g

Weight of empty crucible No. 1 + oven dry soil = W_3 g

Weight of empty crucible No. 2 + furnace dry soil = W_4 g

$$\text{Organic Matter content}(\%) = \frac{(W_3 - W_1) - (W_4 - W_2)}{(W_3 - W_1)} \times 100$$

19.2 Indirect Method of Soil Organic Matter Determination

19.2.1 Determination of Organic Carbon by Dry Combustion Method

The dry combustion method is based on oxidation of organic carbon and thermal decomposition of carbonate minerals in a medium-temperature resistance furnace or in a high temperature induction furnace. The CO_2 liberated is commonly trapped in a suitable reagent and determined titrimetrically or gravimetrically.

Alternatively, the CO_2 released can be reduced to CH_4 and estimated with a gas chromatography fitted with a flame ionization detector.

Principle

In the dry combustion method (Allison *et al*, 1965), the sample is burnt in a stream of purified O_2 and CO_2 in the effluent gas stream is absorbed by Ascarite or some other suitable absorbent and weighed. Other absorbable contaminating gases formed during combustion are removed from the O_2 stream before they could reach the CO_2 absorption bulb. A typical combustion train is comprised of 10 basic components (Fig. 19.1).

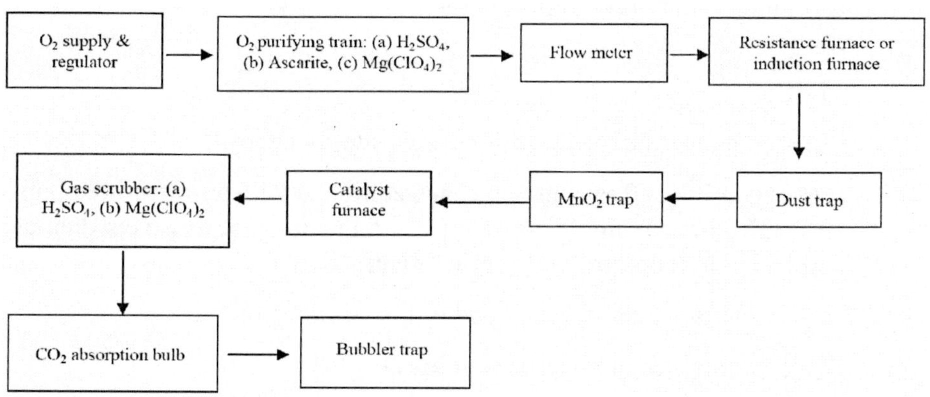

Fig. 19.1:

The O_2 supply is first scrubbed by passage through a train consisting of concentrated H_2SO_4 to remove NH_3 and hydrocarbon, an absorbent like soda lime/ascarite to absorb CO_2, and Mg $(ClO_4)_2$ to remove water vapour. The rate of O_2 flow is regulated by a needle valve and measured by a flow meter.

A furnace provides the heat necessary for combustion of the organic carbon to CO_2 and for decomposition of carbonates. In a resistance furnace, the sample is heated by radiation, conduction and convection in a tube surrounded by heating elements made of high resistance material such as Nichrome. Cupric oxide (CuO) or other accelerator is mixed with soil to aid in combustion of organic matter and elemental C. Catalysts are included in the combustion tube at the rear of the heating zone to ensure complete oxidation of CO or other volatile C compounds. Platinized asbestos or CuO wire may be used as a catalyst.

The gas stream leaving the furnace is made free from particulate matter by a dust trap in the exit end of the combustion tube. Activated MnO_2 appears satisfactory as a dry absorber of oxides of N and S and the halogens. Most of the water vapour formed during combustion is removed by a concentrated

H_2SO_4 tower immediately following the catalyst furnace and rest passing through is trapped by a tower of anhydrous $Mg(ClO_4)_2$ next in the line. The CO_2 is finally absorbed in a suitable bulb containing Ascarite or other absorbent backed by Mg $(ClO_4)_2$ to ensure equal water vapour pressure both in exit gas and entering gas.

Sequential combustion of the same sample at 575°C for 15 minutes followed by combustion at 1000°C for 10 minutes would quantitatively recover organic and inorganic C, respectively from soils (Rabenhorst, 1988).

Special Apparatus

(i) O_2 cylinder and pressure regulator

(ii) O_2 purifying train consisting of concentrated H_2SO_4, Ascarite and $Mg(ClO_4)_2$

(iii) Flow indicator and needle valve for O_2 supply control.

(iv) Furnace unit: (a) Resistance furnace equipped with temperature control and indicator, (b) sample inserter, (c) combustion tube of 2.5 cm diameter and 75 cm length (zircon ceramic or equivalent).

(v) Dust trap inserted in the exit end of the combustion tube.

(vi) Sulphur trap filled with activated MnO_2.

(vii) Catalyst furnace and tube.

(viii) Gas scrubber: (a) H_2SO_4 tower and (b) $Mg(ClO_4)_2$

(ix) CO_2 absorption tube packed with CO_2 absorbent (Ascarite) and anhydrous $Mg(ClO_4)_2$.

(x) Bubbler trap to seal the train from atmosphere and indicate flow of exit gas.

Accessories: (a) combustion boats, ceramic (Alundum, zircon, etc); (b) boat puller with eye shield; (c) combustion tube cleaning brush; (d) plastic tubing for connecting components; (e) analytical balance.

Reagents

- Oxygen gas
- Sulphuric acid concentrated
- MnO_2, activated
- Platinized asbestos, 5%, low in carbon for combustion tube catalyst

- Cupric oxide powder, low in carbon, to serve as an accelerator when mixed with soil in the boat
- Alundum, refractory grade (60- or 90 mesh) size, carbon free
- Anhydrous magnesium perchlorate [Mg $(ClO_4)_2$]
- CO_2 absorbent, 14 to 20 mesh size [Ascarite, Caroxide]
- Standard C source [dextrose ($C_6H_{12}O_6$) or benzoic acid ($C_2H_6O_2$) of reagent grade]

Procedure

- Loosely pack the combustion tube with a 7.5 cm core of platinized asbestos so that it comes within the exit end of the heated zone of the furnace.
- Bring the furnace temperature to 950°C to 1000°C.
- Connect the train, and sweep the apparatus with O_2 @ 100 ml/min for 10 minutes.
- Remove and weigh the CO_2 combustion bulb.
- Repeat the step until CO_2 combustion bulb attains a constant weight (±0.2 mg).
- Replace the CO_2 combustion bulb in the train, and introduce well within the heated zone of the combustion tube, a ceramic boat containing 1.0g of finely divided CuO.
- Admit O_2 and continue the flow @ 100 ml/min for 10 minutes.
- Close the stop cocks of the CO_2 combustion bulb, disconnect the bulb and weigh it. The increase in weight of the bulb represents the blank.
- Remove the boat from the combustion tube. Repeat the determination until a representative blank (negligible in weight) is attained.
- Mix 1.000 g of mineral soil (100 or 140 mesh) with 1.0 g of finely divided CuO in a combustion boat. Cover the mixture lightly with Alundum and follow the steps used for blank. The increase in weight, corrected for blank represents the CO_2 evolved from the sample.

Calculation

$$\text{Total Carbon}(\%) = \frac{[g\ CO2, sample - g\ CO2, blank]}{g, water\ free\ soil} \times 0.2727 \times 100$$

[as 1 g $CO_2 \equiv$ 12/44 or 0.2727 g C]

[*Note*: Carbon laboratory should be free from dust, dirt or fumes. It is advisable habitually to ignite all boats at 900°C before use. Handle boats and covers only with tong.]

To make absorption bulb free from static charge:

- Before each weighing wipe absorption bulb slowly with lint free paper tissue. Handle the absorption bulb with clean cotton gloves.
- After wiping the CO_2 combustion bulb, it should be touched to the grounded plate before weighing.

19.2.2 Determination of Oxidizable Organic Carbon by Wet Oxidation Method (Walkley and Black, 1934)

Principle

Soil organic carbon is oxidized to CO_2 by chromic acid formed by reaction of potassium dichromate ($K_2Cr_2O_7$) and concentrated sulphuric acid (H_2SO_4) utilizing the heat of dilution of H_2SO_4. The excess $K_2Cr_2O_7$ is estimated by back titration with standard ferrous ammonium sulphate and the amount of $K_2Cr_2O_7$ actually consumed for oxidation of organic carbon is determined as a difference of ferrous ammonium sulphate required between blank and sample titration. Schollenberger method (1927) involves same procedure as that of Walkley and Black method except more heating in former case applied through external source. The lesser heating in case of Walkley and Black method largely helps to differentiate soil humus from elemental carbon such as graphite and charcoal.

Oxidation of carbon

$$2K_2Cr_2O_7 + 8H_2SO_4 \rightarrow 2K_2SO_4 + 2Cr_2\ (SO_4)_3 + 8H_2O + 6O$$

$$3C + 6O \rightarrow 3CO_2$$

$$2K_2Cr_2O_7 + 8H_2SO_4 + 3C \rightarrow 2K_2SO_4 + 2Cr_2\ (SO_4)_3 + 8H_2O + 3CO_2$$

Or

$$4Cr^{6+} + 3C^{o} \xrightarrow{\text{Acidic medium}} 4Cr^{3+} + 3C^{4+}$$

Titration

$$K_2Cr_2O_7 + 4H_2SO_4 \rightarrow K_2SO_4 + Cr_2\ (SO_4)_3 + 4H_2O + 3O$$

$$6FeSO_4 + 3H_2SO_4 + 3O \rightarrow 3Fe_2\ (SO_4)_3 + 3\ H_2O$$

$$K_2Cr_2O_7 + 6FeSO_4 + 7H_2SO_4 \rightarrow K_2SO_4 + Cr_2\ (SO_4)_3 + 3Fe_2\ (SO_4)_3 + 7H_2O$$

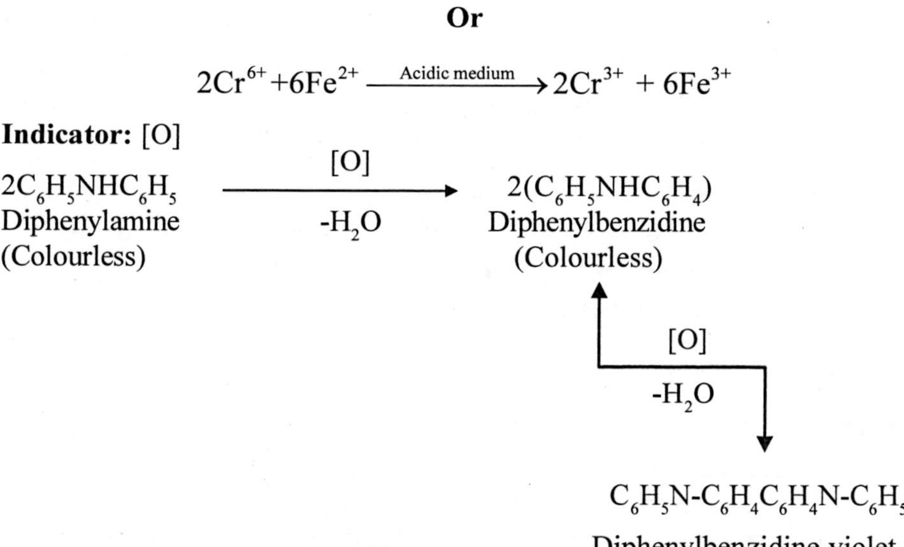

The Cr^{3+} ions formed by the reduction of $K_2Cr_2O_7$ are green in colour; but by simple visual inspection it is difficult to ascertain the end point of this dichromate titration. Thus, a redox indicator must be employed which gives strong and unmistakable colour change. At the beginning the colour of the solution is deep violet or dull green with the oxidized form of diphenylamine (diphenylbenzidine violet), then shifts to a turbid blue as the titration proceeds. At the end point the colour sharply shifts to a brilliant green due to presence of reduced form of chromium ion (Cr^{3+}). The reduced form of diphenylamine is colourless.

Equipments and Materials

Analytical balance, 500 ml conical flask, 10 ml single mark pipette, measuring cylinder of 25 and 250 ml capacity, burette.

Reagents

- Potassium dichromate ($K_2Cr_2O_7$) solution, 1 **N**: Dissolve 49.04 g of AR grade $K_2Cr_2O_7$ (oven dry at 105°C) in distilled water and dilute to 1 litre.
- Ferrous ammonium sulphate [$FeSO_4(NH_4)_2SO_4 \cdot 6H_2O$] or Mohr's solution, 0.5 **N**: Dissolve 196.1 g of AR grade ferrous ammonium sulphate in about 800 ml of distilled water and add 20 ml concentrated H_2SO_4 and make up volume to 1 litre.
- Diphenylamine indicator: Dissolve 0.5 g indicator in a mixture of 20 ml water and 100 ml concentrated H_2SO_4.

Or

- Ferroin or o-phenanthroline ferrous sulphate indicator: Dissolve 1.485g o-phenanthroline in about 80 ml distilled water. Add 0.69g ferrous sulphate heptahydrate ($FeSO_4.7H_2O$) and dilute to 100 ml.
- Concentrated H_2SO_4(sp. gravity. 1.84): In case of analyzing soil containing chloride add 15 g silver sulphate [$Ag_2(SO_4)$] per litre of H_2SO_4, otherwise use pure concentrated H_2SO_4.
- Concentrated ortho phosphoric acid (H_3PO_4), 85%
- Sodium fluoride (NaF)

Procedure

- Take accurately weighed 1.00 g air dry processed soil sample (pass through 0.2 mm nonferrous sieve) in a 500 ml conical flask. The amount of soil varies from 0.5 - 2.0 g for mineral soil and 0.05 – 0.05 g for organic soil.
- Add 10 ml of 1 **N** $K_2Cr_2O_7$ solution with the pipette and swirl gently.
- Add 20 ml concentrated H_2SO_4 [or H_2SO_4 containing Ag_2 (SO_4)] rapidly and mix thoroughly by swirling the flask for about a minute.
- Keep the flask on an asbestos sheet for 30 minutes.
- Add about 200 ml of distilled water and 10 ml H_3PO_4, 0.02 g NaF and 1 ml of diphenylamine indicator. A blue violet colour will appear. In case of ferroin indicator there is no need to add H_3PO_4 and NaF.
- Titrate with 0.5 **N** ferrous ammonium sulphate solution until colour changes from blue violet to brilliant green. In case of ferroin indicator colour changes from blue to red at the end point.

  [*Note:* If burette reading is less than 4 ml (≈ more than 8 ml $K_2Cr_2O_7$ consumed for organic carbon oxidation) repeat with less amount of soil and if it is more than 16 ml (≈ less than 2 ml $K_2Cr_2O_7$ consumed for oxidation) repeat with more amount of soil].
- Simultaneously run a blank in a similar way without soil.

Calculation

Say,

Weight of soil taken = W g

Volume of ferrous ammonium sulphate solution required for blank titration = B ml

Volume of ferrous ammonium sulphate solution required for sample titration = T ml

Volume of $K_2Cr_2O_7$ solution added = V_A ml

Volume of ferrous ammonium sulphate solution required to titrate the volume of $K_2Cr_2O_7$ consumed for organic carbon oxidation = (B - T) ml

$$\text{Oxidizable organic carbon (\%)} = V_A X \left(\frac{B-T}{B}\right) X\ 0.003 X \frac{100}{W} = Z$$

[*Note 1:* As the normality of $K_2Cr_2O_7$ (1 **N**) is twice the normality of ferrous ammonium sulphate (0.5**N**), the volume of $K_2Cr_2O_7$ consumed for oxidation of organic carbon should be (B - T) / 2. But to avoid any change in the strength of ferrous ammonium sulphate during storing it is advisable to consider it to V_A x (B - T) / B, where B / V_A is the ratio of strength between $K_2Cr_2O_7$ and ferrous ammonium sulphate.]

[*Note 2:* 1 gram equivalent weight of $K_2Cr_2O_7$ a" 1 gram equivalent weight of carbon = 12 / 4 = 3 g carbon, as the change in oxidation No. of carbon during oxidation reaction is 4.]
1 equivalent weight of $K_2Cr_2O_7$ = 1000 ml of 1 **N** $K_2Cr_2O_7$
So, 1000 ml of 1 **N** $K_2Cr_2O_7 \equiv$ 3 g carbon
or, 1 ml of 1 **N** $K_2Cr_2O_7 \equiv$ 0.003 g carbon].

$$\text{Total organic carbon}(\%) = \frac{Z \times 100}{77}$$

[Considering recovery factor of this method as 77%]

$$\text{Organic matter}(\%) = \frac{Z \times 100}{77} \times \frac{100}{58}$$

[Considering organic matter contains 58% carbon]

19.2.3 Determination of Total Organic Carbon by Modified Wet Oxidation Method (Nelson and Sommers, 1982)

Principle

The procedure involves the same principle of oxidation of organic carbon with potassium dichromate in strongly acid solution as in Walkley and Black method or Tyurin method. The amount of oxygen consumed during oxidation of organic carbon is calculated from the difference between the amount of dichromate added and the amount remaining after oxidation. In modified wet oxidation method by Nelson and Sommers, the concentration of potassium dichromate and sulphuric acid remain same as was used by Tyurin (1931). Tyurin method recovers about 90 per cent of the organic carbon compared to the carbon recovered by dry combustion method, while Nelson and Sommers claim total recovery of organic carbon. However, carbonates are not included in this estimation.

Equipments and Materials

Analytical balance, 250 ml digestion tube, block digester, 50 ml measuring cylinder, 5 ml and 10 ml pipette, burette.

Reagents

- Potassium dichromate solution, 0.4 **N**: Dissolve 39.22g potassium dichromate ($K_2Cr_2O_7$) in 1 litre of distilled water. Slowly add 1 litre of concentrated sulphuric acid (H_2SO_4) to it avoiding excessive evolution of heat. This results 0.4 **N** (0.066 **M**) potassium dichromate solution in 9 **M** H_2SO_4.
- Ferrous ammonium sulphate, 0.2 **N**: Dissolve 157 g ferrous ammonium sulphate [$Fe(NH_4)_2(SO_4)_2.6H_2O$] in about 1000 ml distilled water containing 100 ml concentrated H_2SO_4. Make up volume to 2 litres with distilled water.
- Phosphoric acid (H_3PO_4), 85%
- N-phenylanthranilic acid indicator: Dissolve 0.1g of N-phenylanthranilic acid and 0.1 g of sodium carbonate (Na_2CO_3) in 100 ml of distilled water.

Procedure

- Take 0.5 to1.0 g soil (pass through 0.2 mm nonferrous sieve) containing between 1 and 10 mg C in a 250 ml digestion tube.
- Add 15 ml of potassium dichromate solution and place on a 150°C preheated block digester for 45 minutes.
- Cool, add 50 ml of water, 5 ml of concentrated H_3PO_4 and four drops of indicator.
- Titrate with ferrous ammonium sulphate to a colour change from dark violet green to light green.
- Run a blank in the same way except soil sample.

Calculation

Say,

Weight of soil taken = W g

Strength (N) of ferrous ammonium sulphate solution = A

Volume of ferrous ammonium sulphate solution required for blank titration = B ml

Volume of ferrous ammonium sulphate solution required for sample titration = T ml Volume of ferrous ammonium sulphate solution required to titrate the volume of $K_2Cr_2O_7$ consumed for organic carbon oxidation = (B - T) ml

$$\text{Total organic carbon, g/100 g soil} = (B\text{-}T) \times A \times 0.003 \times \frac{100}{W}$$

19.3 Characterization of Soil Organic Matter

In mineral soils organic components are remained closely associated with the mineral components. Therefore, one cannot think about the characterization of organic component without separating it from mineral parts of the soil. Numerous extractants had been tried in search of ideal one for this purpose. The ideal extractant should be such that it should remove all organic matter from soil without producing any artifact.

19.3.1 Extraction of Soil Organic Matter

Humic substances are soluble in alkali, and thus use of dilute sodium hydroxide as one of the oldest and most frequently used extractant is quite logical. With this extractant many of the polyvalent cations are replaced by Na and are removed from solution through formation of their insoluble hydroxide at high pH. On the other hand, action of sodium pyrophosphate ($Na_4P_2O_7$) depends on the ability of pyrophosphate ion to interact with polyvalent cations bound to the soil organic matter and to form their insoluble precipitate or soluble complex.

$$R(COO)_4Ca_2 + Na_4P_2O_7 \rightarrow \underset{\text{(soluble)}}{R(COONa)_4} + Ca_2P_2O_7 \downarrow$$

Reagents

- Sodium hydroxide (NaOH), 0.1N and 0.5N: Dissolve 4 g or 20 g of sodium hydroxide pellet in 1 litre of distilled water to get 0.1N or 0.5N NaOH solutions, respectively.
- Sodium pyrophosphate, 0.1M: Dissolve 4.46 g of sodium pyrophosphate decahydrate ($Na_4P_2O_7.10\ H_2O$) 1 litre of distilled water to obtain 0.1M solution.

- Hydrochloric acid, 0.05N and 2.0N: Dilute 4 ml or 167 ml of concentrated hydrochloric acid (HCl) to 1 litre to get 0.05N or 2.0N HCl solutions, respectively.
- Sulphuric acid, 0.05N: Dilute 1.4 ml of concentrated sulphuric acid (H_2SO_4) to 1 litre to get 0.05N H_2SO_4 solution.

Procedure

- If the soil under investigation contains free carbonates as evidenced by effervescence with dilute HCl, allow the soil to stand with an excess amount of 0.05N HCl or H_2SO_4 at room temperature until effervescence stops.
- Wash out excess acid with distilled water, and then thinly spread the soil on a glass plate to dry at room temperature.
- Take 10 g of air dry soil in a 200 ml polypropylene flask and add 100 ml of either 0.1N (or 0.5N) NaOH, 0.1 M $Na_4P_2O_7$ or a mixture of 0.1 N NaOH and 0.1 M $Na_4P_2O_7$ and displace the air in the flask by N_2 gas.
- Stopper the flask and keep it for 24 hours at room temperature.
- Separate dark coloured supernatant solution from the residual soil by centrifugation at 10,000 rpm for 10 minutes.
- Suspend the soil residue in 50 ml distilled water and separate the phases by centrifugation as before. Add solution to the supernatant.

19.3.2 Fractionation of Soil Extract

Classical method of soil organic matter fractionation is based on the solubility of soil extract in aqueous solution at different pH levels (Figure.19.2)

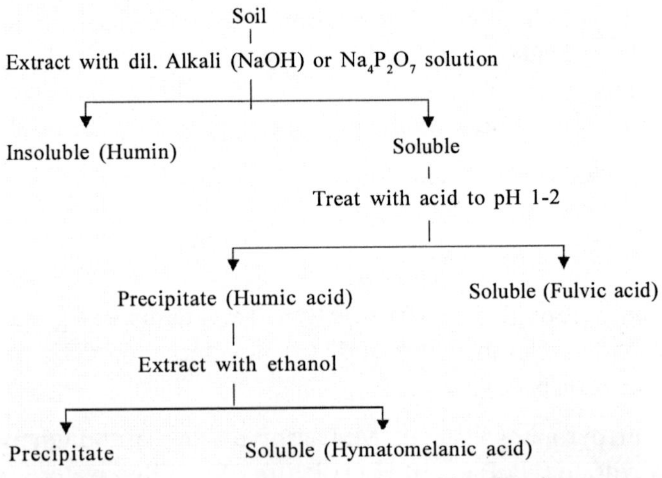

Fig. 19.2: Extraction and fractionation of humic substance

The three principal fractions, *i.e.* humic acid, fulvic acid and humin are not chemically discrete, but differ widely in chemical and physical characteristics.

Procedure

- Acidify the alkali extract to pH 2 with 2.0 N HCl and allow it to stand for 24 hours at room temperature.
- Separate the soluble part (Fulvic acid) from coagulate (Humic acid) by centrifugation.
- Freeze dry both the fractions or take them to dryness on a rotator evaporator at about 40°C.

19.3.3 Purification of Humic Acid and Fulvic Acid

Objectives behind purification are to minimize ash content and to remove low molecular weight organic molecules that are not structural humic acid or fulvic acid constituent. Humic caid and fulvic acid can be purified by repeating the extraction and fractionation process of the dried materials as discussed above.

Reagents

- Hydrochloric acid and hydrofluoric acid (HCl-HF): Dissolve 5 ml of concentrated HCl and 5 ml of 52% HF in 990 ml of distilled water.
- Amberlite IR-120 or Dowex-50 cation exchange resin (H-form)

Procedure

- Shake 1.0 g of humic acid in a polypropylene bottle with 100 ml of HCl-HF mixture and allow it to stand for 24 hours in room temperature.
- Remove the acid mixture by filtration and treat the residue again with a fresh 100 ml portion of HCl-HF mixture.
- Repeat the steps 3 or 4 times in successions and finally wash the residue thoroughly with distilled water and dry at room temperature.
- Fulvic acid is purified by passing aqueous solution 2 or 3 times in successions over Amberlite IR-120 or Dowex-50 resin (H-form) and freeze dry the eluate.

19.3.4 Characterization of Humic Substances

Many chemical, physical and spectroscopic methods have been used to determine the composition and general structure of major components of humic substance. Three general characteristics of a chemical compound are elemental composition, arrangement of these elements in the chemical structure, and nature and location of the functional groups in the compound. Methods determing the nature of various functional groups in the chemical structure of humic substances are discussed in section 21.7. In this section only quantification of various functional groups will be dealt.

General methods of organic analysis revealed that major functional groups in humic substances are carboxylic acids, carbonyls, phenols, alcohols, with N and S located in minor functional groups.

19.3.4.1 Determination of Acidic Functional Groups Concentration

Acidic behaviour of humic substances is due to the presence of a mixture of both stronger (mostly carboxylic acid) and weaker (mostly phenolic acid) organic acids. The pK_a of these two groups of acid are around pH = 4.5 and pH = 10, respectively.

A. Total Acidity: Total acidity of a humic sample encompasses all the acidic (replaceable) hydrogens present. Therefore, determibnation of total acidity requires the use of basic reagent at high pH, sothat even the weakest acids (higher pK_a values) would take part in the neutralization reaction.

Principle

The humic substance is allowed to equilibrate in an excess of barium hydroxide and the amount of barium hydroxide actually comsumed by humic substance is calculated by back-titration with a standard acid.

Equipments and Materials

pH-meter, Shaker, N_2 gas, Erlenmeyer flask (100 ml), measuring cylinder (50 ml), membrane (0.45 μm) and burette.

Reagents

- Barium hydroxide, 0.1M: Dissolve 17.1 g barium hydroxide [$Ba(OH)_2$] and make up volume to 1 liter with distilled water.
- Hydrochloric acid, 0.5 M: Prepare 0.5 M HCl by diluting 41 ml of concentrated HCl in 1 liter and standardize with standard alkali (see section 17.3.5).

Procedure

- Take 50 mg of humic substance (in H^+-form) in a 100 ml ground-glass stoppered Erlenmeyer flask.
- Add 20 ml of 0.1 M $Ba(OH)_2$ solution to the flask containg humic sample and same amount to another flask containing no humic sample (blank).

- Displace the air of the flasks by N_2, stopper and then shake in a mechanical shaker for 24 hours at room temperature.
- Filter the solutions through 0.45 μm membrane and wash the residues troroughly with CO_2-free distilled water.
- Titrate the filtrate plus washings of both sample and blank to pH 8.4 using standard 0.5 M HCl using pH-meter.

Calculation

Let,

Weight (g) of humic sample taken = W

Volume of standard HCl required for titration of sample = V_s ml

Volume of standard HCl required for titration of blank = V_b ml

Strength (M) of standard HCl = S_a

$$\text{Total acidity, } [\text{cmol}_c\ \text{kg}^{-1*}] = \frac{(V_b - V_s) \times S_a \times 1000}{W \times 10} \text{ or } \frac{(V_b - V_s) \times S_a \times 100}{W}$$

[*$\text{cmol}_c\ \text{kg}^{-1}$ designates centimole of charge per kg of sample]

B. Stronger Organic Acid (Carboxylic Groups)

As there is no distinctive equivalence point in the titration curves of humic substances, determination of exact pH at which a particular group of acid neutralized is quite difficult. Again, since weak acid by definition has pK_a value above pH = 10, this value can be used as a mean of differentiating weak acid from strong organic acid.

Principle

The most common method used to determine the acidity associated with carboxylic groups is calcium acetate method. In this method acetic acid liberated from the reaction of calcium acetate with carboxylic acid is estimated by titrating with standard sodium hydroxide solution.

Equipments and Materials

pH-meter, Shaker, Erlenmeyer flask (100 ml), measuring cylinder (50 ml), membrane (0.45 μm) and burette.

Reagents

- Calcium acetate, 0.5 M: Dissolve 77 g of calcium acetate $[(CH_3COO)_2Ca]$ and make up volume to 1 liter with distilled water.

- Sodium hydroxide, 0.1 M: Prepare 0.1 M NaOH solution by dissolving 40 g of NaOH pellets in 1 liter and standardize with standard acid (see section 17.3.1).

Procedure

- Take 50 mg of humic substance (in H^+-form) in a 100 ml ground-glass stoppered Erlenmeyer flask.
- Add 10 ml of 0.5 M calcium acetate solution and 40 ml of CO_2-free distilled water to the flask containing humic sample and same amounts to another flask containing no humic sample (blank).
- Shake the flasks in a mechanical shaker for 24 hours at room temperature.
- Filter the solutions through 0.45 μm membrane and wash the residues thoroughly with CO_2-free distilled water.
- Titrate the filtrate plus washings of both sample and blank to pH 9.8 with standard 0.1 M NaOH using pH-meter.

Calculation

Let,

Weight (g) of humic sample taken = W

Volume of standard NaOH required for titration of sample = V_s ml

Volume of standard NaOH required for titration of blank = V_b ml

Strength (M) of standard NaOH = S_b

$$\text{Carboxylic acidity, } [\text{cmol}_c\ \text{kg}^{-1}] = \frac{(V_s - V_b) \times S_b \times 100}{W}$$

C. Weaker Organic Acids (Phenolic Groups)

Acidity associated with pheolic groups is considered as the difference between the total acidity and acidity attributed to the stronger organic acids (carboxylic groups) of humic substance.

Therefore, Phenolic OH groups, $[\text{cmol}_c\ \text{kg}^{-1}]$

$$= \text{Total acidity } [\text{cmol}_c\ \text{kg}^{-1}] - \text{Carboxylic acidity, } [\text{cmol}_c\ \text{kg}^{-1}]$$

19.3.4.2 Determination of Hydroxyl (OH) Groups Concentration

A. Total Hydroxyl Groups

Principle

Total hydroxyl groups in humic substance accounts for both aromatic (phenolic) and aliphatic (alcoholic) hydroxyl groups. For determination of total hydroxyl group humic substances are acetylated with acetic anhydride in pyridine. Acetyl groups are then hydrolysed to acetic acid, which is distilled and titrated with standard base.

$$ROH + \underset{\text{(Acetic anhydride)}}{(CH_3CO)_2O} \rightarrow CH_3COOR + CH_3COOH$$

Equipments and Materials

Distillation set, Reflux condenser, Shaker, N_2 gas, measuring cylinder (50 ml), membrane (0.45 μm) and burette.

Reagents

- Pyridine (C_5H_5N), 95% purity
- Acetic anhydride [$(CH_3CO)_2O$], 97% purity
- Sodium hydroxide, 3 M, 0.1 M: Dissolve 12 g of NaOH pellets in distilled water and make up the volume to 100 ml to get 3 M NaOH solution, and dissolve 2 g of NaOH pellets in distilled water and make up the volume to 500 ml to get 0.1 M NaOH solution and standardize with standard acid (see section 17.3.1).
- Sulphuric acid, 3M: Dilute 167 ml of concentrated H_2SO_4 to 1 liter with distilled water.
- Phenolphthalein indicator solution, 0.5%: Dissolve 500 mg of Phenolphthalein in 95% ethanol and make up volume to 100 ml.

Procedure

- Reflux 50 mg of humic substance with 5 ml of equal parts of pyridine and acetic anhydride for 2 to 3 hours under N_2 atmosphere.
- Cool the mixture, pour it into distilled water and collect the precipitate by filtration through 0.45 μm membrane.
- Wash the precipitate thoroughly with distilled water and dry it under a vacuum in presence of P_2O_5.
- Again reflux the acetylated sample with 25 ml of 3 M NaOH solution for 2 to 3 hours under N_2 atmosphere.

- Cool the mixrture, add 25 ml of 3 M H_2SO_4 and 25 ml of distilled water, and distill through a splash head.
- Add 25ml portion of distilled water to the distillation mixture and continue distillation. Collect the distillate and titrate with standard 0.1 M NaOH.
- Repeat addition of 25ml portions of distilled water, collection of distillate and titration until sample and reagent-blank titrate equally.

Calculation

Let,

Weight (g) of humic sample taken = W

Volume of standard NaOH required for titration of sample = V_s ml

Volume of standard NaOH required for titration of reagent-blank = V_b ml

Strength (M) of standard NaOH = S_b

$$\text{Acetyl content, } [\text{cmol}_f\text{kg}^{-1}] = \frac{(V_s - V_b) \times S_b \times 100}{W}$$

$$\text{Total hydroxyl content } [\text{cmol}_f\text{kg}^{-1}] = \frac{\text{Acetyl content}}{(100 - 0.042 \times \text{Acetyl content})}$$

The factor of 0.042 is a consequence of the difference in molecular weight between ROH and CH_3COOR.

B. Alcoholic Hydroxyl Groups

Alcoholic OH groups, [cmol_c kg^{-1}]

= Total hydroxyl, [cmol_c kg^{-1}] – Phenolic hydroxyl, [cmol_c kg^{-1}]

19.3.4.3 Determination of Carbonyl (C=O) Groups Concentration

A. Total Carbonyl (>C=O) Groups

Principle

The carbonyl groups in humic substance are allowed to react with an excess of hydroxylamine hydrochloride in methanol/2-propanol. The unreacted hydroxylanine is titrated with standard perchlorie acid.

$$\begin{matrix} R_1 \\ R_2 \end{matrix} \rangle C{=}O + NH_2OH.HCl \longrightarrow \begin{matrix} R_1 \\ R_2 \end{matrix} \rangle C\text{-}NOH + H_2O + HCl$$

Equipments and Materials

pH-meter, water bath, Erlenmeyer flask (50 ml), measuring cylinder (10 ml), and burette.

Reagents

- Dimethyaminoethanol [$(CH_3)_2NCH_2CH_2OH$], 99% purity, 0.25 M: Dissolve 22.5g of Dimethyaminoethanol in 2-propanol and dilute to 1 liter.
- Hydroxylamine hydrochloride [$NH_2OH.HCl$], 0.4 M: Dissolve 27.8 g hydroxylamine hydrochloride in 300 ml absolute methol and dilute to 1 liter with 2-propanol.
- Perchloric acid ($HClO_4$), 0.2 M: Dilute 21.7 ml of concentrated $HClO_4$ to 1 liter with distilled water and standardize it with standard Na_2CO_3.

Procedure

- Place 50 mg of humic substance in a 50 ml ground-glass stoppered Erlenmeyer flask.
- Add 5 ml of 0.25 M dimethylaminoethanol solution and 6.3 ml of 0.1 M hydroxylamine hydrochloride solution to the flask containing humic substance and to another flask containing no humic substance (as blank).
- Heat the flask on steam bath for 15 minutes. Cool the solution and back-titrate the untreated hydroxylamine hydrochloride potentiometrically with standard perchloric acid solution.
- Determine the end point by plotting e.m.f (Volts) vs. volume (ml) of acid used (Fig.19.3).

Calculation

Let,

Weight (g) of humic sample taken = W

Volume of standard $HClO_4$ required for back-titration of sample = V_s ml

Volume of standard $HClO_4$ required for back-titration of blank = V_b ml

Strength (M) of standard $HClO_4$ = S_a

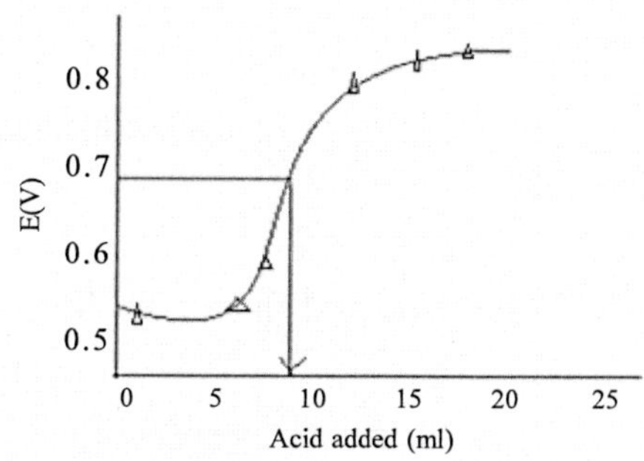

Fig. 19.3: Titration curve of potentiometric titration

$$\textbf{Total carbonyl content [cmol}_f\textbf{ kg}^{-1}\textbf{]} = \frac{(V_b - V_s) \times S_a \times 100}{W}$$

B. Quinone (O=⟨⟩=O) Groups

Principle

Quinone groups undergo redox type interaction with a number of soil constituents including iron. This type of interaction can be used to determine the concentration of this functional group in humic substance. Quinoids present in humic substance are reduced by Fe2+ in alkaline triethanolamine (TEA) solution. The excess reductant (Fe^{2+}) is back-titrated amperometrically with standard potassium dichromate.

O=⟨⟩=O + $2Fe^{2+}$-TEA $\rightarrow$ HO–⟨⟩–OH + $2Fe^{3+}$

Equipments and Materials

Polarograph, tall-form titration cell (200 ml), N_2 gas, measuring cylinder (10 ml), and burette.

Reagents

- Sodium hydroxide (NaOH), 2M: Dissolve 80 g of sodium hydroxide pellets in distilled water and make up volume to 1 liter.
- Triethanolamine (TEA) 97% purity, 2M
- Ferrous ammonium sulphate hexahydrate [$FeSO_4.(NH_4)_2SO_4.6H_2O$], 0.05 M: Dissolve 19.6 g of AR grade ferrous ammonium sulphate hexahydrate in about 800 ml of distilled water and add 2 ml of concentrated H_2SO_4 and make up volume to 1 liter.
- Potassium dichromate ($K_2Cr_2O_7$), 0.004 M: Dissolve 1.177 g of $K_2Cr_2O_7$ in distilled water and make up volume to 1 liter.

Procedure

- Place 20 mg of humic substance in a 200 ml tall form titration cell and add 45 ml of distilled water, 25 ml of 2 M NaOH solution and 25 ml of 2 M TEA.
- Fit the lid and flush with N_2 gas continuously during the analysis.
- Stir the solution with magnetic stirrer for 30 minutes, then add 5 ml of 0.05 M ferrous ammonium sulphate solution and leave for 30 minutes.

- Back-titrate excess ferrous ammonium sulphate using standard 0.004 M $K_2Cr_2O_7$ solution at a constant potential of –80 mV, determined using a platinum wire electrode system connected to a polarograph.
- Carry out blank titration under similar condition without using humic substance.

Calculation

Let,

Weight (g) of humic sample taken = W

Volume of standard NaOH required for titration of sample = V_s ml

Volume of standard NaOH required for titration of blank = V_b ml

Strength (M) of standard $K_2Cr_2O_7$ = S_k

$$\text{Quinoid}\ (O{=}C_6H_4{=}O),\ [\text{cmol}_f\,\text{kg}^{-1}] = \frac{(V_b - V_s)\times S_k \times 6 \times 100}{W}$$

The factor of 6 in the above is a consequence of change of oxidation No. per molecule of $K_2Cr_2O_7$.

C. Ketonic (C=O) Group

Ketonic C=O groups, $[\text{cmol}_f\,\text{kg}^{-1}]$

= Total C=O groups, $[\text{cmol}_f\,\text{kg}^{-1}]$ – Quinone groups, $[\text{cmol}_f\,\text{kg}^{-1}]$

19.4 Determination of Soil Microbial Biomass Carbon

The soil microbial biomass is an important component of the soil organic matter that regulates the transformation and storage of nutrients. It is a labile component of soil organic fraction contributing 1 to 3% of the total soil C and up to 5% of the total soil N. The methods of determination include (i) the chloroform fumigation incubation method, (ii) chloroform fumigation extraction method, (iii) substrate induced respiration and (iv) adenosine triphosphate analysis. The former two methods are also useful to recover tracers (e.g., ^{14}C and ^{15}N). Chloroform fumigation extraction method (Joergensen, 1996) is discussed here.

Principle

Microbial constituents released by chloroform fumigation and extracted can directly be used to determine the size of the soil biomass. Chloroform fumigation is done for overnight to kill and lyse microbial cells with release of cytoplasm into the soil environment, following which the amount of organic C in the sample can be measured by standard chemical method.

Equipments and Materials

Moisture box, separatory funnel, glass vacuum desiccator and vacuum pump, shaker, glass beads, hot plate, beaker (50 and 100 ml capacity), conical flask (250 and 500 ml capacity), measuring cylinder, pipette, and burette.

Reagents

- Distilled chloroform
- Sulphuric acid, concentrated
- Potassium sulphate, 0.5 **M**: Dissolve 87.128 g potassium sulphate (K_2SO_4) in distilled water and make up volume to 1 litre.
- Potassium dichromate, 0.2 **N**: Dissolve 0.9808 g potassium dichromate ($K_2Cr_2O_7$) in distilled water and make up volume to 100 ml.
- Orthophosphoric acid, concentrated
- Ferrous ammonium sulphate, 0.005 **N**: Dissolve 3.92 g ferrous ammonium sulphate [$FeSO_4$ $(NH_4)_2SO_4.6H_2O$] in distilled water; add 0.15 ml of concentrated H_2SO_4 and make up volume to 2 litre.
- Ferroin indicator: Dissolve 1.485 g of orthophenanthraline monohydrate in about 80 ml of distilled water and add 0.69 g of ferrous sulphate heptahydrate and dilute to 100 ml.

Procedure

- Put the field moist soil sample in plastic bag to prevent drying. Pass the soil through 2 mm sieve and analyze on the same day.
- Weigh 5 sets of 10 g soil for each sample.
- Put one set in a previously weighed moisture box and keep in oven at 105°C for 24 hours. Cool in desiccator and weigh to determine moisture content of the soil.
- Keep two sets in 50 ml beaker for fumigation and rest two sets for immediate extraction with 0.5 **M** K_2SO_4.
- For immediate extraction take 10 g unfumigated soil sample in 250 ml conical flasks. Add 25 ml of 0.5 **M** K_2SO_4 and shake for 30 minutes on a mechanical shaker. Filter through Whatman No. 1 filter paper.
- Take 20 ml of chloroform in a separatory funnel for each 10 g soil sample. Wash the chloroform twice with concentrated H_2SO_4 (10 ml for each time). Shake thoroughly and discard the acid (lower phase) after phase separation. After each shaking carefully release the pressure of the funnel by opening the stop cock.

- Wash twice with distilled water (10 ml for each time) in similar way to make the chloroform free from ethanol and collect the whitish bottom phase (chloroform).
- Place the beaker containing 10 g soil sample in the vacuum dessicator lined inner surface with moist filter paper.
- Take 20 ml ethanol free chloroform in a 100 ml beaker containing some glass beads and place it in the desiccator.
- Seal the lid-joint with high density vacuum grease. Use rubber tube to direct the exhaust through water.
- Run the vacuum pump till chloroform boils for 5 minutes. Close the outlet and keep desiccator in dark for overnight at 25°C.
- Next day release the vacuum slowly, takes out the beaker containing chloroform and lining filter paper. Back suck for at least 5 times to remove the excess /adhered chloroform and release the vacuum slowly.
- Transfer fumigated soil sample in 250 ml conical flasks. Add 25 ml of 0.5 **M** K_2SO_4 and shake for 30 minutes on a mechanical shaker. Filter through Whatman No. 1 filter paper.
- Pipette out 10 ml of filtrate in a 500 conical flask, add 2 ml of 0.2 **N** $K_2Cr_2O_7$, 10 ml concentrated H_2SO_4 and 5 ml orthophosphoric acid.
- Run blank with 10 ml distilled water and with all the above reagents.
- Place the flasks on the hot plate and reflux for 30 minutes.
- Remove the flask from the hot plate and immediately add about 250 ml of distilled water. Cool it to room temperature.
- Add 2 to 3 drops of ferroin indicator and titrate against ferrous ammonium sulpahte till brick-red end point appears.

Calculation

(a) Moisture content of soil (M_S)

$$M_S\,(\%) = \frac{\text{Wt. of moist soil} - \text{Wt. of oven dry soil}}{\text{Wt. of oven dry soil}} \times 100$$

(b) Weight of oven dry soil taken for microbial biomass carbon calculation (W_{DS})

$$W_{DS} = \frac{\text{Wt.of moist soil}}{(100 + M_S\%)} \times 100$$

(c) Actual volume of solution in the extracted soil (V_S)

V_S (ml) = Wt. of the moist soil (g) – Wt. of the oven dry soil (g) + extractant volume (ml)

d) Determination of carbon in the extract (C_E), mg/ ml

Say,

Volume of ferrous ammonium sulphate solution required for blank titration = B ml

Volume of ferrous ammonium sulphate solution required for sample titration = T ml

Volume of $K_2Cr_2O_7$ solution added = V_A ml

Volume of the filtrate used for titration = V_C ml

Volume of $K_2Cr_2O_7$ consumed for oxidizing mineralisable C = $V_A \times \left(1 - \frac{T}{B}\right)$

Therefore, the amount of carbon in the extract (C_E), µg/ml =

$$V_A \times \left(1 - \frac{T}{B}\right) \times \frac{\text{Strength of } K_2Cr_2O_7}{V_C} \times 0.003$$

[As 1 ml 1 N $K_2Cr_2O_7$0.003g of C]

Amount of C extracted in fumigated (C_{EF}) or unfumigated (C_{EUF}) sample

$$C_{EF} \text{or } C_{EUF}\text{, in } \mu g / g \text{ soil} = \frac{C_E \times V_S}{W_{DS}}$$

$$\text{Microbial biomass Carbon (MBC), in } \mu g / g \text{ soil} = \frac{C_{EF} - C_{EUF}}{K_{EC}}$$

Where, K_{EC} is the efficiency of extraction of organic carbon and its value depends on physical and chemical properties of soil; usually considered as 0.45 (Joergensen, 1996).

Questions and Hints

Q.1 How will you calculate the equivalent weight of $K_2Cr_2O_7$?

Hint: $4Cr^{6+} + 3C^0 \rightarrow 4Cr^{3+} + 3C^{4+}$. In this redox reaction the change in oxidation No. of chromium ion per molecule of $K_2Cr_2O_7$ is (6 - 3) 2 = 6 [The change in oxidation No. per atom multiplied by number of atom per molecule]. So, the equivalent weight of $K_2Cr_2O_7$ = Molecular weight of $K_2Cr_2O_7$ / 6 = 294.22 / 6 = 49.04.

Q.2 What is the role of NaF in oxidizable organic carbon determination by Walkley and Black method?

Hint: NaF is added to eliminate the interference of Al^{3+} in soil (particularly acidic soil) by forming stable complex.

$NaF + Al_2(SO_4)_3 \rightarrow 2AlF_3 + 3Na_2(SO_4)$

Q.3 What function does ortho phosphoric perform in Walkley and Black method?

Hint: To get a sharp colour change at the end point H_3PO_4 plays very important role which can be discussed in the following ways:

a) During titration of $K_2Cr_2O_7$ with ferrous ammonium sulphate, Fe^{2+} is oxidized to Fe^{3+} form. As the reaction proceeds, more and more yellow coloured Fe^{3+} will be accumulated in the system which will not only interfere the detection of correct end point but also slows down the speed of forward reaction. H_3PO_4 forms a colourless complex $[Fe(HPO_4)]^+$ with Fe^{3+} and thus eliminates the effect of Fe^{3+} ions in the solution.

$Fe^{3+} + HPO_4^{2-} \rightarrow [Fe(HPO_4)]^+$

b) In this analysis two redox systems are operating: one where Cr^{6+} is reduced to Cr^{3+} and other where Fe^{2+} is oxidized to Fe^{3+}. The ideal redox indicator will be one with reduction potential intermediate between the redox potentials of these two systems and at least 0.15 volts away from the two standard redox potentials of the system. The standard reduction potential of Cr^{6+} / Cr^{3+} system is + 1.33 volts and that of Fe^{3+} / Fe^{2+} system is + 0.77 volts. Thus the equivalence potential of these two redox systems is (1.33 + 0.77) / 2 = + 1.05 volts. But the standard (strictly speaking the formal) potential of the diphenylamine indicator indicator is +0.76 volts which does not satisfy the criteria of an ideal redox indicator in this titration. Addition of H_3PO_4 which forms stable complex Fe^{3+} drops down the formal potential of Fe^{3+} / Fe^{2+} system to + 0.61 volts in presence of H_2SO_4 and satisfy the criteria of an ideal redox indicator.

Q.4 What necessary correction or precaution to be adopted in Walkley and Black method in case of saline soil?

Hint: In case of saline soil or soil containing chloride ions, organic carbon may be overestimated in this method as $K_2Cr_2O_7$ will also be consumed for oxidation of chloride ions.

$$Cr_2O_7^{2-} + 6Cl^- + 14H^+ \rightarrow 2Cr^{3+} + 3Cl_2 + 7H_2O$$

Acidic medium

$$4Cr^{6+} + 3C^0 \xrightarrow{\text{Acidic medium}} 4Cr^{3+} + 3C^{4+}$$

From the above two reactions, we can write:

$1K_2Cr_2O_7 \equiv 6Cl^-$ and $2K_2Cr_2O_7 \equiv 3C \equiv 12Cl^-$

Thus, $4Cl^- \equiv 1\ C \equiv \frac{12}{4 \times 35.5} = \frac{1}{12} g\ C(\text{approx})$

4×35.5 g $Cl^- \equiv 12$ g C, or 1 g $Cl^- \equiv \frac{12}{4 \times 35.5} = \frac{1}{12} g\ C(\text{approx})$

Corrected soil carbon (%) = Uncorrected soil carbon (%) $- \frac{\textit{Soil chloride}(\%)}{12}$

However, this correction factor has been found to be valid up to Cl: C ratio of 5:1.

Alternately, chloride interference can be overcome by leaching out soil on an asbestos filter and placing asbestos filter and sample together for determination.

Consumption of $K_2Cr_2O_7$ for oxidation of Cl^- can be prevented by precipitating Cl^- as AgCl through addition of Ag_2SO_4 to concentrated H_2SO_4 @ 15 g/litre of acid.

Q.5 Whether iron can any way interfere in the determination of organic carbon by Walkley and Black method?

Hint: Yes; if soil contains iron in Fe^{2+} form (reduced soil) it will consume $K_2Cr_2O_7$ and thus organic carbon in this soil will be overestimated. To eliminate the effect of Fe^{2+}, the soil must be dried for 1-2 days before analysis so that Fe^{2+} will be converted to non interfering Fe^{3+} form.

Q.6 Why the strength of ferrous ammonium sulphate is not considered in the calculation of organic carbon in Walkley and Black method?

Hint: Ferrous ammonium sulphate is not a primary standard but $K_2Cr_2O_7$ is a primary standard. In this method the amount of organic carbon in soil is calculated according to the formula [V × (B-T)/B] in terms of the amount of 1**N** $K_2Cr_2O_7$ required for its oxidation by transforming the volume and the strength of ferrous ammonium sulphate in terms of $K_2Cr_2O_7$. Thus, further mentioning of strength of ferrous ammonium sulphate is immaterial.

Q.7 How MnO_2 interference can be prevented in Walkley and Black method?

Hint: Highly reactive (freshly precipitated) higher oxide of Mn which interferes in the estimation of soil organic carbon by $K_2Cr_2O_7$ can be nullified by pretreatment of soil with $FeSO_4$ before analysis. As active MnO_2 reacts with cold acidified $FeSO_4$, the amount of $FeSO_4$ required for this reaction can be calculated by titration. Two ml of H_3PO_4, 5 ml of water and 1 ml of diphenyamine indicator are added to the soil followed by addition of sufficient $FeSO_4$ (1 **N**). The mixture is allowed to stand for 10 minutes with occasional shaking, and then excess $FeSO_4$ is back titrated with $K_2Cr_2O_7$. The amount of $FeSO_4$ actually oxidized by MnO_2 is determined by subtracting sample test value from the blank test value (without soil).

Now, the amount of $FeSO_4$ required for MnO_2 reduction is added to the fresh soil sample, together with 2 ml H_3PO_4. The mixture is allowed to stand for 5 minutes and proceed for organic carbon estimation following usual procedure.

Q.8 Whether HNO_3 can be used in place of H_2SO_4 in Walkley and Black method?

Hint: No. The standard oxidation potential of NO - NO_3^- system (-0.96V) is higher than that of Cr^{3+}- Cr^{6+} system (-1.33V) which means Fe^{2+} will reduce NO_3^- first and then Cr^{6+}. Therefore, more amount of ferrous ammonium sulphate will be consumed to titer the unreacted $K_2Cr_2O_7$ than the actual.

Q.9 Whether $KMnO_4$ can be used instead of $K_2Cr_2O_7$ in wet digestion method?

Hint: Although $KMnO_4$ is more powerful oxiding agent (std. red. potential +1.52V) than $K_2Cr_2O_7$, it has several disadvantages- (i) $KMnO_4$ is not usually available in chemically pure form, thus it can not be used as primary standard solution, (ii) its aqueous solution is unstable, it undergoes auto decomposition in presence of MnO_2 which often remains present as impurity, (iii) in bright sunlight it undergoes photo-decomposition, thus it should always be kept in dark or kept in dark brown glass bottle.

Q.10 Whether any approximation can be done regarding total N from organic matter content of soil?

Hint: Yes: conventionally organic matter contains 58 % carbon and bears a C: N ratio of 11.6.

Thus, N content (%) in organic matter = 58/11.6 = 5%

Total N (%) = Organic matter (%) X 0.05.

Or, organic matter (%) = Total N (%) X 20

Loss in ignition (%) at 400°C×(0.022 to 0.03) = Total N

However, these approximation factors are subject to vary widely in different soils.

Q.11 Why purification of humic acid and fulvic acid is necessary?

Hint: Purification is required to minimize the ash content of humic acid and fulvic acid below 1% and to remove low molecular weight organic molecules that are not structural humic acid or fulvic acid constituent.

20

Ion Exchange Capacity and Exchangeable Bases

20.1 Exchange Capacity

Soil colloidal particles carry both negative and positive charges on their surface which attract and hold equivalent amount of oppositely charged ions to their surface. The total number of negative charges present on a given mass of soil or capacity of a given mass of soil to exchange cations or total number of cation held by a given mass of soil is termed as **cation exchange capacity** (CEC). Similarly, the capacity of a soil to retain anions is known as **anion exchange capacity** (AEC). The ion exchange capacity of a soil depends upon the type and amount of colloidal particle (clay and organic matter) present in the soil. Ion exchange capacity (CEC/AEC) is expressed as milliequivalent ion (cation/ anion, respectively) per 100 g of soil (meq/100g). It is also expressed as centimoles of charge (positive/negative) per kg of soil [cmol (±) / kg]. Meq/ 100g = cmol (±) / kg of soil.

20.1.1 Determination of Cation Exchange Capacity of Soil

Principle

Soil is first saturated with normal ammonium acetate (pH 7.0). Excess ammonium (NH_4^+) ions and displaced cations are then removed by washing with alcohol. The amount of NH_4^+ ions retained by the soil is measured by steam distillation of NH_4-saturated soil with MgO. During distillation evolved ammonia is absorbed in a boric acid solution as ammonium borate and the amount ammonium borate formed is determined by titration with standard sulphuric acid in presence of mixed indicator.

NH_4 saturation:

Soil	K K + $2NH_4OAc$	⟶	Soil	NH_4 + 2KOAc NH_4

Distillation:

Soil	NH_4 + MgO + H_2O NH_4	⟶		Mg + $2NH_3$ + $2H_2O$

$NH_3 + H_3BO_3 \rightarrow NH_4H_2BO_3$

Titration:

$2NH_4H_2BO_3 + H_2SO_4 \rightarrow (NH_4)_2SO_4 + 2H_3BO_3$

Equipments and Materials

Kjeldahl distillation set, shaker, Buchner funnel, Whatman No. 42 filter paper, balance, 800 ml Kjeldahl flask, 250 ml conical flask, 50 ml measuring cylinder.

Reagents

- 1 **N** ammonium acetate (pH 7.0): Either dilute 57 ml of glacial acetic acid (99.5%) to 800 ml with water containing 70 ml concentrated ammonia solution or dissolve 77.08g ammonium acetate (NH_4OOCCH_3) crystal in about 800 ml water. Cool and adjust pH to 7.0 with dilute acetic acid or ammonia solution. Make up volume to 1 litre.
- Ethanol, 60%: Dilute 630 ml of absolute alcohol (95%) to 1 litre with distilled water.
- Ammonium chloride crystal
- Magnesium oxide powder (MgO)
- Mixed indicator: Dissolve 0.099 g bromo cresol green and 0.066 g methyl red in 100 ml of 95% ethanol. Adjust the solution to the bluish purple midcolour at pH 4.5 with dilute NaOH or HCl. This indicator is pink at pH 4.2 or lower and bluish green as pH rises to 4.9 and above.
- Boric acid- indicator solution, 2%: Dissolve 20 g boric acid in about 800 ml hot distilled water. Cool, then add 20 ml of mixed indicator and adjust the pH of the solution by dilute NaOH or HCl until the bluish colour of the indicator weakens toward pink. Make up volume to 1 litre.
- Standard sulphuric acid, 0.1N: Dilute 2.8 ml of concentrated sulphuric acid to 1 litre with distilled water and standardize against standard sodium carbonate (0.1 **N**) using methyl red indicator.

- Methyl red indicator, 0.5%: Dissolve 0.5 g indicator in 100 ml of 95% ethanol.
- Silver nitrate solution, 0.1 **M**: Dissolve 8.5 g silver nitrate ($AgNO_3$) in 500 ml distilled water and add 2 ml concentrated HNO_3 to it.

Procedure

- Take 10 g air dry processed soil in a 250 ml conical flask and add 50 ml of 1 **N** ammonium acetate solution.
- Shake the suspension for an hour and leave it for overnight.
- Filter the content through Whatman No. 42 filter paper fitted on Buchner funnel under suction. Transfer the soil completely to the filter paper with ammonium acetate solution.
- Leach the soil with neutral normal ammonium acetate solution for about ten times. Before adding fresh aliquot (20 ml) allow the leachate to be drained out completely.
- Leachate is preserved for determination of individual exchangeable cations.The soil left on the filter paper will be required for CEC determination.
- Wash the soil with 60% ethanol to remove excess ammonium acetate. To be confirmed add a pinch of ammonium chloride to the soil on the filter paper and continue washing until the leachate becomes free from chloride (test with $AgNO_3$).
- Transfer the soil with filter paper to an 800 ml distillation flask; add about 200 ml of water, little liquid paraffin and one spoonful of MgO.
- Distill and collect ammonia by absorbing in 50 ml of 2% boric acid mixed indicator solution taken in a 250 ml conical flask. Collect about 150 ml distillate.
- Titrate the distillate with standard sulphuric acid (0.1**N**) till pink colour appears.
- Run a blank distillation in a similar way without soil.

Calculation

Say,

Weight of soil taken = W g

Volume of standard sulphuric acid required for sample titration = V_S ml

Volume of standard sulphuric acid required for blank titration = V_B ml

Strength (N) of standard sulphuric acid = S

$$\text{Cation exchange capacity,meq/ 100g} = (V_S - V_B) \times S \times \frac{100}{W}$$

[As 1000 ml 1 **N** of any acid or alkali ≡ 1 equivalent of any cation

Or 1 ml 1 **N** of any acid or alkali ≡ 1 milliequivalent of any cation]

Rating

Soil Class on CEC	CEC (meq / 100 g soil)
Low	< 10
Medium	10 – 25
High	25 – 45
Very high	> 45

20.1.2 Determination of Anion Exchange Capacity of Soil

Principle

Soil is first leached with barium chloride-triethanolamine buffered at pH 8.1, then with calcium chloride to saturate the soil with calcium. Calcium saturated soil is equilibrated with known amount of phosphoric acid solution and the amount of phosphous adsorbed is evaluated from the phosphorus concentration in the equilibrated solution and the added solution. Initial phosphous concentration of the soil is also determined from a separate sample.

Anion exchange capacity, meq/100 g = (Initial P concentration, meq/100 g soil + Adsorbed P concentration, meq/100 g soil).

Reagents

- Barium chloride-triethanolamine ($BaCl_2$-TEA) solution: Dilute 90 ml of triethanolamine to about 150 ml with distilled water. Adjust pH to 8.1 with dilute HCl and dilute to 200 ml with water. Mix it with 200 ml of distilled water containing 100 g of barium chloride ($BaCl_2$, $2H_2O$).
- Calcium chloride solution: Dissolve 50 g calcium chloride ($CaCl_2$, $2H_2O$) in 100 ml of distilled water and adjust pH to 8.0 with saturated Ca $(OH)_2$ solution.
- Ethanol (95%)
- Phosphoric acid (H_3PO_4), 0.01 **M**: Dilute 1.36 ml of concentrated phosphoric acid to 2 litre with distilled water.

- Bray (I) reagent for P extraction: 0.025 **N** NH_4F in 0.03 **N** HCl (see section 22.1.2).
- Dickman and Bray reagent for colour development: (see section 22.1.2).
- Standard P solution for standard curve preparation: (see section 22.1.2).

Equipment and Materials

Balance, shaker, centrifuge, oven, colorimeter, Buchner funnel, Whatman No. 42 filter paper, 25 ml measuring cylinder.

Procedure

- Take 10 g air dry processed (<2 mm) soil sample on a Whatman No. 42 filter paper fitted on a Buchner funnel and leach with 100 ml $BaCl_2$-TEA solution.
- Wash about 6 times with 95% ethanol to remove excess $BaCl_2$-TEA.
- Leach the soil with 100 ml $CaCl_2$ solution and again wash at least 6 times with 95% ethanol to remove excess $CaCl_2$.
- Dry calcium saturated soil at 45°C temperature in oven.
- Take some amount of dry soil (having CEC 0.2 meq/ 100 g of soil) in a centrifuge tube and add 20 ml of 0.01 **M** H_3PO_4 solution.
- Shake for half an hour and allow it to stand for 24 hours.
- Again shake for half an hour and centrifuge.
- Take 1 ml clear extract and determine P concentration colorimetrically (see section 22.1.2).
- Taking suitable aliquot determine P concentration of 0.01 **M** H_3PO_4 solution also.
- With a separate soil sample determine initial P concentration after extracting the soil with Bray (I) extractant (see section 22.1.2).

Calculation

Say,

Adsoption study:

Weight of soil taken = W g

Volume of 0.01**M** H_3PO_4 solution added = V_P ml

Volume of aliquot taken for P concentration determination = V_A ml

Final volume of coloured P solution = V_B ml

Concentration of P obtained from standard curve = C_A ppm

Thus,concentration of P in equilibrated solution = $C_A \times \frac{V_B}{V_A} \times \frac{V_P}{W} = X$ ppm

Concentration of P in added H3PO4 solution = $C_C \times \frac{V_D}{V_C} = Y$ ppm

Where, C_C is the concentration of P obtained from standard curve for the added H_3PO_4 solution, ppm

V_C is the volume of aliquot taken from the added H_3PO_4 solution for coloured development, ml

V_D is the final volume of coloured P solution for the added H_3PO_4 solution, ml

P adsorbed $= [Y - X] = A$ ppm $= \frac{A}{6.2 \times 10}$ meq / 100g soil

[As equivalent weight of P (v) is 31 / 5 = 6.2]

Initial P concentration study

Weight of soil taken = W g

Volume of Bray (I) solution added = V_E ml

Volume of aliquot taken for P concentration determination = V_M ml

Final volume of coloured P solution = V_N ml

Concentration of P obtained from standard curve = C_B ppm

Thus,concentration of P in soil $= C_B \times \frac{V_N}{V_M} \times \frac{V_E}{W} = B$ ppm

Initial P concentration = B ppm $= \frac{B}{6.2 \times 10}$ meq / 100g soil

Anion exchange capacity, meq / 100 g $= \frac{A+B}{6.2 \times 10} = \frac{A+B}{62}$

20.2 Determination of Exchangeable Bases

Exchangeable basic cations which are measured for general purposes includes Na^+, K^+, Ca^{2+} and Mg^{2+}. Other cations present in soil are acidic cations like H_3O^+, Al^{3+}, and Fe^{3+} which may be dominant in acid soils.

20.2.1 Determination of Exchangeable Na and K

Exchangeable Na and K can be measured directly from the leachate preserved for exchangeable cation estimation (see section 20.1.1) using flame photometer. Only modification of the procedures discussed for water soluble Na (see section 18.2.5) and K determination (see section 18.2.6) is respective standard solutions for the preparation of standard curves should be prepared in neutral normal ammonium acetate solution and flame photometer zero setting also be made with neutral normal ammonium acetate solution instead of distilled water.

20.2.2 Determination of Exchangeable Ca and Mg

Where atomic absorption spectrophotometer (AAS) is available exchangeable Ca and Mg can be determined directly from the leachate preparing standards in 1**N** ammonium acetate solution and considering 1 **N** ammonium acetate solution as blank. But where AAS is not available exchangeable Ca and Mg can be estimated by complexometric titration with EDTA. But presence of ammonium acetate and dispersed organic matter may interfere in EDTA titration. Thus, complete removal of these interfering agents (pretreatment) is essential before EDTA titration.

Pretreatment

- Pipette out 100 ml of ammonium acetate solution leachate in a 500 ml beaker and evaporate to dryness.
- Cool and add 5 ml of aqua regia (3 parts concentrated HCl + 1 part concentrated HNO_3) and cover with the watch glass immediately.
- Again evaporate the solution to dryness in a fume hood. Very dark coloured leachate requires additional treatment with aqua regia.
- Add 1 ml of concentrated HCl followed by about 20 ml distilled water, stir and filter through Whatman No. 42 filter paper into a 100 ml volumetric flask.
- Wash the beaker 3-4 times with distilled water and collect the washings in the volumetric flask through filter paper.
- Make up volume to the mark with distilled water.
- Prepare a blank in the same manner taking 100 ml 1 **N** ammonium acetate solution.

Analysis

Determine exchangeable Ca and Mg following the procedures as described for water soluble Ca and Mg (see section 18.2.4).

Questions and Hints

Q.1 Why alcohol is used to remove excess NH_4OAc but not water?

Hint: Use of water as washing agent may cause hydrolysis of ammonium complex. Secondly, ammonium saturated soil in contact with water may disperse and colloidal particles will pass through the filter paper.

Q.2 What is the basic principle of determining CEC of a soil?

Hint: The principle behind the determination of CEC of a soil is first the saturation of the exchange sites with any index cation (NH_4^+, Na^+, Ca^{2+}, Ba^{2+}, K^+, etc), then removal of excess cations (in solution phase) and finally estimation of the amount of cation actually retained by the exchange complex.

Q.3 During alcohol washing why a pinch of NH_4Cl is added?

Hint: To ascertain whether any amount of excess NH_4^+ (solution phase) is still present in the soil. If the chloride test of the leachate is negative, be sure that the soil is free from excess NH_4^+ ions.

Q.4 What are the precautions needed during washing of excess NH_4^+ ions from the soil?

Hint: (a) During washing soil must not be dried at any moment, otherwise some NH_3 will be lost by volatilization.

(b) Be sure about complete removal of excess NH_4^+ ions from the soil, otherwise there will be overestimation of CEC.

Q.5 How can you get the measure of exchange capacity of soil due to organic matter?

Hint: The difference beween the exchange capacities of the soil before and after the removal of organic matter (with H_2O_2) are the measure of exchange capacity only due to organic matter.

Q.6 Is there any relationship between CEC and total exchangeable metallic cation?

Hint: CEC (approx) = Total exchangeable metallic cation + Exchangeable H^+

Q.7 Establish the relationship between cmol (±)/kg and meq/100g.

Hint: $$\frac{cmol(\pm)}{kg} = \frac{10 \times mmol(\pm)}{kg} = \frac{mmol(\pm)}{\frac{kg}{10}} = \frac{meq}{100g}$$

21

Nitrogen in Soil

21.1 Determination of Total Nitrogen in Soil (Modified Kjeldahl Method)

Nitrogen mostly present in soil in organic form. Relatively small amount of nitrogen usually occurs in ammonium and nitrate form, the available form. The Kjeldahl method of total N determination includes both organic and ammonium forms and with modification nitrate form of N can also be included.

Principle

When soil is digested with concentrated sulphuric acid organic nitrogen is converted to ammonium sulphate. To speed up the reaction a digestion mixture composed of sodium (or potassium) sulphate (to raise the boiling point of H_2SO_4) and cupric sulphate ($CuSO_4$) - selenium powder (catalyst) is used.

$$\text{Organic N} + H_2SO_4 \xrightarrow{\text{Catalyst}} (NH_4)_2SO_4$$

In modified method concentrated H_2SO_4 – salicylic acid is used along with sodium thiosulphate ($Na_2S_2O_3$) to trap nitrate N (otherwise lost by volatilization) as nitro salicylic acid.

$$\underset{\text{Salicylic acid}}{C_6H_4OHCOOH} + HNO_3 \rightarrow \underset{\text{Nitro salicylic acid}}{C_6H_3OH.NO_2.COOH} + H_2O$$

Addition of sodium thiosulphate (or Zn dust) reduces nitro salicylic acid to amino salicylic acid which on digestion with H_2SO_4 is converted to ammonium sulphate.

$$Na_2S_2O_3 + H_2SO_4 \longrightarrow Na_2SO_4 + H_2SO_3 + S$$

$$C_6H_3OH.NO_2.COOH + 3\ H_2SO_3 + H_2O \rightarrow \underset{\text{Amino salicylic acid}}{C_6H_3OH.NH_2.COOH} + 3H_2SO_4$$

$$C_6H_3OH.NH_2.COOH + 27\ H_2SO_4 \rightarrow (NH_4)_2SO_4 + 26SO_2 + 14CO_2 + 3H_2O$$

The ammonium sulphate on distillation with strong alkali (NaOH) liberates NH_3.

$(NH_4)_2SO_4 + NaOH \rightarrow Na_2SO_4 + 2NH_3 + 2H_2O$

The liberated NH_3 is absorbed in boric acid and absorbed ammonia is determined by titration with standard acid (HCl or H_2SO_4) in presence of suitable indicator.

$$NH_3 + H_3BO_3 \rightarrow NH_4H_2BO_3$$

Ammonium borate

$$NH_4H_2BO_3 + H_2SO_4 \rightarrow (NH_4)_2SO_4 + 2\ H_3BO_3$$

Equipments and Materials

Kjeldahl distillation flask of 800 ml capacity, 150 ml microKjeldahl flask, digestion chamber with heaters and fume exhaust facility, steam distillation set, balance, glass beads, measuring cylinder, burette.

Reagents

- Sulphuric acid-salicylic acid mixture: Dissolve 1 g salicylic acid in 30 ml concentrated H_2SO_4.
- Digestion mixture: Mix 1 part dry cupric sulphate ($CuSO_4.5H_2O$), 10 parts potassium sulphate (K_2SO_4) and 0.5 part of selenium powder. Grind to powder in the mortar.
- Sodium thiosulphate ($Na_2S_2O_3$. $5H_2O$)
- Mixed indicator: Dissolve 0.099 g bromo cresol green and 0.066 g methyl red in 100 ml of 95% ethanol. Adjust the solution to the bluish purple midcolour at pH 4.5 with dilute NaOH or HCl.
- Boric acid- indicator solution, 2%: Dissolve 20 g boric acid in about 800 ml hot distilled water. Cool, then add 20 ml of mixed indicator and adjust by dilute NaOH or HCl until the bluish colour of the indicator weakens toward pink. Make up volume to 1 litre.
- Standard sulphuric acid, 0.1 **N**: Dilute 2.8 ml of concentrated sulphuric acid to 1 litre with distilled water and standardize against standard sodium carbonate (0.1 **N**) using methyl red indicator until colour changes from yellow to red.
- NaOH, 40%: Dissolve 400 g NaOH pellets in about 500 ml distilled water in a thick walled 2 litre beaker. Allow it to cool and make up volume in a 1 litre volumetric flask. [Note: Place a small piece of paper at the neck before putting the stopper, otherwise you may face difficulty in opening the flask]

Procedure

Digestion

- Take 1.0 g air dry processed (passed through 0.2 mm sieve) medium texture soil (0.5 g for clayey soil, 2.0 g for sandy soil, 0.1 g for muck or peat) in an 150 ml microKjeldahl flask.
- Add about 2 g digestion mixture and about 5 ml of water. Allow the soil sample to soak for 30 minutes.
- Add 5 ml of sulphuric acid-salicylic acid mixture. Swirl the flask until the acid is thoroughly mixed with the soil. Allow it to stand for 20 minutes for nitrate to react with salicylic acid.
- Then add 5 g sodium thiosulphate and heat the flask on digestion rack for first 10-30 minutes slowly until frothing stops.
- Gradually raise the temperature until the acid reaches a boil but never allow to rise the temperature above 410°C as evidenced by condensation of acid reaches approximately 1/3rd the way up the neck of flask.
- Rotate the flask at intervals and continue heating until the organic matter completely destroyed which is best judged by the clear grayish blue or greenish colour of the solution (1±0.25 hour).
- Cool and slowly add 30 ml of distilled water. Further cool the solution (heat of dilution). [If large quantities of sand are present, to avoid bumping during distillation discard sand particles during transfer of the solution by repeated washing with distilled water to an 800 ml Kjeldahl flask].

Distillation

- Take approximately 10 ml of boric acid-indicator solution in a 100 ml conical flask and place in such a way that the end of delivery tube dips into the boric acid – indicator solution.
- Add few pieces of glass beads to the Kjeldahl distillation flask containing digested materials to prevent bumping.
- Start flowing of cold water in the condenser.
- Connect the distillation flask to the distillation apparatus.
- To make the solution alkaline carefully pour 18 ml of 40% NaOH so that it runs down the neck to the bottom of the flask without mixing.
- The mixture is heated to boil by steam distillation.

- Continue distillation until NH_3 evolution ceases (test by placing red filter paper at the delivery end) and collect about 25 ml of distillate. Disconnect the flask to prevent sucking back.
- Run a blank in a similar way without soil sample.

Titration

- Titrate the distillate - boric acid solution with standard H_2SO_4. At the end point blue colour just disappears and one drop in excess turns the solution pink.

Calculation

Say,

Weight of soil taken = W g

Volume of H_2SO_4 required for sample titration = V_S ml

Volume of H_2SO_4 required for blank titration = V_B ml

Strength (N) of H_2SO_4 = S

$$\text{Total N in soil (\%)} = (V_S - V_B) \times S \times 0.014 \times \frac{100}{W}$$

[$2NH_3 + H_2SO_4 \rightarrow (NH_4)_2SO_4$. It means 1 mole of $H_2SO_4 \equiv$ 2 moles of NH_3.

½ mole of $H_2SO_4 \equiv$ 1 mole of NH_3

1 g equivalent weight (equivalent weight of H_2SO_4 is ½ mole) of $H_2SO_4 \equiv$ 1 mole of NH_3 = 17 g $NH_3 \equiv$ 14 g N

So, 1000 ml of 1**N** $H_2SO_4 \equiv$ 14 g N

1 ml of 1 **N** $H_2SO_4 \equiv$ 0.014 g N]

21.2 Determination of Mineralizable (Available) Nitrogen in Soil

The term available nitrogen in soil can be divided into two categories one is immediate available nitrogen which includes ammonium (both soluble and exchangeable), nitrite and nitrate nitrogen, another is potential available nitrogen or the mineralizable nitrogen which includes easily oxidizable organic fraction of nitrogen. Soil nitrogen which mostly present in organic form becomes available to the crop plant through the process of mineralization. Mineralization process is mostly biological in nature. Incubation method of determination of mineralizable nitrogen in soil which is more reliable than any other methods requires incubation of soil under aerobic and anaerobic condition and estimation of NH_4 and NO_3– N produced through mineralization by microbes. Chemical method of determination of available nitrogen which is very well correlated with the biological estimate involves measurement of a fraction of easily hydrolysable nitrogen by using dilute acid or alkali. Alkaline permanganate method (Subbiah and Asija, 1956) which involves distillation of soil with alkaline potassium permanganate estimates readily oxidizable as well as reactive forms of soil nitrogen ignoring the fraction present as nitrite and nitrate form. Ammonia thus formed during distillation is absorbed in boric acid and the amount of absorbed ammonia is determined by titration with standard acid as in usual ammonia distillation.

Distillation

$$2KMnO_4 + H_2O \xrightarrow{\text{alkaline medium}} MnO_2 + 2KOH + 3O$$

$$RCH.NH_2.COOH + O \longrightarrow RCO.COOH + NH_3$$

Equipments and Materials

Kjeldahl digestion flask of 800 ml capacity, steam distillation set, balance, glass beads, pure liquid paraffin, 250 ml conical flask, measuring cylinder, burette.

Reagents

- Potassium permanganate solution, 0.32%: Dissolve 3.2g potassium permanganate ($KMnO_4$) in distilled water and make up volume to 1 litre.
- Sodium hydroxide solution, 2.5%: Dissolve 25 g sodium hydroxide (NaOH) pellets in distilled water and make up volume to 1 litre.
- Mixed indicator: As in section 20.1
- Boric acid- indicator solution, 2%: As in section 21.1
- Standard sulphuric acid, 0.02**N**: Dilute 0.1 **N** H_2SO_4 (*see section 15.1 page no 157*) five times with distilled water [take 100 ml of 0.1 N H_2SO_4 and make up the volume to 500 ml] and standardize with standard 0.02 **N** Na_2CO_3 using methyl red indicator until colour changes from yellow to red.

Procedure

Distillation

- Pipette out approximately 25 ml of boric acid-indicator solution in a 250 ml conical flask and place in such a way that the end of delivery tube dips into the boric acid- indicator solution.

- Take accurately weighed 20g of processed soil sample in a Kjeldahl distillation flask.
- Add 20 ml of distilled water, 100 ml of 0.32% potassium permanganate solution and 100 ml of 2.5% sodium hydroxide solution.
- Add 1 ml of pure paraffin (to prevent frothing) and few pieces of glass beads (to check bumping).
- Determine the liberated NH_3 by steam distillation as described in section 21.1, but collect the distillate in a 250 ml conical flask which contains 25 ml of boric acid -indicator solution.
- Stop distillation when 100 ml of distillate will be collected.
- Run a blank in a similar way without soil sample.

Titration

- As described in the section 21.1

Calculation

Say,

Weight of soil taken = W g

Volume of H_2SO_4 required for sample titration = V_S ml

Volume of H_2SO_4 required for blank titration = V_B ml

Strength (N) of H_2SO_4 = S

Available N in soil (%)=$(V_S\text{-}V_B)\times S\times 0.014\times \frac{100}{W} = Z$ (*See section 21.1 page no 293*)

Available N, ppm = $\frac{Z\times 10^6}{10^2} = Z\times 10^4$

Available N, kg/ha = $Z \times 10^4 \times 2.24$ [Considering weight of 1 hectare 15 cm soil = 2.24×10^6 kg]

Rating

Soil Class on Available N	Kg / ha
Low	< 280
Medium	280 – 450
High	> 450

21.3 Determination of Exchangeable Ammonium Nitrogen in Soil

As ammonium ion is subjected to oxidation to nitrite and then nitrate in soils stored under warm and moist condition, its determination should be done immediately after the collection of sample. If it is not possible, ammonification (conversion of amine or amide nitrogen to ammonium form) and nitrification (conversion of ammonium form to nitrite and then nitrate form) may be arrested temporarily by rapid drying in an oven at 50°C after spreading out in a thin layer, or by freezing.

Principle

Exchangeable ammonium nitrogen can easily be extracted by potassium or sodium ion. The extract containing ammonium ion (NH_4^+) is then distilled with alkali and liberated ammonia is absorbed in boric acid and titrated as in usual ammonia distillation (*see section 21.1 page no 293*).

Equipments and Materials

Distillation flask of 800 ml capacity, steam distillation set, shaker, balance, Buchner funnel, 250 ml conical flask, Whatman No. 42 filter paper, desiccator.

Reagents

- Potassium chloride, 2**M**: Dissolve 150 g potassium chloride (KCl) in about 800 ml of water, adjust pH to 2.5 with HCl and make up volume to 1 litre.
- Magnesium oxide powder: Magnesium oxide is heated to 600°C for 2 hours in muffle furnace, cool in desiccator and store in air tight container.
- Mixed indicator: See section 21.1
- Boric acid - indicator solution, 2%: See section 21.1
- Standard sulphuric acid, 0.02**N**: See section 21.2.

Procedure

Extraction

- Weigh 20 g freshly collected soil sample in a 250 ml conical flask and moisten with distilled water. Add 100 ml of acidified potassium chloride solution.
- Shake on a mechanical shaker for half an hour.
- Filter it through Whatman No. 42 filter paper fitted on a Buchner funnel under suction.

- Wash the soil on the filter paper with about 200 ml of acidified KCl solution in increments using first few increments to rinse out the conical flask.

Distillation

- Transfer the entire leachate to an 800 ml distillation flask.
- Now add 0.5 g MgO powder to the distillation flask and connect the flask to the steam distillation set.
- Determine the liberated NH_3 by steam distillation as described in section 20.1, but collect the distillate in a 250 ml conical flask which contains 20 ml of boric acid -indicator solution.
- Stop distillation when 50 ml of distillate will be collected.
- Carry out a blank distillation with 300 ml KCl solution but without soil.
- Determine the moisture content by drying a known weight of fresh soil sample in an oven at 105°C for about 24 hours, cooling in desiccator and weighing.
- Run a blank in a similar way with equal volume of distilled water in place of leachate.

Titration

- As described in the (*section 21.1 page no 293*)

Calculation

Say,

Weight of oven dry soil = W g

Volume of H_2SO_4 required for sample titration = V_S ml

Volume of H_2SO_4 required for blank titration = V_B ml

Strength (N) of H_2SO_4 = S

Exchangeable NH_4- N in soil (%) = $(V_S - V_B) \times S \times 0.014 \times \frac{100}{W} = Z$ (*See section 20.1 page no 285*)

$$\text{Exchangeable } NH_4\text{-N, ppm} = \frac{Z \times 10^6}{10^2} = Z \times 10^4$$

Exchangeable NH_4-N, kg/ha = $Z \times 10^4 \times 2.24$

[Considering weight of 1 hectare 15 cm soil = 2.24×10^6 kg]

21.4 Determination of Nitrate Nitrogen in Soil

Like exchangeable ammonium, nitrate nitrogen content of soil should be determined immediately after sampling or may be temporarily stored as described in (*See section 21.3 page no 299*).

Principle

Nitrate N content of the soil can be estimated from the residue left after the distillation of exchangeable ammonium N after reduction of nitrate to ammonia by Davarda's alloy in alkaline condition.

$$3NO_3^- + 8Al + 5OH^- + 2H_2O \rightarrow 8AlO_2^- + 3NH_3$$

Evolved NH_3 is then estimated as described (in section 20.1 Page no 285).

Reagents

- Davarda's alloy (an alloy containing Cu: Al: Zn: 50:45:5)
- Sodium hydroxide, 1%: Dissolve 10 g NaOH in distilled water and make up volume to 1 litre.
- Mixed indicator: See section 21.1
- Boric acid- indicator solution, 2%: (*See section 21.1 page no 293*)
- Standard sulphuric acid, 0.02 N: (*See section 21.2 oage no 296*)

Procedure

Distillation

- Cool the residue left in the distillation flask after distillation of exchangeable NH_4 –N.
- Add about 1 g Davarda's alloy and 25 ml of 1% of NaOH solution.
- Determine the liberated NH_3 by steam distillation as described in (*See section 21.3 page no 293*).
- Carry out a blank distillation in the same way with the residue left after distillation of exchangeable NH_4^+ -N from blank sample.

Titration

- As described in the (*See section 21.1 page no. 293*)

Calculation

Same as *(See Section 21.3 page no. 299)*

21.5 Determination of Organic Forms of Nitrogen in Soil (Stevenson, 1996)

More than 95% of the total N in most of the surface soil is organically combined. Hydrolysis studies with hot acid have shown that 20-40% of the total N in most of the surface soils is in the form of bound amino acid and that 4-10% is in the form of combined hexosamine. Account of purine and pyrimidine derivatives is not exceeding 1% of the total N. Some unidentified organic N are in the form of lignin-ammonia, quinine- ammonia, quinine –amino acid condensation products, but the chemical nature of about one half of the N in the soils still remains obscure.

Principle

In the procedure generally adopted for preparation of soil hydrolysates, the soil sample is heated with 6 **N** HCl or 6 **N** H_2SO_4 for several hours and filtered hydrolysate is prepared for analysis by removal of acid used for hydrolysis. The hydrolysis with acid is usually conducted by boiling the soil-acid mixture under reflux.

Equipments and Materials

Micro Kjeldahl digestion stand, steam distillation set, distillation flask of 50 ml and 500-600 ml capacities, micro burette of 5 ml graduated with 0.01 ml intervals, electric hot plate with heat control device, Whatman No. 42 filter paper

Reagents

- Hydrochloric acid, 6 **N**: Add 490 ml of concentrated hydrochloric acid (HCl) to about 300 ml water; dilute the cooled solution to 1 litre.
- n-Octyl alcohol
- Sodium hydroxide, 10 **N**: Dissolve 400 g of sodium hydroxide (NaOH) pellets in about 500 ml of distilled water in a glass beaker. Cool it and make volume to 1 litre.
- Sodium hydroxide, 5 **N**: Dilute 500 ml of 10 **N** NaOH to 1 litre with distilled water.
- Sodium hydroxide, 0.5 **N**: Dilute 50 ml of 10 **N** NaOH to 1 litre with distilled water.
- Digestion mixture: (*See section 21.1 page no. 293)*

- Sulphuric acid, concentrated
- Magnesium oxide powder: *(See section 21.3 page no. 293)*
- Phosphate–borate buffer, pH 11.2: Place 100 g of sodium phosphate ($Na_3PO_4.12H_2O$), 25 g of sodium tetra borate ($Na_2B_4O_7.10H_2O$) and 900 ml of water in 1 litre volumetric flask. Shake to dissolve and dilute to 1 litre.
- Citric acid: Grind 100 g citric acid ($C_6H_8O7.H_2O$) in a mortar and store in a small wide mouth bottle.
- Ninhydrin: Grind 10 g of ninhydrin (triketohydrindene hydrate; indane-trione hydrate) in a mortar and store in a small wide mouth bottle.
- Mixed indicator: (See section 20.1 page no
- Boric acid- indicator solution, 2%: *(See section 21.1 page no.* 293*)*
- Standard sulphuric acid, 0.005 **N**: Dilute 1.4 ml of concentrated H_2SO_4 to 1 litre. Again dilute 100 ml of the above solution to 1 litre. That gives you approx. 0.005 **N** H_2SO_4 solution. Standardize it with standard 0.005 N Na_2CO_3 solution using methyl red as indicator.

Procedure

A) Preparation of Hydrolysate

- Place 5 g sample of fresh ground (< 100 mesh) soil containing about 10 mg of N in 125 ml Erlenmeyer flask fitted with 24/40 ground glass joint.
- Add 2 drops of n-octyl alcohol and 20 ml of 6 **N** HCl. Thoroughly mix the acid with the soil.
- Place the flask on an electric hot plate and connect the flask to a Liebig condenser.
- Then continue the circulation of cold water through the water jacket of the condenser.
- Heat the mixture so that it boils gently under reflux for 12 hours.
- After completion of hydrolysis, allow the flask to cool and filter the hydrolysis mixture through Buchner funnel fitted with Whatman No. 42 filter paper under suction.
- Wash the hydrolysis residue with small (5-10 ml) portion of water for about 5 to 6 times.

- Transfer the filtrate to a 100 ml beaker and neutralize the hydrolysate using pH meter (pH 6.5 ± 0.1) by cautious addition of NaOH. Use 5 **N** NaOH to bring the pH of the hydrolysate to around 5.0, and complete neutralization using 0.5 **N** NaOH.
- Transfer the hydrolysate to 100 ml volumetric flask and dilute it to the volume with washings obtained by rinsing electrodes, glass rod and beaker several times with small quantities of water. Make up the volume to the mark with distilled water.
- Stopper the flask and mix the content by inverting it several times.

B) Analysis of Hydrolysate

21.5.1 Determination of Total Hydrolysable Nitrogen

- Place 5 ml of the neutralized hydrolysate in 50 ml of distillation flask.
- Add 0.5 g of digestion mixture and 2 ml of concentrated H_2SO_4.
- Heat the flask first at low temperature until frothing ceases, then increase the temperature until boiling of acid starts. Continue digestion until the mixture clears.
- After completion of digestion, allow the flask to cool, and add about 10 ml of water.
- Transfer the digested materials to 600 ml Kjeldahl distillation flask by repeated washing.
- Pipette out 5 ml boric acid- indicator solution in 100 ml conical flask and place the flask under the condenser of the steam distillation set so that the end of the condenser is below the surface of the boric acid solution.
- Connect the cooled distillation flask to the distillation set; pour 10 ml of 10**N** NaOH in the funnel of the set. Rinse the funnel rapidly with 5 ml of water and seal the funnel plug.
- Commence steam distillation and collect about 35 ml distillate (approx. 4 minutes).
- Rinse the end of the condenser and determine ammonium-N in the distillate by titration with standard 0.005 **N** H_2SO_4 from the micro burette. The colour change at the end point is from green to faint permanent pink.

21.5.2 Determination of Ammonium Nitrogen

- Pipette out 10 ml of the hydrolysate in 600 ml distillation flask.
- Add about 0.07 g of MgO and connect the flask to steam distillation set.

- Determine the amount of NH_3 liberated by steam distillation as described in section 21.5.1, but collect 20 ml distillate in a 100 ml conical flask containing 5 ml boric acid indicator (approx. 2 minutes).

21.5.3 Determination of Ammonium + Hexosamine Nitrogen

- Pipette out 10 ml of hydrolysate in 600 ml distillation flask.
- Add 10 ml of phosphate–borate buffer.
- Connect the flask to the distillation set
- Determine the amount of NH_3 liberated by steam distillation as described in section 21.5.1, and collect 35 ml distillate in a 100 ml conical flask containing 5 ml boric acid indicator (approx. 4 minutes).

20.5.4 Determination of Amino Acid Nitrogen

- Place 5 ml of hydrolysate in the 50 ml distillation flask and add 1 ml of 0.5 **N** NaOH, heat the flask in boiling water until the volume reduced to half (20 minutes).
- Allow the flask to cool, add 500 mg of citric acid and 100 mg ninhydrin and place the flask in a boiling water bath, so that its bulb is completely immersed in boiling water.
- Swirl the flask intermittently without removing it from the water bath. Keep it for 10 minutes.
- Cool it and transfer the content completely to a 600 ml Kjeldahl distillation flask.
- Add 10 ml of phosphate buffer solution and 1 ml of 5 **N** NaOH.
- Connect the flask to the distillation set and determine the amount of NH_3 liberated by steam distillation as described in section 21.5.1, and collect 35 ml distillate in a 100 ml conical flask containing 5 ml boric acid indicator (approx. 4 minutes).

21.5.5 Determination of Hydrolysable Unknown Nitrogen

This form of soil nitrogen can be calculated as the difference between the estimate of total hydrolysable nitrogen and nitrogen present in the form of ammonium, amino acid and amino sugar in the acid hydrolysate.

Hydrolysable Unknown N = Total Hydrolysable N - N accounted for as (NH_3 + amino Acid + Amino sugar) N

21.5.6 Determination of Acid-Insoluble Nitrogen

This form of soil nitrogen can be calculated as the difference between the estimate of total soil nitrogen (modified Kjeldahl method) and total hydrolysable nitrogen.

Acid -Insoluble (Non-hydrolysable) N = Total Soil N – Total Hydrolysable N

Calculation

Say,

Weight of dry soil taken = W g

Volume of the hydrolysate = V_H ml

Volume of the aliquot taken for analysis = V_A ml

Volume of H_2SO_4 required for sample titration = V_S ml

Strength (N) of H_2SO_4 = S_N

Form of N in soil (%) = $V_S \times S_N \; 0.014 \times \frac{V_H}{V_A} \times \frac{100}{W}$ = Z *(See section 21.1 page no 293)*

Form of-N, ppm = $(Z \times 10^6/ 10^2) = Z \times 10^4$

Form of-N, kg/ha = $(Z \times 10^4) \times 2.24$

[Considering weight of 1 hectare 15 cm soil = 2.24×10^6]

21.6 Determination of Soil Microbial Biomass Nitrogen

Equipments and Materials

Moisture box, separatory funnel, glass vacuum desiccator and vacuum pump, shaker, digestion chamber with heaters, glass beads, beaker (50 and 100 ml capacity), 250 ml conical flask, Kjeldahl digestion flask of 800 ml capacity, steam distillation set.

Reagents

- Distilled chloroform
- Concentrated sulphuric acid
- K_2SO_4, 0.5M: As in section 19.2.4
- Plus reagents as in section 21.1

Procedure

- Follow the same steps as were followed during determination of soil microbial biomass carbon (Section 19.2.4) up to 30 minutes shaking of fumigated and unfumigated soil samples with 25 ml of 0.5 **N** K_2SO_4.
- Determine total N in the extract by Modified Kjeldahl digestion method as outlined in the section 21.1.

Calculation

$$\text{Microbial biomass Nitrogen (MBN), in } \mu g / g \text{ soil} = \frac{N_{EF} - N_{EUF}}{K_{EN}}$$

Where, N_{EF} = Total N in the fumigated soil extract, µg/g soil

N_{EUF} = Total N in the unfumigated soil extract, µg/g soil

K_{EN} = Efficiency of extraction of microbial N plus inorganic N from soil; usually the value ranges from 0.54 to 0.62 (Jenkinson, 1988). However, 0.54 is considered ideal for many soils.

Questions and Hints

Q.1 Which forms of nitrogen in the soil are determined by the Kjeldahl method of total N determination?

Hint: The Kjeldahl method includes both organic and ammonium forms, and with modifications (inclusion of salicylic acid and sodium thiosulphate) it includes the nitrate form also.

Q.2 What is the role of each of the following in the digestion for total N determination?

(a) $CuSO_4$, (b) K_2SO_4/Na_2SO_4, (c) Se powder, (d) Salicylic acid and (e) sodium thiosulphate.

Hint: (a) $CuSO_4$ in the digestion mixture acts as catalyst to accelerate the digestion rate.

(b) K_2SO_4/Na_2SO_4 is used to raise the boiling temperature of the H_2SO_4 in the digestion.

(c) Se powder accelerates the digestion rate further over the $CuSO_4$ alone.

(d) Salicylic acid in H_2SO_4 helps to trap NO_3-N as nitro salicylic acid.

$$\underset{\text{Salicylic acid}}{C_6H_4OHCOOH} + HNO_3 \rightarrow \underset{\text{Nitro salicylic acid}}{C_6H_3OH.NO_2.COOH} + H_2O$$

(e) Sodium thiosulphate reduces nitro salicylic acid to amino salicylic acid which on digestion with H_2SO_4 is converted to $(NH_4)_2SO_4$.

$$Na_2S_2O_3 + H_2SO_4 \longrightarrow Na_2SO_4 + H_2SO_3 + S$$

$$C_6H_3OH.NO_2.COOH + 3\ H_2SO_3 \rightarrow C_6H_3OH.NH_2.COOH + 3H_2SO_4$$

Amino salicylic acid

$$C_6H_3OH.NH_2.COOH + 27\ H_2SO_4 \rightarrow (NH_4)_2SO_4 + 26SO_2 + 14CO_2 + 3H_2O$$

Q.3 Why digestion temperature is so important in total N determination?

Hint: If the digestion temperature is less than 360°C, the release of N from organic source is slow or incomplete and if the temperature is more than 410°C, some volatilization loss of NH_3 from the mixture results. In both the cases the recovery is less.

Q.4 Why exchangeable NH_4- and NO_3-N should be estimated from the freshly collected field moist soil sample?

Hint: NH_4 ions are subjected to oxidation to nitrite and then nitrate when soil is kept in warm moist condition through microbial activity (nitrification), that's why immediate analysis of field moist soil sample is needed. If immediate analysis is no way possible then either rapidly dry the soil at 50°C temperature in oven or preserve in refrigerator to temporarily arrest the microbial conversion of NH_4 to NO_3.

Q.5 Whether use of accurate amount of digestion mixture or MgO powder is essential?

Hint: No; however it is convenient to use a calibrated spoon or scoop for addition of these reagents.

Q.6 Whether during distillation of exchangeable NH_4-N in place of MgO powder NaOH can be used?

Hint: Yes; but it will be required in very less amount as the extractant is a neutral salt solution and should be added carefully down the side of the flask so as to collect at the bottom of the flask without much agitation, otherwise volatilization of NH_3 will start before connecting the flask to the distillation set.

Q.7 What is role of n-Octyl alcohol in acid hydrolysis of soil?

Hint: Octyl alcohol prevents the troublesome frothing which sometimes encountered during hydrolysis of calcareous or dolomitic soils and it increases the precision of the results obtained with soils that do not wet readily on treatment with acid.

Q.8 Why hydrolysate is neutralized before distillation?

Hint: Neutralization increases the precision of analysis and decreases bumping during Kjeldahl digestion for total hydrolysable N.

22

Phosphorus in Soil

Phosphorus is present in soil both in inorganic form and organic form. The inorganic P can be divided into different pools. The major active inorganic forms are: P bound to aluminium (Al-P), iron (Fe-P), calcium (Ca-P) and silicate minerals and relatively less active forms are occluded and reductant soluble forms of P. On the other hand, the principal organic P compounds present in soil are (i) inositol phosphate, (ii) phospholipids, (iii) nucleic acid and other unidentified esters and phosphoproteins.

22.1 Determination of Available Soil P

The availability of soil P to plant varies greatly depending upon the reaction, minerological composition and amount of the soil colloid. Very little P is present in the soil solution except in recently fertilized or sodic soils. The type of phosphate ions present in the soil solution depends on the soil pH. In soils of neutral to slightly alkaline pH, HPO_4^{2-} ion is the dominant form. In slightly or moderately acidic soils, both $H_2PO_4^-$ and HPO_4^{2-} ions prevail. At strongly acidic soil $H_2PO_4^-$ ion tends to dominate, while above pH 9.0, the PO_4^{3-} ion becomes more important than $H_2PO_4^-$.

As the P is drawn from soil solution largely through diffusion by the plant, it is replenished from solid phase pool of available P. P fertilizer is added to recharge this pool.

$$\text{Solid phase P} \rightleftharpoons \text{Solution P} \rightleftharpoons \text{Precipitated P}$$

The Al-P and Fe-P fractions are the major contributor to plant available P under most of the situations. Calcium P also contributes to plant available P, but its contribution is usually less than Al-P and Fe-P even where Ca-P is much more abundant.

Several chemical extractants has been developed over the years to simulate the chemically extracted P with biologically available P. None of these methods is equally suitable for all types of soils. Accordingly depending mostly upon the

Some common methods for extracting available P in soil

Extractant	Extractant Composition	Soil : Extractantratio	Shaking time	Suitability
Olsen (Olsen *et al.*, 1954)	0.5 **M** $NaHCO_3$ (pH 8.5): Dissolve 41 g $NaHCO_3$ in 1 litre and adjust pH to 8.5 with NaOH.	1:20	30 minutes	Neutral to alkaline soil
Bray I(Bray and Kurtz, 1945)	0.03**N** NH_4F in 0.025**N** HCl (pH 3.5): Dissolve 1.11 g NH_4F and 2.1 ml of conc. HCl in 1 litre.	1:10	5 minutes	Acid soil
Bray II	0.03 **N** NH_4F in 0.1 **N** HCl (pH 1.0): Dissolve 1.11 g NH_4F and 8.5 ml of conc. HCl in 1 litre.	1:20	40 seconds	Acid soil
Troug (Troug, 1930)	0.002 **N** H_2SO_4 (pH 3.0): Dilute 0.2 **N** H_2SO_4 100 times and add 3 g K_2SO_4 per litre	1:100	30 minutes	Acid soil
Morgan (Morgan, 1941)	10 ml of glacial acetic acid in 1 litre of 0.5% NaOH adjusted to pH 8.4.	1:5	30 minutes	Acid soil
Mehlich-I(Mehlich, 1953)	0.05 **N** HCl + 0.025 **N** H_2SO_4 (pH 1.2): Dilute 4.3 ml conc. HCl and 0.7 ml conc. H_2SO_4 in 1 litre.	1:4	5 minutes	Acid soil
Mehlich 3(Mehlich, 1984)	0.2 **M** CH_3COOH + 0.25**M** NH_4NO_3 + 0.015**M** NH_4F + 0.013**M** HNO_3 + 0.001**M** EDTA	1:10	15 minutes	
AB-DTPA(Soltanpour and Schwab, 1977)	1 **M** NH_4HCO_3 + 0.05**M** DTPA (pH 7.6): Dissolve 79 g NH_4HCO_3 and 19.67 g AR grade DTPA in 1 litre.	1:2	15 minutes	Neutral to alkaline soil

pH, the amount of Fe and Al compound and calcite and dolomite content, one particular method should be selected.

22.1.1 Determination of Available Soil P using Olsen's Extractant (Olsen *et al.*, 1954)

Principle

A) Extraction

Olsen's extractant is found to be suitable for soils having pH more than 5.5. The 0.5 **M** $NaHCO_3$ (pH 8.5) solution employs for extraction of P from alkaline and calcareous soils actually designed to control the activity of Ca, through the solubility product of $CaCO_3$. As the carbonate activity in the soil is raised by this extractant, the Ca activity is decreased. Again, as the Ca activity decreases phosphate activity increases, thus through solubility product of calcium phosphate some phosphate from the surface of calcium is extracted.

The importance of buffering of carbonate during extraction in calcareous soils can be explained by two way role of carbonic acid:

(a) It increases the solubility of calcium phosphate as expected by any acid.

(b) It decreases solubility of calcium phosphate due to increased activity of Ca as $CaCO_3$ is dissolved in H_2CO_3.

I. Exchange Reaction

$$\boxed{\text{Soil}}\, H_2PO_4 + NaHCO_3 \rightarrow \boxed{\text{Soil}}\ HCO_3 + NaH_2PO_4$$

II. Chemical Reaction

$$Ca_3\ (PO_4)_2 + 6NaHCO_3 \rightarrow 3Ca\ (HCO_3)_2 + 2Na_3PO_4$$

$$3Ca\ (HCO_3)_2 \rightarrow 3\ CaCO_3 + 3H_2CO_3$$

$$2Na_3PO_4 + 3H_2CO_3 \rightarrow 2H_3PO_4 + 3Na_2CO_3$$

$$Ca_3\ (PO_4)_2 + 6NaHCO_3 \rightarrow 3\ CaCO_3 + 2H_3PO_4 + 3Na_2CO_3$$

The reagent also extracts some phosphate from the surface of Al and Fe phosphates by repression of the Al and Fe activities (Al as aluminate complex formation and Fe by precipitation as oxide). The carbonate ions added through the reagent, maintain Ca activity in low enough in all soil (acid, neutral and alkaline) by the solubility product of $CaCO_3$, to prevent reprecipitation of liberated phosphate as calcium phosphate.

B) Colour development

The heteropoly complexes are thought to be formed when extracted H_3PO_4 is reacted with ammonium molybdate in acid medium by coordination of molybdate ions, with phosphate as the central coordinating atom, the oxygen of the molybdate radicals being substituted for that PO_4.

$$(NH_4)_6Mo_7O_{24}.4H_2O + 6HCl \longrightarrow 7H_2MoO_4 + 6NH_4Cl$$

(Ammonium molybdate) (Molybdic acid)

$$H_3PO_4 + 12H_2MoO_4 \longrightarrow H_3P\ (Mo_3O_{10})_4 + 12H_2O$$

Phosphomolydate complex (Yellow colour)

Besides P^{5+} ions which act as the central coordinating atom to form 12 fold heteropoly acid with molybdate, arsenic (As^{5+}), silicon (Si^{4+}), germanium (Ge^{4+}) and even under some conditions molybdenum (Mo^{6+}) and boron (B^{3+}) also act as a central ion. Before reduction the heteropoly complexes give yellow colour to their water solution. In soils of low enough P concentration to be suitable for determination by reduction to form the blue colour, the yellow colour is so faint that it is not noticed, but spectrophotometric measurement is done without any problem. The molybdenum blue colour is produced when either molybdate or its heteropoly complexes are partially reduced (Mo^{6+} reduced to Mo^{3+} or Mo^{5+} state) during reaction with ascorbic acid (Murphy and Riley, 1962). Presence of unpaired electrons is responsible for spectrophotometric resonance (blue colouration). The intensity of the colour is measured at 730 nm wavelength in the spectrophotometer.

Alkaline extractant solubilizes some organic matter resulting coloured extract which interferes in colorimetric estimation. Activated charcoal is used with soil to adsorb soluble organic matter.

Equipments and Materials

Colorimeter or spectrophotometer, 50 ml volumetric flasks, 100 ml conical flask, measuring cylinder, mechanical shaker, pipette, Whatman No. 42 filter paper, balance.

Reagents

- Olsen's extractant, 0.5 **M** sodium bicarbonate solution: Dissolve 42.0 g sodium bicarbonate ($NaHCO_3$) in about 800 ml distilled water. Adjust pH to 8.5 with 1**M** NaOH (4 g NaOH / 100 ml of water). Usually 10-12 ml is required per litre of $NaHCO_3$ solution.

- Phosphorus free activated charcoal or Dargo G-60: To make the charcoal free from P repeatedly wash with Olsen's extractant followed by warm distilled water.
- Ammonium molybdate solution: Dissolve 40 g ammonium molybdate [$(NH_4)_6Mo7O_{24}.4H_2O$] in water and make up volume to1 litre.
- Ascorbic acid solution: Dissolve 26.4 g of L-ascorbic acid ($C_6H_8O_6$) and make up volume to 500 ml with distilled water.
- Antimony potassium tartarate or potassium antimonyl tartarate solution: Dissolve 1.454 g of antimony potassium tartarate [$K(SbO)C_4H_4O_6.½H_2O$] and make up volume to 500 ml with distilled water.
- Sulphuric acid, 2.5**M**: Dilute 140 ml of concentrated H_2SO_4 to 1 litre.
- Murphy-Riley colour developing solution: Add 250 ml of 2.5 **M** H_2SO_4, followed by 75 ml of ammonium molybdate solution, 50 ml of ascorbic acid and 25 ml of antimony potassium tartarate in a 500 ml volumetric flask. Make up volume with distilled water and store in an amber bottle in a dark place. Prepare the reagent mixture for daily use.
- p-nitrophenol indicator, 0.25%: Dissolve 0.25 g of indicator and make up volume to 100 ml with distilled water.
- Standard P solution: Dissolve 0.439 g of potassium dihydrogen phosphate (KH_2PO_4) in about 500 ml of distilled water. Add 25 ml of 7 **N** H_2SO_4 to the solution and make up volume to 1 litre. It gives 100 ppm P stock solution. From 100 ppm P solution prepare 5 ppm P solution by taking 5 ml of 100 ppm P solution in a 100 ml volumetric flask and making up volume with distilled water.

Procedure

A) Preparation of standard curve

- Take 1, 2, 3, 4 and 5 ml of 5 ppm P solution in a series of 50 ml volumetric flasks.
- Add 5 ml of Olsen's extractant to each flask.
- Add 10 ml of distilled water and five drops of p-nitrophenol indicator.
- Add 2.5 **M** H_2SO_4 drop by drop until the solution becomes clear. The point at which yellow colour just disappears, indicates the correct pH (5.0) for colour development. [If the pH of the solution exceeds to the desired point bring back by adding NaOH.]

- Add 8 ml Murphy-Riley solution to each of the flask and make up volume to the mark which will give 0.1, 0.2, 0.3, 0.4 and 0.5 ppm P solution, respectively.
- Run a blank without P solution in the same way.
- After waiting for 15 minutes, read the intensity of the blue colour on a colorimeter or spectrophotometer at 730 nm wavelength, setting spectrophotometric zero with the blank solution.
- Prepare the standard curve by plotting absorbance or optical density in the Y axis (ordinate) against corresponding P concentration in the X axis (abscissa).

B) Extraction of P from soil

- Take 2.5 g air dry processed (passed through 2 mm sieve) soil in a 100 ml conical flask.
- Add 50 ml of Olsen's extractant and one teaspoonful of P free activated charcoal.
- Stopper the flask with rubber cork and shake the suspension for 30 minutes on a mechanical shaker.
- Filter the suspension through Whatman No. 42 filter paper.
- Run a blank without soil in the same manner.

C) Colour development of test solution

- Pipette out 10 ml aliquot of the filtrate in a 50 ml volumetric flask.
- Add 10 ml of distilled water and five drops of p-nitrophenol indicator.
- Add 2.5 M H_2SO_4 dropwise to bring down the pH at 5.0 till yellow colour disappears.
- Add 8 ml of Murphy-Riley solution and make up volume to the mark.
- Read the intensity of the colour at 730 nm wavelength after 15 minutes waiting. The colour remains stable for 24 hours.

Calculation

Say,

Weight of the soil taken = W g

Volume of the Olsen's extractant added to the soil = V_D ml

Volume of the aliquot taken for colour development = V_A ml

Final volume of the coloured solution = V_C ml

Absorbance reading of the test solution = X

Absorbance reading of the blank solution = Y

Concentration of P corresponding to X absorbance (from standard curve) = A ppm

Concentration of P corresponding to Y absorbance (from standard curve) = B ppm

$$\text{Available soil P,ppm} = (A\text{-}B) \times \frac{V_C}{V_A} \times \frac{V_D}{W} = Z$$

Available soil P, kg / ha = Z × 2.24

Available soil P_2O_5, kg / ha = Z×2.24×2.29

Rating

P fertility Class	Available P_2O_5, kg / ha
Very low	< 7
Low	8 - 16
Medium	17 – 25
High	26 – 45
Very high	> 45

22.1.2 Determination of Available Soil P using Bray-I Extractant

Principle

The extraction of available P from soils with 0.03**N** NH_4F in 0.025 **N** HCl (Bray and Kurtz, 1945) or Bray I extractant is suitable for soils having pH less than 5.5. Fluoride ion (F^-) has the property to form complex with Al^{3+} and Fe^{3+} ions in acid medium with consequent release of phosphate held by these trivalent ions in the soils.

$$3NH_4F + 3HF + AlPO_4 \rightarrow H_3PO_4 + (NH_4)_3AlF_6$$

$$3NH_4F + 3HF + FePO_4 \rightarrow H_3PO_4 + (NH_4)_3FeF_6$$

The formulae $AlPO_4$ and $FePO_4$ represent various hydrated and hydroxyl phosphate, including any adsorbed or precipitated surface layers on oxides and silicates of Al and Fe, respectively.

The extraction of available P from soils with 0.03**N** NH_4F in 0.1 **N** HCl or Bray II extractant includes more of the soil apatite (for example, that in near neutral or calcareous soils or that remaining from rock phosphate additions) as a higher concentration of HCl is employed. The procedure is same as that of Bray I extraction except 0.1**N** HCl and 40 seconds shaking period are employed.

Equipments and Materials

Same as required in *(See section 22.1.1. page no. 311)*

Reagents

- Bray I extractant, 0.03**N** NH_4F in 0.025 **N** HCl: Dissolve 1.11 g of AR grade NH_4F in about 20 ml of distilled water and filter. Dilute 2.1 ml of concentrated HCl in 500 ml of distilled water. Add NH_4F solution to the HCl solution and make up the volume to 1 litre.
- Ammonium molybdate solution: As mentioned in *(See section 22.1.1. page no. 311).*
- Ascorbic acid solution: As mentioned in *(See section 22.1.1. page no. 311)*
- Antimony potassium tartarate or potassium antimonyl tartarate solution: As mentioned in *(See section 22.1.1. page no. 311).*
- Sulphuric acid, 2.5**M**: As mentioned in *(See section 22.1.1. page no. 311).*
- Murphy-Riley colour developing solution: As mentioned in *(See section 22.1.1. page no. 311).*
- Standard P solution: As mentioned in *(See section 22.1.1. page no. 311).*

Procedure

A) Preparation of standard curve

- Follow the same procedure as in *(See section 22.1.1. page no. 311)* except instead of using Olsen's extractant use Bray I extractant.

B) Extraction of P from soil

- Take 5 g air dry processed (passed through 2 mm sieve) soil in a 100 ml conical flask.
- Add 50 ml of Bray I extractant and stopper the flask with rubber cork.
- Shake the suspension for 5 minutes on a mechanical shaker.
- Filter the suspension through Whatman No. 42 filter paper.
- Run a blank without soil in the same manner.

C) Colour development of test solution

- Pipette out 10 ml aliquot of the filtrate in a 50 ml volumetric flask.[If necessary, add 15 ml of 0.8 **M** boric acid (50 g of H_3BO_3 per litre) to the aliquot to avoid the interference of fluoride ions].
- Add 15 ml of distilled water and 8 ml of Murphy-Riley solution and make up volume to the mark.
- Read the intensity of the colour at 730 nm wavelength after 15 minutes waiting.

Calculation

Same as followed in section 22.1.1

Rating

P fertility Class	Available P_2O_5, kg / ha
Very low	< 11
Low	12 - 27
Medium	28 – 56
High	57 – 90
Very high	> 90

22.2 Determination of Total Soil P

The total content of elemental P in silicates and other solids can be extracted and determined by several methods: (a) by Na_2CO_3 fusion, (b) by perchloric acid ($HClO_4$) digestion followed by colorimetric or titrimetric determination or (c) by digestion in HF followed by gravimetric determination as $Mg_2P_2O_7$. The perchloric acid method has been widely used for the determination of total P in soil, although recovery is sometimes less than obtained in the Na_2CO_3 or HF method.

Equipments and Materials

Colorimeter or spectrophometer, mechanical shaker, balance, hot plate, 250 ml volumetric flasks, 250 ml conical flask, measuring cylinder, pipette, Whatman No. 42 filter paper,

Reagents

- Perchloric acid ($HClO_4$), 60%: Dilute 857 ml of concentrated (70%) perchloric acid to1 litre.
- HNO_3, concentrated

- Sodium hydroxide, 5**M**: Dissolve 200 g of sodium hydroxide (NaOH) pellets in distilled water and make up volume to 1 litre.
- Ammonium molybdate solution: As mentioned in section 22.1.1.
- Ascorbic acid solution: As mentioned in section 22.1.1.
- Antimony potassium tartarate or potassium antimonyl tartarate solution: As mentioned in section 22.1.1.
- Sulphuric acid, 2.5**M**: As mentioned in section 22.1.1.
- Murphy-Riley colour developing solution: As mentioned in section 22.1.1.
- p-nitrophenol indicator: As mentioned in section 22.1.1.
- Standard P solution: As mentioned in section 22.1.1.

Procedure

A) Preparation of standard curve

- As followed in section 22.1.2 except the addition of Bray I's extractant

B) Extraction

- Take 2 g (5 g for soil low in phosphorus) of air dry processed (passed through 0.5 mm sieve) in a 250 ml conical flask. [If the sample is high in organic matter, add 20 ml of concentrated HNO_3 for preliminary oxidation on a steam plate.]
- Then add 30 ml of 60% $HClO_4$ and carry out digestion at 130°C in digestion chamber designed to remove $HClO_4$ fumes. A funnel can be used to reflux the $HClO_4$ during digestion in the flask.
- Carry out digestion until the solution appears colourless with a slight increase in temperature, if necessary. Usually about 40 minutes digestion is sufficient. As the digestion is completed, dense white fumes of $HClO_4$ appear and the silica particles become white. Additional $HClO_4$ may be employed to wash down any dark particles that stick to the sides of the flask.
- When the digestion is completed, remove the flask, cool it sufficiently and add 50 ml distilled water to it.
- Transfer the solution through a Whatman No. 42 filter paper to a 250 ml volumetric flask.
- Wash the residue several times and make the volume up to the mark.
- Run a blank without soil in the same way.

C) Colour development of sample solution

- Pipette out 10 ml aliquot of the filtrate in a 50 ml volumetric flask.
- Add 15 ml of distilled water and adjust pH at 5.0 with dropwise addition of 5 **M** NaOH until the colour of the solution changes from colourless to yellow.
- Add 8 ml of Murphy-Riley solution and make up volume to the mark.
- Read the intensity of the colour at 730 nm wavelength after 15 minutes waiting.

Calculation

Say,

Weight of the soil taken = W g

Volume of the digested solution = V_D ml

Volume of the aliquot taken for colour development = V_A ml

Final volume of the coloured solution = V_C ml

Absorbance reading of the test solution = X

Absorbance reading of the blank solution = Y

Concentration of P corresponding to X absorbance (from standard curve) = A ppm

Concentration of P corresponding to Y absorbance (from standard curve) = B ppm

$$\text{Total soil P, ppm} = (A-B)\times\frac{V_C}{V_A}\times\frac{V_D}{W} = Z$$

Total soil P, kg / ha = Z x 2.24

22.3 Determination of Different Fractions of Inorganic Soil P

As discussed earlier inorganic P in soil may present either in ionic or in combined form. The combined form may be of with Ca, Fe and Al as well as may be adsorbed on the colloidal surfaces. The P adsorbed on the soil colloid surface through polyvalent cations is termed as saloid bound P. The Al-P and Fe-P generally constitute 8-10% of total P while Ca-P constitutes 40-50% or more of total P. In some soils (old, highly weathered acid soils) a significant amount of Al-P and Fe-P may remain occluded in the oxides of Fe. The occluded Al-P is termed here as **occluded P** (OC-P) and occluded Fe-P is termed here as

reductant soluble P (RS-P). While Ca-P is the dominant fraction in young, neutral-alkaline soils of high base saturation, the occluded P and reductant soluble P, the ultimate P sink in soil are dominant in highly weathered acidic soils.

Principle

Soil inorganic P fractionation adopted here is the modified method of Chang and Jackson (1957) and after Petersen and Corey (1966).The soil is first extracted with1**M** NH_4Cl to extract the loosely bound P (Saloid P) to the soil colloids (Clay and organic matter) through polyvalent cations. Controls with different chemical species showed that the NH_4F and NaOH treatments must precede the H_2SO_4 treatment since later removes considerable $AlPO_4$ and $FePO_4$ as well as total $CaPO_4$. In neutral or alkaline solution, the fluoride complex of Al forms, but that of iron does not to any appreciable extent. Neither is $CaPO_4$ appreciably dissolved in this reagent. Therefore, the phosphate extracted by 0.5 **M** NH_4F is mainly $AlPO_4$. After removal of $AlPO_4$, the soil is extracted with 0.1 **N** NaOH to extract $FePO_4$. The decantate is made acidic to floccules organic colloids. To extract occluded $AlPO_4$ and $FePO_4$, soil is treated with Na-citrate-Na dithionite which removes oxides of Fe by reduction chelation technique. During reduction $FePO_4$ is reduced to soluble $Fe_3(PO_4)_2$ and comes in solution phase. Insoluble $AlPO_4$ left in the solution is then extracted with NaOH. The residual soil is now left with none other than $CaPO_4$ which is extracted with 0.25 **M** H_2SO_4. During estimation of any fraction separate standard curve preparation is must with respective extractant *e.g.* for saloid bound P with1**M** NH_4Cl, Al-P with 0.5 **M** NH_4F, Ca-P with 0.25 **M** H_2SO_4, etc.

Equipments and Materials

Colorimeter or spectrophotometer, 50 ml volumetric flasks, 100 ml polypropylene centrifuge tube, measuring cylinder, mechanical shaker, pipette, Whatman No. 42 filter paper, balance, centrifuge, water bath.

Reagents

- Ammonium chloride, 1**M**: Dissolve 53.3 g ammonium chloride (NH_4Cl) in distilled water and make up volume to 1 litre.
- Ammonium fluoride, 0.5 **M** (pH 8.2): Dissolve 18.5 g ammonium fluoride (NH_4F) in about 900 ml of distilled water. Adjust pH of the solution to 8.2 with dilute ammonia solution.
- Sodium chloride, saturated solution: Dissolve sufficient (200 g) amount of NaCl crystal to 500 ml of water. Shake vigorously for 15 minutes on mechanical shaker and filter through Whatman No.1 filter paper.

- Sodium citrate, 0.3**M**: Dissolve 88 g tribasic sodium citrate ($Na_3C_6H_5O_7.2H_2O$) and make up volume to 1 litre with distilled water.
- Sodium dithionite, solid
- Potassium permanganate, 0.25**M**: Dissolve 39.5 g potassium permanganate ($KMnO_4$) in 1 litre.
- Ammonium molybdate solution: As mentioned in section 22.1.1.
- Ascorbic acid solution: As mentioned in section 22.1.1.
- Antimony potassium tartarate or potassium antimonyl tartarate solution: As mentioned in section 22.1.1.
- Sodium hydroxide, 2 **M**: Dissolve 80 g of sodium hydroxide (NaOH) pellets in distilled water and make up volume to 1 litre.
- Hydrochloric acid, 2 **M**: Dilute 168 ml of concentrated hydrochloric acid (HCl) and make up volume to 1 litre with distilled water.
- Sulphuric acid, 0.25**M**: Dilute 14 ml of concentrated H_2SO_4 to 1 litre with distilled water.
- Murphy-Riley colour developing solution: As mentioned in section 22.1.1.
- p-nitrophenol indicator : As mentioned in section 22.1.1.
- Boric acid, 0.8**M**: Dissolve 50 g of boric acid (H_3BO_3) and make up volume to 1 litre with distilled water.
- Sodium hydroxide, 0.1**M**: Dissolve 4 g of sodium hydroxide (NaOH) and make up volume to 1 litre with distilled water.
- Standard P solution: As mentioned in section 22.1.1, but using respective extractant.
- Transfer 10 ml aliquot from each of **solution1, 2, - - - - - 6** to 50 ml volumetric flasks.
- Add 15 ml of water and five drops of indicator and adjust pH at 5.0 until indicator colour just changes from colourless to yellow.

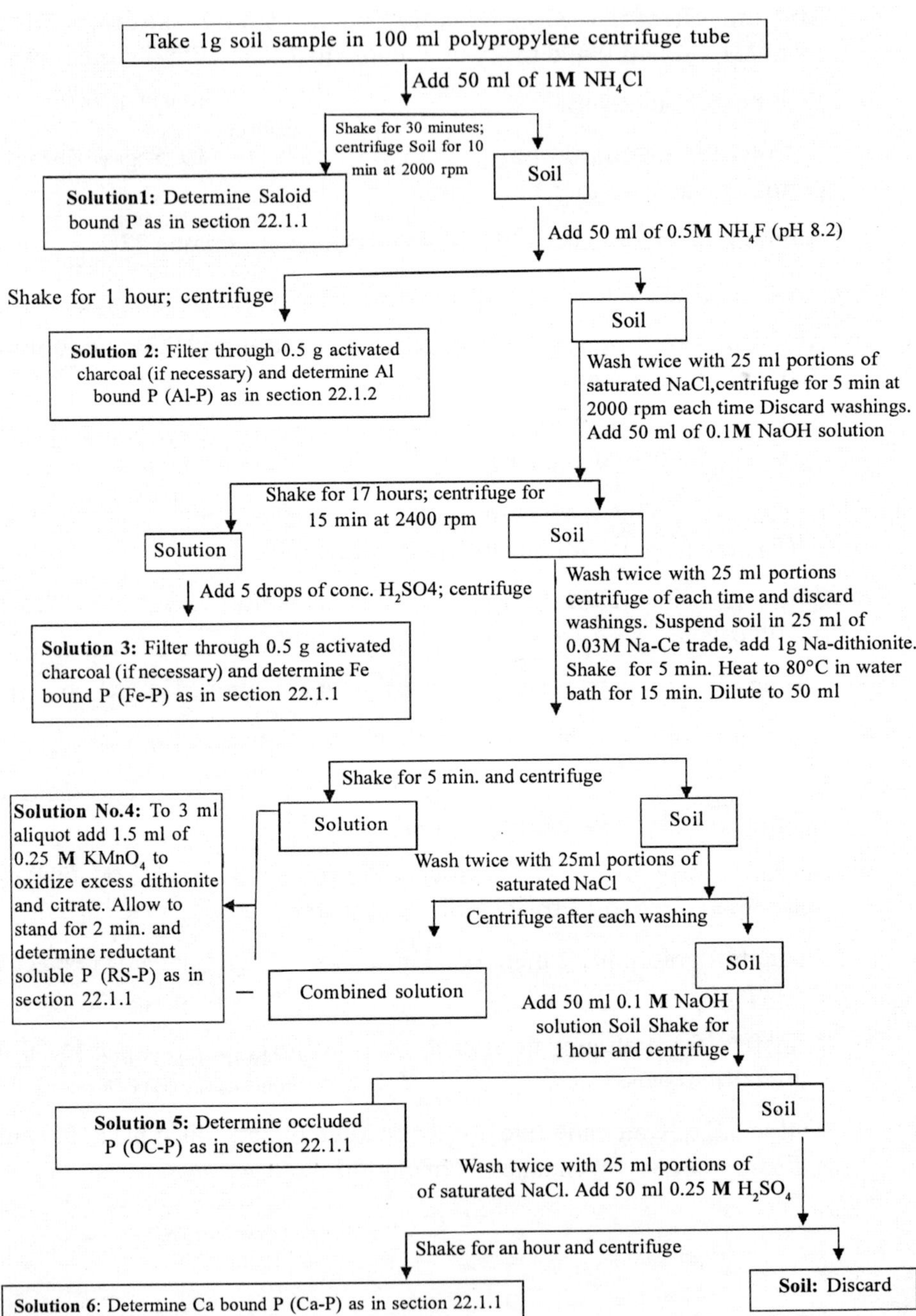
Take 1g soil sample in 100 ml polypropylene centrifuge tube
Add 50 ml of 1M NH_4Cl
Shake for 30 minutes; centrifuge Soil for 10 min at 2000 rpm
Soil
Solution1: Determine Saloid bound P as in section 22.1.1
Add 50 ml of 0.5M NH_4F (pH 8.2)
Shake for 1 hour; centrifuge
Soil
Solution 2: Filter through 0.5 g activated charcoal (if necessary) and determine Al bound P (Al-P) as in section 22.1.2
Wash twice with 25 ml portions of saturated NaCl,centrifuge for 5 min at 2000 rpm each time Discard washings. Add 50 ml of 0.1M NaOH solution
Shake for 17 hours; centrifuge for 15 min at 2400 rpm
Solution
Soil
Add 5 drops of conc. H_2SO4; centrifuge
Wash twice with 25 ml portions centrifuge of each time and discard washings. Suspend soil in 25 ml of 0.03M Na-Ce trade, add 1g Na-dithionite. Shake for 5 min. Heat to 80°C in water bath for 15 min. Dilute to 50 ml
Solution 3: Filter through 0.5 g activated charcoal (if necessary) and determine Fe bound P (Fe-P) as in section 22.1.1
Shake for 5 min. and centrifuge
Solution No.4: To 3 ml aliquot add 1.5 ml of 0.25 M $KMnO_4$ to oxidize excess dithionite and citrate. Allow to stand for 2 min. and determine reductant soluble P (RS-P) as in section 22.1.1
Solution
Soil
Wash twice with 25ml portions of saturated NaCl
Centrifuge after each washing
Soil
Combined solution
Add 50 ml 0.1 M NaOH solution Soil Shake for 1 hour and centrifuge
Soil
Solution 5: Determine occluded P (OC-P) as in section 22.1.1
Wash twice with 25 ml portions of of saturated NaCl. Add 50 ml 0.25 M H_2SO_4
Shake for an hour and centrifuge
Soil: Discard
Solution 6: Determine Ca bound P (Ca-P) as in section 22.1.1

22.4. Determination of Organic P in Soil

Calculation

Say,

Weight of the soil taken = W g

Volume of the extractant = V_D ml

Volume of the aliquot taken for colour development = V_A ml

Final volume of the coloured solution = V_C ml

Absorbance reading of the test solution = X

Concentration of P corresponding to X absorbance (from standard curve) = A ppm

Concentration of **soil P, ppm** = $A \times \frac{V_C}{V_A} \times \frac{V_D}{W}$

22.4. Determination of Organic P in Soil

Organic P of soil can be extracted and determined by several methods, but these have been classified into two general types: (a) alkaline extraction in NaOH or NH_4OH after acid pretreatment and (b) dilute acid extraction after oxidation of organic matter by H_2O_2 or ignition. The difference between the P content in the extract after and before oxidation of organic matter represents organic P.

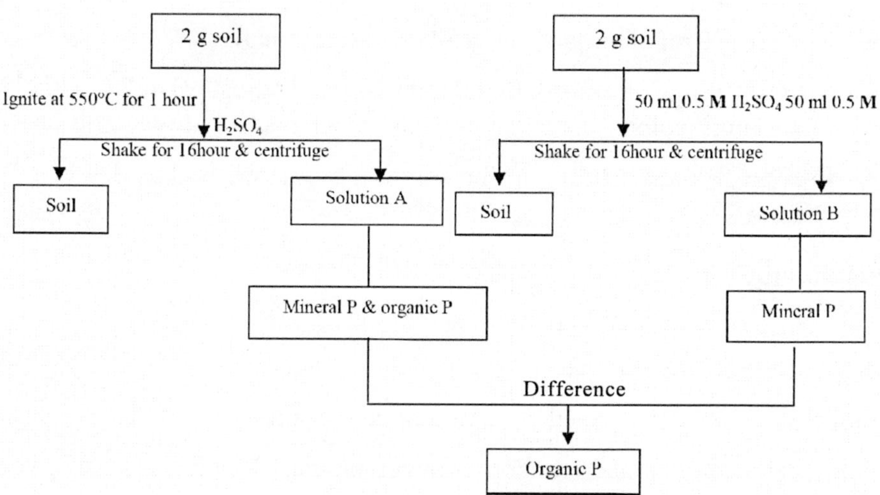

Equipments and Materials

Colorimeter or spectrophotometer, muffle furnace, porcelain crucible, centrifuge, 100 ml centrifuge tube, 50 ml volumetric flasks, measuring cylinder, balance.

Reagents

- Sulphuric acid, 0.5**M**: Dilute 27.8 ml of concentrated sulphuric acid (H_2SO_4) to 1 litre with distilled water.
- Sodium hydroxide, 5**M**: Dissolve 200 g of sodium hydroxide (NaOH) and make up volume to 1 litre with distilled water.
- p-nitrophenol: As mentioned in the section 22.1.1.

Procedure

- Take 2.0 g air dry processed (passed through 0.5 mm sieve) soil sample in a porcelain crucible and place it in a cool muffle furnace.
- Increase the temperature of the furnace gradually to 550°C, and maintain for 1 hour.
- Allow to cool and transfer the ignited soil to a 100 ml centrifuge tube.
- Place 2.0 g of another fresh (unignited) sample in another 100 ml centrifuge tube.
- Add 50 ml of 0.5**M** H_2SO_4 to each centrifuge tube and shake for 16 hours.
- Centrifuge to obtain a clear solution.
- Transfer 10 ml of aliquot into a 50 ml volumetric flask.
- Add five drops of p-nitrophenol indicator and adjust pH at 5.0 with 5 **M** NaOH until colour just changes from colourless to yellow.
- Determine the P content in the solution by the method as described in section 22.1.1.

Calculation

Say,

Weight of the soil taken = W g

Volume of 0.5 **M** H_2SO_4 used for extraction = V_D ml

Volume of the aliquot taken for colour development = V_A ml

Final volume of the coloured solution = V_C ml

Absorbance reading of the test solution = X

Concentration of P corresponding to X absorbance (from standard curve) = A ppm

Concentration of P, ppm = $A \times \frac{V_C}{V_A} \times \frac{V_D}{W}$

Organic P in soil = Total P $_{ignited}$ – Total P $_{unignited}$

22.5 Determination of Phosphate Fixing Capacity of Soil

Principle

The processes whereby readily soluble plant nutrient are changed to less soluble forms by reacting with inorganic or organic components of the soil, with the result that the nutrients become restricted in their mobility in the soil and suffer a decrease in their availability to the plant is called fixation of plant nutrient. Phosphorus use efficiency is generally very low in all soils due to conversion of major part of applied P to unavailable form. The extent of P fixation varies from soil to soil depending on the amount and type of clay minerals present, pH, presence of oxides of Fe and Al, $CaCO_3$, etc. This leads to the necessity of making appropriate correction in the recommended doses of P fertilizer. The P fixing capacity can be determined as per procedure outlined by Waugh and Fitts (1966). In this method soil sample is treated with graded dosed of P and incubated for a period sufficient to allow maximum fixation. The amount of P fixed is then determined by extracting P from the soil and subtracting from the value of amount of P added.

Equipments and Materials

Colorimeter or spectrophotometer, mechanical shaker, 50 ml volumetric flasks, 100 ml conical flasks, measuring cylinder, pipette, Whatman No. 42 filter paper, balance, centrifuge, water bath.

Reagents

- Standard stock P solution (2500 ppm): Dissolve 4.720 g of anhydrous monocalcium phosphate [Ca $(H_2PO_4)_2$] in water and make up volume to 500 ml. This corresponds to 2500 ppm P solution.
- Working standard: Dilute 0, 1, 2, 3, 4, 5, 6, 8, 10 and 15 ml of 2500 ppm P stock in 50 ml volumetric flasks which correspond to 0, 50, 100, 150, 200, 250, 300, 400, 500 and 750 ppm of P solution, respectively.

Procedure

- Accurately weigh 2g of air dry soil sample in a series of 100 ml conical flasks.
- Add 1 ml of each of the working standard P solutions to different flasks. This will give 0, 25, 50, 75, 100, 125, 150, 200 and 275 ìg of P added per g of soil.
- Plug the conical flasks with cotton wool and allow incubating at room temperature for 96 hours.
- Extract the soil by Olsen's extractant (pH > 5.5) or Bray-I extractant (pH < 5.5) of the soil samples.
- Take 5 ml of clear filtrate and determine available P content as described in section 22.1.1 or 22.1.2.
- Plot the amount of P extracted along Y- axis against the amount of P added along X– axis. Find out the inflexion point on the curve where there is an abrupt increase in extractable P.
- From this inflexion point find out the corresponding value of X i.e., approximate amount of P required to be added to more than the crop removal to overcome the effect of fixation.
- Calculate the amount of P fixed as the difference between the P added and amount of P extracted.
- If the above curve is linear, consider the shape of the curve to work out the amount of P to be applied which would take into account the fixing capacity.

Calculation

Say,

Available P (extracted) content of the soil without P treatment, μg of P / g of soil = A

Available P (extracted) content of the P treated soil, μg of P / g of soil = B

Amount of P extracted out of added P, μg of P / g of soil = (B – A)

Amount of P added, μg of P / g of soil = C

Amount of P fixed by soil, μg of P / g of soil = C- (B - A)

$$\text{P fixation,\% of added P} = \frac{C-(B-A)}{C} \times 100$$

22.6 Measurement of Microbial Biomass phosphorus in Soil (Brookes *et al.* 1982)

Equipments and Materials

Moisture box, separatory funnel, glass vacuum desiccator and vacuum pump, shaker, glass beads, beaker (50 and 100 ml capacity), 250 ml conical flask,

Reagents

- Distiiled chloroform
- Concentrated sulphuric acid
- Plus reagents as in section 22.1.1

Procedure

- For each sample, weigh seven sets of 10 g soil.
- Use one set for moisture determination (Section 19.2.4).

A. Fumigation and Extraction

- Take two sets of soil sample in 250 ml conical flask and add 200 ml 0.5 **M** $NaHCO_3$ (pH 8.5). Shake for 30 minutes on a mechanical shaker and filter through Whatman No.42 filter paper.
- For fumigation follow the same steps with two sets of soil sample as were followed during determination of soil microbial biomass carbon (Section 19.2.4).
- Transfer fumigated soil samples in 250 ml conical flask for extraction with 0.5 **M** $NaHCO_3$ as mentioned earlier (for unfumigated soil sample).
- Measure P concentration in neutralised aliquot (10 or 5 ml) of $NaHCO_3$ extracts from both fumigated and unfumigated soil sample following the method as described in the section 22.1.1.

B. Correction of Inorganic P fixed during $NaHCO_3$ extraction

During extraction, some of the inorganic P in the solution can react with the soil colloids. This P (fixed against extraction by 0.5 **M** $NaHCO_3$) was estimated by measuring the recovery of a known amount of orthophosphate P added to the soil in the $NaHCO_3$ extract. 1.0 ml of a solution containing 250 μg P is added to 200 ml 0.5 **M** $NaHCO_3$ (in rest two sets of soil sample) and shaken under same condition as the fumigated and unfumigated samples.

$$\text{Recovery of P (\%)} = \frac{(C-A)}{25} \times 100$$

Where, A = µg inorganic P per g oven dry soil extracted from unfumigated soil by 0.5 **M** $NaHCO_3$

C = µg inorganic P per g oven dry soil extracted from unfumigated soil by 0.5 **M** $NaHCO_3$ spiked with 25 µgP per g soil.

Calculation

$$\text{Microbial biomass Phosphorus (MBP), in µg/g soil} = \frac{P_{EF} - P_{EUF}}{K_{EP} \times R_P}$$

Where, P_{EF} = Inorganic P in the fumigated soil extract, µg/g soil

P_{EUF} = Inorganic P in the unfumigated soil extract, µg/g soil

K_{EP} = Efficiency of extraction of microbial P from soil; usually the value 0.40 is considered as ideal

R_P = Recovery percentage

Questions and Hints

Q.1 What are the different forms of P present in soil?

Hint: Phosphorus is present in soil both in inorganic form and organic form. The inorganic P can be divided into different pools. The major active inorganic forms are: P bound to aluminium (Al-P), iron (Fe-P), calcium (Ca-P) and silicate minerals and relatively less active forms are occluded and reductant soluble form of P. On the other hand the principal organic P compounds present in soil are (i) inositol phosphate, (ii) phospholipids, (iii) nucleic acid and other unidentified esters and phosphoproteins.

Q.2 What are the factors that dictate the choice of available P method?

Hint: (a) Soil pH, (b) amount of Fe and Al compounds, (c) amount of calcite and dolomite and (iv) interfering substances present in the solution to be analyzed.

Q.3 Which forms of soil P are extracted by Bray's extractants?

Hint: Bray's extractants release P from various hydrated and hydroxyl phosphates of Al and Fe including any adsorbed or precipitated surface layers on oxides and aluminosilicates and also from active calcium phosphate.

Q.4 What is the basic principle of extraction of soil P by Olsen extractant?

Hint: See extraction of section 22.1.1.

Q.5 Why ascorbic acid method is preferred over stannous chloride method for colour development?

Hint: The advantages of ascorbic acid method over stannous chloride are the prolonged stability (about 24 hours) of the molybdenum blue colour and tolerance of high salt and Fe^{3+} (upto 2.5 ppm) concentration.

Q.6 Why activated charcoal or Darco G is used during extraction of soil P by Olsen extractant?

Hint: To adsorb the coloured soluble organic matter dissolved by the $NaHCO_3$ so that it would not interfere in the colour development for colorimetric determination of soil P.

Q.7 Which chemical form of soil P is extrcated by neutral fluoride solution?

Hint: Neutral fluoride solution extracts soil P mostly from $AlPO_4$ as in neutral or alkaline solution fluoride complex of Al forms, but that of Fe does not at any appreciable extent. Dissolution of $CaPO_4$ is also negligible.

Q.8 Why boric acid is used during estimation of available P by Bray's extractants?

Hint: To eliminate the interference of fluoride ions sometimes present in sandy acid soils during colour developemnt.If the concentration of fluoride ions is greater than 5 ppm, it may produce negative interference with molybdenum blue reaction in the determination of P. Possible role of boric acid is:

$$4F^- + H_3BO_3 + 3H^+ \longrightarrow (BF_4)^- + 3H_2O$$

Neither boric acid nor fluoroborate interferes in the colour development.

Q.9 Which form of soil P is extracted by citrate-dithionite extractant?

Hint: Citrate-dithionite extractant extracts reductant soluble P in soil. Citrate-dithionite extractant removes iron oxide by an effective reduction-chelation process, thus releases P occluded in the iron oxides i.e. reductant soluble P.

Q.10 What chemical principles are applied in the extraction of total organic P in soils?

Hint: (*See the section 22.4 page no 323*).

Q.11 If you are asked to determine the available P content of a soil, what will you do?

Hint: First determine the pH of the soil and then choose the appropriate method suitable for that soil.

Q.12 In which form P is taken up by the plant?

Hint: The primary ($H_2PO_4^-$) and secondary (HPO_4^{2-}) orthophosphate ions are most commonly taken up by the plants; of which primary orthophosphate ion is the preferred form.

23

Potassium in Soil

23.1 Forms of Soil K

Ignoring potassium (K) in soil organism, total K in soil may be classified as (i) water soluble K, (ii) readily exchangeable K (exchangeable K), (iii) slowly exchangeable K (non- exchangeable or fixed K) and (iv) mineral or lattice K. The different forms are in dynamic equilibrium with one another.

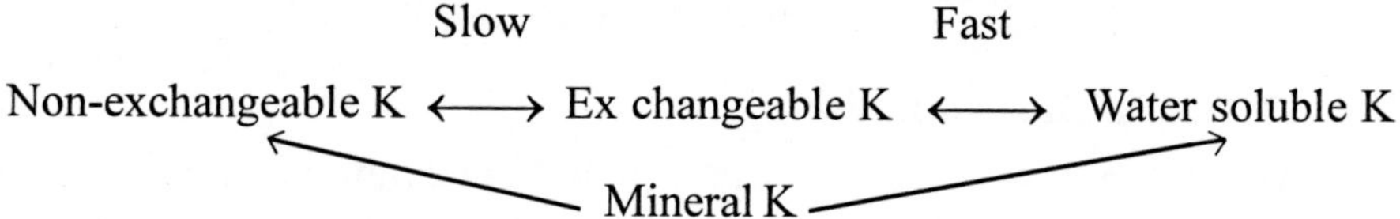

23.1.1 Determination of Water Soluble K in Soil (USSLS, 1954)

Potassium is present as soluble cation in soil solution. Its concentration is dependent on number of factors such as: type of clay, water content, intensity of leaching, the amount of exchangeable K, etc. The dilution of soil increases the concentration of water soluble K and drying decreases it still further. As this form of K is in fast equilibrium with exchangeable K, hence it is difficult to distinguish from the later.

Equipments and Materials

Flame photometer, mechanical shaker, balance, 100 ml conical flask, 50 ml volumetric flasks, graduated pipette, Whatman No. 42 filter paper.

Reagents

- Standard potassium chloride solution: Dissolve 1.907 g of A.R. grade potassium chloride (KCl) (dried at 60°C for 1 hour) and make up volume to 1 litre. It gives 1000 ppm K solution. It can be preserved as stock solution. Dilute it 10 times (dilute 100 ml of 1000 ppm K solution to 1 litre) to get 100 ppm K solution.

Procedure

Preparation of Standard Curve

- Pipette out 1, 2, 5, 8, 10 ml of 100 ppm K solution in a series of 50 ml volumetric flasks and make up volume to the mark with distilled water. These solutions correspond to 2, 4, 10, 16 and 20 ppm K solution, respectively.
- Adjust air and gas supply as per the direction in the flame photometer manual.
- Set K filter.
- Set zero and 100 scale of flame photometer with distilled water and highest concentration of standard K solution (here, 20 ppm).
- Note flame photometer readings against different standard K solutions and construct the standard curve plotting concentration (ppm K) along X axis (abscissa) and flame photo meter reading along Y axis (ordinate).
- Draw an average line passing through the origin.

Extraction and Determination

- Weigh 5 g air dry processed (passed through 2 mm sieve) soil sample in a 100 ml conical flask.
- Add 25 ml of distilled water to it and stopper the conical flask with rubber cork.
- Shake for 5 minutes on a mechanical shaker.
- Filter through Whatman No. 42 filter paper.
- Determine K concentration in the extract in the flame photometer.

Calculation

Say,

Weight of the soil taken = W g

Volume of water added for extraction = V_W ml

Concentration of K in the water extract obtained from standard curve = A ppm.

$$\text{So, Water soluble K in soil, ppm} = A \times \frac{V_W}{W}$$

23.1.2 Determination of Available K in Soil

Principle

The term available K includes both readily exchangeable K plus water soluble K. Except in recently fertilized soils, usually exchangeable K constitutes the major potion of available K which in turn constitutes about 1–2% of total K. Over the years researchers tested number of chemical extractant in different soils to simulate the amount of K extracted with the plant uptake. They opined that no single method could cover all the soils investigated by them. The common methods for extracting available K in soils are:

Extractant	Extractant Composition	Soil : Extractant ratio	Shaking time
Mehlich 1 (Mehlich, 1953)	0.05**M** H_2SO_4 + 0.05**M** HCl	1:5	5 minutes
Mehlich 3 (Mehlich, 1984)	0.2 **M** CH_3COOH + 0.25**M** NH_4NO_3 + 0.015**M** NH_4F + 0.013**M** HNO_3 + 0.001**M** EDTA	1:10	15 minutes
NH_4OAc (Hanway and Heidel, 1952).	1**N** NH_4OAc (pH 7.0)	1:5	5 minutes
Morgan (Morgan, 1941)	0.72**M** NaOAc + 0.52 **M** CH_3COOH	1:5	15 minutes
Modified Morgan (McIntosh, 1969)	0.62 **M** NH_4OH + 1.25**M** CH_3COOH	1:5	15 minutes
AB-DTPA (Soltanpour and Schwab, 1977)	1**M** NH_4HCO_3 + 0.005 **M** DTPA	1:2	15 minutes

Among the different extractants tried for this purpose, extraction with neutral normal ammonium acetate (NH_4OAc) is the most widely used method (Hanway and Heidel, 1952). Use of NH_4OAc is considered as standard because the level of exchangeable K in soils rises rapidly within an hour after extraction with H or Ca ions, but does not happen with NH_4OAc extraction. Since NH_4^+ holds highly charged layers of 2:1 layer silicates (particularly of vermiculite) together just as K, thus the release of non-exchangeable K from interlayer space to exchangeable form is retarded during NH_4OAc extraction.

Equipments and Materials

Flame photometer, mechanical shaker, balance, 100 ml conical flask, 50 ml volumetric flasks, graduated pipette, measuring cylinder, Whatman No. 1 filter paper.

Reagents

- Ammonium acetate, 1**N** (pH 7.0): Either dissolve 77.09 g of ammonium acetate (NH_4OOCCH_3) in about 800 ml of distilled water or dilute 57 ml glacial acetic acid (99.5%) to 800 ml distilled water containing 69 ml of concentrated ammonia solution and cool. Adjust pH of the solution at 7.0 with dilute acetic acid or ammonia solution and make up the volume to 1 litre.
- Standard potassium chloride solution: Prepare working standards of 2, 4, 10, 16 and 20 ppm K from 100 ppm K solution as in the section 23.1.1, but make up the volume with neutral normal ammonium acetate solution.

Procedure

Preparation of Standard Curve

- Prepare standard curve as described in the section 23.1.1 except during setting zero instead of using distilled water use neutral normal ammonium acetate.

Extraction and Determination

- Weigh 5 g air dry processed (passed through 2 mm sieve) soil sample in a 100 ml conical flask.
- Add 25 ml of neutral normal ammonium acetate to it and stopper the conical flask with rubber cork.
- Shake for 5 minutes on a mechanical shaker.
- Filter through Whatman No. 1filter paper.
- Determine K concentration in the NH_4OAc extract in the flame photometer.

Calculation

Say,

Weight of the soil taken = W g

Volume of extractant (NH_4OAc) added for extraction = V_W ml
Concentration of K in the NH_4OAc extract obtained from standard curve = A ppm.

So,1N NH4OAc extractable K or Available K in soil, ppm $= A \times \frac{V_W}{W}$

$$\text{Available K in soil, kg / ha} = A \times \frac{V_W}{W} \times 2.24$$

$$\text{Available } K_2O \text{ in soil, kg / ha} = A \times \frac{V_W}{W} \times 2.24 \times 1.20$$

[As % K x 1.20 = % K_2O]

Exchangeable K, ppm = (Available K, ppm – Water soluble K, ppm)

Rating

K Fertility Class	Available K_2O (kg / ha)
Low	< 150
Medium	150– 340
High	> 340

23.1.3 Determination of Non-exchangeable K in Soil (Wood and DeTurk, 1940)

Principle

Non-exchangeable or fixed K or reserve K or potentially available K is that fraction of soil K which is firmly bound by the soil and not immediately replaceable with neutral salt. The non-exchangeable K is bound between the basal planes of chiefly micaceous mineral.

Non-exchangeable K has been found to contribute appreciably towards K availability to crops. Usually this form of K constitutes 1-10% of the total K. The procedure for determination of non-exchangeable K involves extraction of soil with boiling 1**N** HNO_3 for 10 minutes. The amount of K extracted is found to be insensitive to soil: extractant ratio, but sensitive to time of boiling. The amount extracted by this acid treatment is found to correlate significantly with the amount removed by exhaustive cropping in many soils. The non-exchangeable K of soil is in fact the difference between the amount of K extracted by 1**N** HNO_3 and 1**N** NH_4OAc (pH 7.0).

Equipments and Materials

Flame photometer, hot plate, balance, 100 ml conical flask, 100 ml volumetric flasks, graduated pipette, measuring cylinder, Whatman No. 42 filter paper.

Reagents

- Nitric acid, 1**N**: Dilute 62.5 ml of concentrated nitric acid (HNO_3) to 1 litre with distilled water.
- Nitric acid, 0.1**N**: Dilute 100 ml of 1**N** nitric acid to 1 litre with distilled water

- Standard potassium chloride solution: From 1000 ppm K stock solution prepare 100 ppm K solution in 0.33**N** HNO_3 and then working standards of 2, 4, 10, 16 and 20 ppm K as in the section 23.1.1, but make up the volume with 0.33**N** HNO_3 solution.

Procedure

Preparation of Standard Curve

- Prepare standard curve as described in the section 23.1.1 except during setting zero instead of using distilled water use 0.33**N** HNO_3 solution.

Extraction and Determination

- Weigh 2.5 g air dry processed (passed through 2 mm sieve) soil sample in a 100 ml conical flask.
- Add 25 ml of 1**N** HNO_3 solution to it and heat to boil (220°C) the suspension for 10 minutes on a hot plate.
- After sufficiently cooling filter the suspension through Whatman No. 42 filter paper into a 100 ml volumetric flask.
- Wash the conical flask and the soil on the filter paper for 3–4 times with 0.1**N** HNO_3 solution and make up the volume with 0.1**N** HNO_3 solution.
- Determine K concentration in the acid extract in the flame photometer.

Calculation

Say,

Weight of the soil taken = W g

Volume of the filtrate is made up after filtration = V_W

Concentration of K in the acid extract obtained from standard curve = A ppm.

$$\text{So, 1N HNO3 extractable K in soil, ppm} = A \times \frac{V_W}{W} \times D$$

[D is the aliquot dilution factor of acid extract (if required) for flame photometric estimation]

Non-exchangeable K, ppm = (1N HNO_3 - K, ppm – 1N NH_4OAc-K, ppm)

23.1.4 Determination of Total K in Soil (Jackson, 1973)

Principle

As mineral K is the major contributor to the total soil K, soil containing more amount of K bearing primary minerals like mica (muscovite and biotite) and feldspars (orthoclase and microcline) and the clay minerals like illite results more total K in soils or minerals. Total K is best determined by decomposition of the sample by means of HF followed by estimation by emission spectrophotometric method.

Equipments and Materials

Flame photometer, sand bath, analytical balance, 30 ml platinum crucible with lid, 100 ml conical flask, 100 ml volumetric flasks, graduated pipette, measuring cylinder, Whatman No. 1 filter paper.

Reagents

- Hydrofluoric acid (HF), 48%
- Perchloric acid ($HClO_4$), 60%
- Hydrochloric acid, 6**N**: Dilute 490 ml of concentrated hydrochloric acid (HCl) with about 300 ml of water, cool the solution and make up volume to 1 litre.
- Standard potassium chloride solution: From 1000 ppm K stock solution prepare 100 ppm K solution and then working standards of 2, 4, 10, 16 and 20 ppm K as in the section 23.1.1, but make up the volume with 0.1**N** HCl solution

Procedure

a) Preparation of Standard Curve

- Prepare standard curve as described in the section 23.1.1 except during setting zero instead of using distilled water use 0.1**N** HCl solution.

b) Digestion and Determination of Total K

- Weigh 0.1000 g finely ground (passed through 0.16 mm sieve) dry soil sample (dry at 100°C for 2 hours or determine moisture content in a separate sample) in a 30 ml platinum crucible.
- Moist the sample with few drops of water.
- Then add 0.5 ml $HClO_4$ and 5 ml of 48% HF.
- Place the crucible covering 9/10th of the top with lid on a sand bath at a temperature of 200° to 225°C and evaporate the acid to dryness. [The solution must not boil vigorously or any spattering may occur].
- Ensure complete digestion of organic matter (no dark colour on the cover or sides of the crucible).

- Remove the crucible from the sand bath, cool and add 5 ml of 6**N** HCl.
- Dilute the suspension to 2/3rd of the volume of the crucible with water (15 ml).
- Cover the crucible and heat with relatively low flame for 5 minutes, so that the solution boils gently without bumping to dissolve the residues completely.
- Cool and transfer the solution of the crucible through a Whatman No. 1 filter paper to a 100 ml volumetric flask, cool and dilute to the mark with distilled water.
- Determine the K content in the digest after necessary dilution.

Calculation

Say,

Weight of the soil taken = W g

Volume of solution after digestion = V_W ml

Concentration of K in the final solution obtained from standard curve = A ppm.

$$\text{So, Total K in soil, ppm} = A \times \frac{V_W}{W} \times D$$

[**D** is the aliquot dilution factor of the acid extract for flame photometric estimation]

[*Note*: Platinum crucible is indispensible for many chemical analyses, but must be used with care for their expensiveness].

*Mineral K, ppm = (Total K, ppm – 1N HNO_3 K, ppm)

23.2 Determination of K Fixation Capacity of Soil (Jackson, 1973)

Principle

The fixation of K by clay minerals is due to the collapse of 2:1 micaceous lattice to approximate 10A^0 spacing. The nature of clay fraction is an important factor in the fixation of K. The intensity of fixation is zero with kaolinite, chlorite and micas; slight with montmorillonite; variable with illite according to their degree of alteration and strong with vermiculite. The importance of K fixation in nature lies with the fact that it not only regulates the supply of soil K to plants but also protects it against loss through leaching. The procedure for determining K fixation capacity of soil involves alternate wetting and drying of soil with sufficient amount of added K as suggested by N. J. Volk.

Equipments and Materials

Flame photometer, steam plate, balance, 100 ml conical flask, 100 ml volumetric flasks, measuring cylinder, Whatman No. 1 filter paper.

Reagents

- Standard potassium chloride solution: As in (*See section 23.1.2 page no. 312*)
- Ammonium acetate, 1**N** (pH 7.0): As in (*See section 23.1.2 page no. 312*)

Procedure

Preparation of Standard Curve

- Prepare standard curve as described in the (*See section 23.1.2 page no. 312*)

Extraction and Determination

- Weigh 10 g air dry processed soil sample in 100 ml conical flask.
- Add 10 ml of KCl solution containing 1000 ppm K (a"10 mg K addition).
- Evaporate the suspension to dryness on a steam plate and wet it with distilled water.
- Repeat the process of alternate wetting and drying for 10 times.
- Extract K with 1**N** NH_4OAc solution as in (*See section 23.1.2 page no. 312*) and determine the K concentration of the extract.
- Perform extraction of K with 1**N** NH_4OAc solution from a separate soil sample immediately after addition of KCl solution (without wetting-drying cycle) and determine K concentration in the extract.
- Calculate K fixation capacity as the difference in the K concentration between 2nd extract (without wetting-drying) and 1st extract (with wetting-drying) expressed in meq of K per 100 g of soil.

Calculation

Say,

Weight of the soil taken = W g

Volume of extractant (1**N** NH_4OAc) used for extraction = V_W ml

Concentration of K in the 1st extract (with wetting-drying) obtained from standard curve = A ppm.

Concentration of K in the 2nd extract (without wetting-drying) obtained from standard curve = B ppm.

$$\text{Amount of K fixed, ppm} = (B - A) \times \frac{V_W}{W}$$

$$\text{So, K fixation capacity, meq/100g of soil} = (B - A) \times \frac{V_W}{W} \times \frac{1}{10 \times 39.1}$$

[Equivalent weight of K = 39.1]

23.3 Determination of Quantity/ Intensity Relationship of K in Soil (Hesse, 1971)

The concept of quantity, intensity and buffering capacity proposed by Beckett (1964) are useful in describing and measuring the potassium supplying power of soils.The intensity factor (I) is a measure of K in soil solution i.e. immediately available for absorption by plant. Since this absorption of K is influenced by the activity of the other cations like Ca^{2+} and Mg^{2+} in the soil solution, the potassium activity ratio (AR_K) is used to indicate the intensity factor. K intensity in soil solution is thus a function of behavior of labile pool, the rate of release of fixed K and rate of diffusion and transport of K ions in the soil solution. The quantity factor (Q) is a measure of the capacity of the soil to maintain the level of K in soil solution over a long period or the duration of crop growth. The potential buffering capacity of soil K is reported to be an intrinsic property which is not usually affected by manuring and cropping. It indicates how the potassium level in the soil solution (I) varies with the amount of labile form (Q). In quantitative terms, the buffering capacity is expressed as the ratio of ΔQ/ΔI. The wider the ratio ΔQ/ΔI, more buffered is the soil.

Samples of soil are equilibrated with solution containing varying concentration of K in $CaCl_2$ solution of known strength in different soil: solution ratios for a given length of time followed by the determination of K, Ca and Mg in the filtrate. For each suspension the amount (± K in meq/100 g soil) by which the exchangeable K content has changed is calculated as the difference between the K concentration of the initial and final (equilibrated) solution. The activity ratio, $AR_K = a_K/\sqrt{(a_{Ca} + a_{Mg})}$ in the equilibrium solution is calculated after determining activity coefficient of the ion from the ionic strength using modified Debye-Huckel equation:

$$\log f_i = - AZ^2 \sqrt{I};\ a = M.f$$

Where, **f** is the activity coefficient, **A** is a constant, **Z** is the valence, **I** is the ionic strength and **a** is the activity of the ion, **M** is the molal concentration. Again, $I = \frac{1}{2}\Sigma M_i Z_i^2$.

The activity ratio is calculated by multiplying concentration ratio by activity coefficient ratio

$$AR_K = \frac{a_K}{\sqrt{a_{(Ca+Mg)}}} = \frac{[K]}{\sqrt{([Ca]+[Mg])}} \times \frac{f_K}{\sqrt{f_{(Ca+Mg)}}}$$

However, Tinker (1964) found that from 0.021 to 0.001 **M** $CaCl_2$ + $MgCl_2$ solution the activity coefficient ratio usually approximates to 1.10. If the soluble salts are low then [Ca] + [Mg] may be close to the original Ca concentration and very rarely changes with K level change.

Equipments and Materials

Flame photometer, Conductivity Bridge, balance, mechanical shaker, centrifuge or Whatman No. 1 filter paper, 100 ml conical flasks or centrifuge tubes, 100 ml volumetric flasks, pipette, burette

Reagents

- Calcium chloride solution, 0.002**M**: Dissolve 2.22 g of anhydrous calcium chloride ($CaCl_2$) and make volume to 10 litre.
- Standard potassium chloride solution, 1000 ppm K: Dissolve 1.9078 g of potassium chloride (KCl) and make up to 1 litre with 0.002 **M** $CaCl_2$ solution
- Working standard solution: Prepare 1 litre of each of 0, 5, 10, 20, 40 and 80 ppm K solution by diluting 1000 ppm K with 0.002 **M** $CaCl_2$
- Ammonium chloride – ammonium hydroxide buffer solution: As section 18.2.4.
- Standard EDTA (0.01**N**) solution: As section 18.2.4.
- Standard calcium chloride (0.01**N**) solution: As section 18.2.4.
- Hydroxylamine hydrochloride ($NH_2OH.HCl$), 5% (w/v): As section 18.2.4.
- Eriochrome black T(EBT) or solochrome black indicator: As section 18.2.4.
- Potassium (or sodium) cyanide solution: (1% w/v): As section 18.2.4.
- Potassium ferro cyanide (4%) solution: As section 18.2.4.

Procedure

- Weigh 6 soil samples of 5 g each in 100 ml conical flasks or centrifuge tubes and add 50 ml of 0, 5, 10, 20, 40 and 80 ppm K solution.

- Also weigh 0.5, 1.0, 2.0, 4.0 g of soil and add 50 ml of 0 ppm of K solution (0.002 **M** $CaCl_2$).
- Shake the contents for half an hour on a mechanical shaker and filter or centrifuge.
- Determine K content in the filtrate as well as in the original working standard solution as described in the section 23.1.1. Also determine Ca plus Mg content in the filtrate as described in the section 18.2.4.
- Express K concentration in the initial (original) and final (after equilibration) solutions in meq/litre (ppm/39.1). The difference between the K concentration in the initial and final solution in terms of meq/100 g or cmol/kg of soil will give ΔK values.
- Express the K and Ca plus Mg concentration in the final solutions in moles/litre. Then calculate activity ratio (AR) by using the formula for each concentration:

$$AR_K = \frac{[K]}{\sqrt{([Ca]+[Mg])}} \times 1.1$$

Or, calculate by calculating activity coeffeicent using Dibye - Huckel equation as shown below.

- Plot ± ΔK values in Y axis against AR_K values in X axis. The intercept of the Q/I curve on the AR_K axis gives the AR_eK (equilibrium activity ratio).
- Draw a tangent from the point on the Q/I curve where ΔK = 0 to get the K_0 (nonspecific bound K) value. The extrapolation of the lower curved portion of the Q/I curve into ÄK line is the K_L (labile K). The difference between K_L and K_0 is K_X (specific bound K).
- Calculate the potential buffering capacity for K (PBC_K) by dividing K_0 by AR_eK in terms of cmolkg^{-1}/(M/L)^{½}.

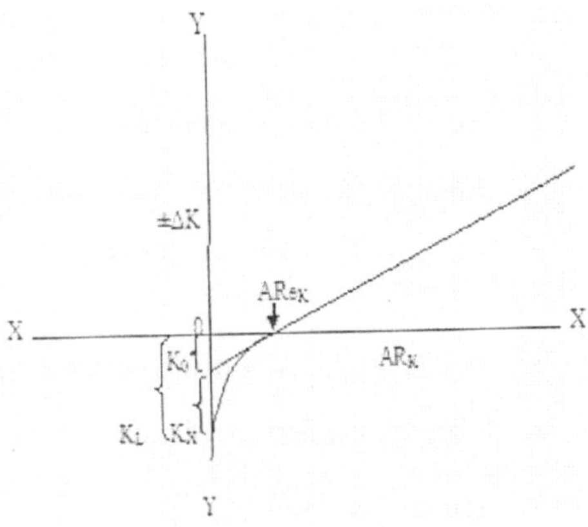

Fig. 23.1: A typical Q/I curve of K

Calculation

Say,

a) Calculation for ±ΔK

Weight of soil taken = W g

Volume of extractant used = V_T ml

Concentration of K in the initial solution = X_i ppm = X_i / 39.1 meq/L

[As ppm = mg /L]

Concentration of K in the final (equilibrated) solution = Y_i ppm = Y_i / 39.1 meq/L

$$\pm\Delta K = \frac{X_i}{39.1} - \frac{Y_i}{39.1} = \frac{X_i - Y_i}{39.1} = B\ meq / L$$

$$= B \times \frac{V_T}{1000} \times \frac{100}{W} \text{ meq / 100g or cmol / kg soil}$$

b) Calulation for a_K

K conc. in the final solution (Y_i ppm)	K conc (M /L)$M_i = Y_i$ /(1000x 39.1)	Ionic strength (I)I_i = ½$M_iZ_i^2$	Activity coefficient (f_K) f_{Ki}= -antilog ($AZ_i^2\sqrt{I_i}$)	Activity of K (a_K) $a_{Ki} = M_2 x f_{Ki}$

According to Dibye-Huckel equation A= 0.5042 at 20°C, Z_i for K = 1

c) Calulation for $a_{(Ca+Mg)}$

Volume of aliquot taken for Ca plus Mg determination = V_A ml

Strength of EDTA (N) = R

Burette reading (ml)P_i	Conc. of (Ca +Mg) (M /L)*M_i = $\frac{P_i \times R}{2\times1000}\times\frac{1000}{V_A}=\frac{P_i \times R}{2\times V_A}$	Ionic strength (I) $I_i = ½M_iZ_i^2$	Activity coefficient $f_{(Ca+Mg)}$ $f_{(Ca+Mg)i}$ = -antilog($AZ_i^2\sqrt{I_i}$)	Activity of (Ca+Mg) $a_{(Ca+Mg)i}$ = M_i x $f_{(Ca+Mg)i}$

*1 ml 1 N EDTA = 1 meq of (Ca+Mg); Z_i = 2 (both are divalent), A= 0.5042 at 20°C

d) Calculation of AR_K

$AR_K=a_K/\sqrt{a_{(Ca+Mg)}}$

Questions and Hints

Q 1. Why K is called a 'Master cation'?

Hint: This univalent cation is present in the largest concentration (100-200 mM) in the plant cell sap and regulates the activity of as many as 60 enzymes in plant, that's why it is called a master cation.

Q 2. In which form K presents in plant?

Hint: Potassium present as free (not bound to any constituent) and ionic (K^+) form in plant.

Q3. Why 1N NH_4OAc extraction is considered as standard for extraction of exchangeable form of soil K?

Hint: NH_4^+ holds highly charged vermiculite interlayers together just as K, thus the release of non-exchangeable K to exchangeable form is retarded during NH_4OAc extraction, while the level of exchangeable K rises rapidly within an hour after replacement with H^+ or Ca^{2+} due to release of K from interlayer spaces of 2:1 layer silicates.

Q 4. Give an account of the contribution of different forms of soil K towards total K.

Hint: Mineral K or lattice K constitutes the largest (92-98%) fraction of total soil K. About 1-10% of total K in soil is present as non-exchangeable form and 1-2% as readily available K. of this readily available K about 98% is constituted by exchangeable K and only 2% by solution K.

Q 5. Arrange clay minerals on the basis of K fixing power?

Hint: The K fixing power of the clay minerals usually follow the order: Vermiculite > Illite > Smectite > Kaolinite.

Q 6. Replenishment of soil solution K from the solid phase is of great practical significance-explain.

Hint: The total demand for K by crop is considerably higher than its presence in solution at any given time. For adequate nutrition, the concentration of K in soil solution must be maintained at a satisfactory level throughout the growth period through replenishment from solid phase (exchangeable phase).

Q 7. PBC_K of smectite dominant soil is higher than that of kaolinite soil - what comment can you make?

Hint: The smectite dominant soil will supply K to plant over a longer period than kaolinite dominant soil.

Q 8. (a)Arrange the texturally similar soils but dominated by different types of clay minerals on the basis of PBC_K.

(b) What is the general relationship between AR_eK and PBC_K among the different clay dominant soil?

Hint: (a) The PBC_K of the texturally similar soils but dominated by different types of clay minerals follow the general order: Smectite > Illite > Kaolinite.

(b) AR_eK and PBC_K among the different clay dominant soils are inversely related.

24

Sulphur in Soil

Sulphur (S) occurs in soil in both organic and inorganic forms. Inorganic S fraction may occurs as sulphate (SO_4^{2-}) and compounds of lower oxidation state such as sulphide (S^{2-}), polysulphide (S_n^{2-}), thiosulphate ($S_2O_3^{2-}$) and elemental sulphur (S^0). In well drained soils inorganic S usually present as sulphate. Under anaerobic condition (reduced or waterlogged soil) inorganic S mainly occur as sulphide and under extreme condition as elemental S form. Estimation of S is carried out in two steps:

a) Extraction of different forms of S from soil by suitable extractant.

b) The filtrate is then analysed for S by the turbidimetric method.

The turbidimetric method is the most widely used one because of its simplicity and rapidity.

24.1 Determination of Available S in Soil (Fox *et al.* 1964)

Principle

Plant absorbs sulphur exclusively as sulphate present in soil solution and a portion adsorbed on the exchange surfaces. This method is suitable for extraction of adsorbed plus soluble sulphate. In this method soil is shaken with calcium phosphate. The phosphate ions exchange for sulphate ions adsorbed on the colloidal surface. Calcium ions depress the extraction of soil organic matter, thus eliminating contamination from organically bound extractable S. The filtrate is then treated with $BaCl_2$ in presence of gum acacia which helps to prevent quick settling of sulphate as barium sulphate precipitate. The turbidity of the solution due to presence of barium sulphate precipitate is measured colorimetrically using blue filter or spectrophotometrically at 490 nm wavelength. This method is the most suitable for soils containing appreciable amount of adsorbed S and for acidic soils.

Soil colloid

$$\boxed{\text{Soil colloid}}SO_4 + Ca\,(H_2PO_4)_2 \rightleftharpoons \boxed{\text{Soil colloid}}\begin{matrix}H_2PO_4\\H_2PO_4\end{matrix} + SO_4^{2-} + Ca^{2+}$$

$$SO_4^{2-} + BaCl_2 \longrightarrow \underset{\text{(ppt)}}{BaSO_4} + 2Cl^-$$

Equipments and Materials

Colorimeter with blue filter or spectrophotometer, shaker, balance, 100 ml conical flask, 25 ml volumetric flasks, measuring cylinder, Whatman No. 42 filter paper, graduated pipette.

Reagents

- Monocalcium phosphate monohydrate solution, 500 mgL^{-1} P: Dissolve 2.03g of monocalcium phosphate [$Ca\,(H_2PO_4)_2.\,H_2O$] in distilled water and make up volume to 1 litre.
- Barium chloride: Grind barium chloride ($BaCl_2.\,2H_2O$) crystal in mortar and pass through 20 to 30 mesh sieve.
- Gum acacia solution, 0.25%: Dissolve 0.25 g gum acacia in distilled water and dilute to 100 ml.
- Standard S solution: Dissolve 1.087 g of AR grade potassium sulphate (K_2SO_4) in distilled water and make up volume to 1 litre. It gives 200 ppm standard stock solution of S. Dilute 10 times (dilute 100 ml of 200 ppm S solution to 1 litre) to get 20 ppm working standard solution of S.

Procedure

Preparation of Standard Curve

- Pipette out 0, 2.5, 5, 7.5 and 10 ml of 20 ppm S solution in a series of 25 ml volumetric flasks.
- Add 10 ml of extracting solution (monocalcium phosphate solution) and 1 g of $BaCl_2$ powder to each flask.
- Shake the flask for 1 minute.
- Add 1 ml of 0.25% gum acacia, make up the volume to the mark with distilled water and again shake for 1 minute. These solutions correspond to 0, 2, 4, 6 and 8 ppm S solution, respectively.
- Measure the turbidity after 1 to 3 minutes in the colorimeter with blue filter or spectrophotometer at 490 nm wavelength. Complete measurement of turbidity by 10 minutes.

- Prepare a standard curve by plotting concentration along X axis (abscissa) and absorbance along Y axis (ordinate).

Extraction of Available S from Soil and Estimation

- Weigh some amount (approximately 10 g dry weight) of fresh field soil sample in 100 ml conical flask. Determine moisture content in a separate sample at 105°C for 24 hours in an oven.
- Add 50 ml of monocalcium phosphate solution to it and shake for 30 minutes on a reciprocating shaker.
- Filter the suspension through Whatman No. 42 filter paper.
- Pipette out 10 ml aliquot in a 25 ml volumetric flask and proceed as described in the preparation of standard curve.

Calculation

Say,

Weight of the oven dry soil taken = W g

Volume of extractant (monocalcium phosphate solution) = V_C ml

Volume of aliquot taken for turbidity development = V_A ml

Final volume of the turbid solution = V_F ml

Concentration of S in the final solution obtained from standard curve = A ppm.

$$\text{Available S in soil, ppm} = A \times \frac{V_F}{V_A} \times \frac{V_c}{W}$$

$$\text{Available S in soil, kg/ha} = A \times \frac{V_F}{V_A} \times \frac{V_c}{W} \times 2.24$$

Rating

S Fertility Class	Available S (ppm)
Low	< 10
Medium	10 – 15
High	> 15

24.2 Forms of Sulphur in Soil

Sulphur (S) occurs in soil in both organic and inorganic forms. Most of the soil sulphur is present in organic forms. Inorganic S fraction may occur as sulphate and non-sulphate S comprised of compounds of lower oxidation state such as sulphide, polysulphide, thiosulphate and even as elemental sulphur.

24.2.1 Determination of Total S Content in Soil (Arkley, 1961)

Principle

The method available for accurate determination of total S in soil involves two steps: (i) conversion of the various S compounds in the soils to one form, either by oxidation to sulphate (dry or wet procedure) or by reduction to sulphide and (ii) determination of the sulphate or sulphide produced. Conversion to sulphate is more common than to sulphide.

Digestion with perchloric acid or a mixture of perchloric acid and nitric acid has been to be found the most popular but perchloric acid digestion may not decompose all the mineral constituents of some soils. In all perchloric acid digestion procedures, predigestion of soils high in organic matter with nitric acid before addition of perchloric acid is necessary to avoid the danger of explosion and fire. Arkley (1961) recommended an acid oxidation procedure involving nitric acid, perchloric acid, orthophosphoric acid and hydrochloric acid. He claims that inclusion of phosphoric acid with nitric acid plus perchloric acid in the digestion mixture increased ease of handling and decreased the amount of perchloric acid consumption.

Equipments and Materials

Spectrophotometer or colorimeter with blue filter, hot plate, hot water bath, watch glass, volumetric flask (250, 100 and 50 ml capacity), and single mark pipettes (5, 10, 15, 20 and 25 ml).

Reagents

- Concentrated nitric acid (HNO_3), 69%
- Concentrated perchloric acid ($HClO_4$), 60%
- Concentrated orthophosphoric acid (H_3PO_4), 85%
- Concentrated hydrochloric acid (HCl), 37%
- Hydrochloric acid, 1**N**: Dilute 83 ml of concentrated HCl to about 800 of water and make up volume to 1 litre.
- Hydrochloric acid, 6**N**: Dilute 490 ml of concentrated HCl to about 800 of water and make up volume to 1 litre.
- Salt buffer solution: Dissolve 50 g magnesium chloride ($MgCl_2.2H_2O$), 4.1 g potassium nitrate (KNO_3) and 28 ml ethanol per litre.
- Barium chloride: As in section 24.1.
- Gum acacia solution, 0.25%: As in section 24.1.
- Standard S solution: As in section 24.1.

Procedure

a) Digestion

- Place 2.0 g of finely ground (< 40 mesh) soil sample in a 100 ml beaker.
- Add 3 ml of concentrated HNO_3, swirl the beaker to mix the content, cover the beaker with watch glass and heat on a water bath for 1 hour.
- Remove the beaker from the water bath, uncover and add 3 ml of concentrated $HClO_4$ and 7 ml of H_3PO_4 (these acids may be premixed for ease of handling).
- Heat the beaker on a hot plate at 190 to 210°C until dense white fume of $HClO_4$ is visible.
- Place the watch glass on the beaker and continue heating for 30 minutes.
- Remove the beaker from hot plate, cool, uncover and add 2 ml of concentrated HCl.
- Heat again until white fume of $HClO_4$ is again visible.
- Transfer the digest quantitatively into 100 ml volumetric flask and make up volume with 1 **N** HCl.

b) Determination

(i) Preparation of Standard Curve

- Pipette 0, 5, 10, 15, 20 and 25 ml aliquot of 200 ppm standard S solution into 250 ml volumetric flasks.
- Add 25 ml salt buffer to each flask, make up volume and mix thoroughly. The resulting solutions are of 0, 4, 8, 12, 16 and 20 ppm S solution, respectively.
- Pipette out 10 ml each of above solution into 50 ml volumetric flasks.
- Add 1.0 ml 6 **N** HCl and 2 ml of 0.25% gum acacia. Mix the content by swirling and add 0.5 g $BaCl_2$ and make up volume to the mark.
- Allow to stand for one minute and read the absorbance of the turbidity at 490 nm wavelength.
- Prepare a standard curve by plotting concentrations (0, 0.8, 1.6, 2.4, 3.2 and 4.0 ppm) along X axis (abscissa) and absorbance along Y axis (ordinate).

(ii) Estimation of S in the Digest

- Take 10 ml of aliquot after proper dilution of the digest and follow the steps as described in standard curve preparation.

Calculation

Say,

Weight of the oven dry soil taken = W g

Volume of the digest = V_D ml

Further dilution required for getting required S concentration = D

Volume of the diluted digest taken for salt buffer treatment = V_B ml

Volume of the diluted digest after salt buffer treatment = V_C ml

Volume of aliquot taken for turbidity development = V_A ml

Final volume of the turbid solution = V_F ml

Concentration of S in the final solution obtained from standard curve = A ppm.

$$\text{Total S in soil,ppm} = A \times \frac{V_F}{V_A} \times \frac{V_C}{V_B} \times \frac{V_D}{W} \times D$$

$$\text{Total S in soil,kg / ha} = A \times \frac{V_F}{V_A} \times \frac{V_C}{V_B} \times \frac{V_D}{W} \times D \times 2.24$$

24.2.2 Determination of Organic S Content in Soil (Evans and Rost, 1945)

Principle

The method involving the determination of organic sulphur content of soil is to oxidize the organic matter with hydrogen peroxide. During oxidation the organically bound sulphur transforms into sulphate form which is measured following the procedure for sulphate estimation.

Equipments and Materials

Same as in (*See section 24.1.Page no 347*)

Reagents

- Hydrochloric acid, 1%: Dilute 13.5 ml of concentrated hydrochloric acid to 500 ml with distilled water.

- Hydrogen peroxide, 30%
- Barium chloride: (*See section 24.1.Page no 347*)
- Gum acacia solution, 0.25%: (*See section 24.1.Page no 347*)
- Standard S solution: (*See section 24.1.Page no 347*)

Procedure

- Leach 5-10 g (depending on the amount of organic matter) air dry soil sample through Whatman.No.42 filter paper with distilled water.
- Then leach with 1% HCl and finally with distilled water until no chloride present in the leachate.
- Transfer the soil to a beaker and oxidize with 30% H_2O_2 as described in the section 2.1.1.
- Filter the suspension through Chamberland filter tube to remove the suspended clay.
- Pipette a suitable aliquot (containing 5 ppm S when diluted to 25 ml) in 25 ml volumetric flask.
- Add 1 g $BaCl_2$, mix thoroughly and add 1 to 2 ml of gum acacia (1 ml if sulphate content is < 20 ppm and 2 ml if sulphate content is > 20 ppm).
- Shake for 1 minute and make up volume.
- Read the absorbance of the turbidity at 490 nm wavelength or in colorimeter with blue filter between 3 to 10 minutes.

Calculation

Say,

Weight of the oven dry soil taken = W g

Volume of filtrate after clay removal = V_D ml

Further dilution required for getting required S concentration = D

Volume of aliquot taken for turbidity development = V_A ml

Final volume of the turbid solution = V_F ml

Concentration of S in the final solution obtained from standard curve = A ppm.

$$\text{Organic S in soil,ppm} = A \times \frac{V_F}{V_A} \times \frac{V_D}{W} \times D$$

$$\text{Organic S in soil, kg / ha} = A \times \frac{V_F}{V_A} \times \frac{V_D}{W} \times D \times 2.24$$

*Inorganic S = Total S – Organic S

24.2.3 Determination of 0.15% $CaCl_2$ Soluble Sulphate S in Soil (Williams and Steinbergs, 1959)

Principle

The method involves the estimation of 0.15% $CaCl_2$ soluble sulphate S. The soil extract is treated with $BaCl_2$ in presence of a stabilizing solution which prevents settling of barium sulphate precipitate. The turbidity of the solution due to presence of barium sulphate precipitate is measured colorimetrically using blue filter or spectrophotometrically at 340 nm wavelength. This method is the most suitable for coarse textured soil.

Equipments and Materials

Same as in section 24.1.

Reagents

- Calcium chloride solution, 0.15%: Dissolve 1.5 g of calcium chloride ($CaCl_2 . 2H_2O$) in about 800 ml of distilled water and make up volume to 1 litre.
- Barium chloride: Same as in section 24.1.
- Stabilizing solution: Dissolve 75 g sodium chloride (NaCl) in 250 ml of distilled water in a 500 ml of volumetric flask. Stir well and add 30 ml glycerol. Continue stirring to dissolve NaCl and make volume up to the mark with distilled water.
- Standard S solution: Same as in (*See section 24.1.Page no 347*).

Procedure

Preparation of Standard Curve

- Pipette out 0, 2.5, 5, 7.5 and 10 ml of 20 ppm S solution in a series of 25 ml volumetric flask.
- Add 10 ml of extracting solution (0.15% $CaCl_2$ solution) to each flask.
- Add 2.5 ml of stabilizing solution and 0.2 to 0.3 g of $BaCl_2$ powder.
- Shake the flask for 1 minute.

- Make up the volume to the mark with distilled water and again shake for 1 minute. These solutions correspond to 0,2,4,6 and 8 ppm S solution, respectively.
- Measure the turbidity after 1 to 3 minutes in the colorimeter with blue filter or in spectrophotometer at 340 nm wavelength. Complete measurement of turbidity by 10 minutes.
- Prepare a standard curve by plotting concentration along X axis (abscissa) and absorbance along Y axis (ordinate)

Extraction of 0.15% $CaCl_2$ Soluble SO_4^{2-}- S and Estimation

- Weigh some amount (approximately 5 g dry soil) of fresh field soil sample in 100 ml conical flask. Determine moisture content in a separate sample at 105°C for 24 hours in an oven.
- Add 50 ml of 0.15% $CaCl_2$ solution to it and shake for 30 minutes on a reciprocating shaker.
- Filter the suspension through Whatman No. 42 filter paper.
- Pipette out 10 ml aliquot in a 25 ml volumetric flask and proceed as described in the preparation of standard curve.

Calculation

Say,

Weight of the oven dry soil taken = W g

Volume of extractant (0.15% $CaCl_2$) = V_C ml

Volume of aliquot taken for turbidity development = V_A ml

Final volume of the turbid solution = V_F ml

Concentration of S in the final solution obtained from standard curve = A ppm.

$$\text{Soluble Sulphate S in soil, ppm} = A \times \frac{V_F}{V_A} \times \frac{V_C}{W}$$

$$\text{Soluble Sulphate S in soil, kg / ha} = A \times \frac{V_F}{V_A} \times \frac{V_C}{W} \times 2.24$$

*Adsorbed sulphate S = Ca $(H_2PO_4)_2$ extractable S – 0.15% $CaCl_2$ extractable S

** Non-sulphate S = (Total S - Organic S) - Ca $(H_2PO_4)_2$ extractable S

24.3 Measurement of Microbial Sulphur in Soil (Saggar *et al.* 1981)

The biomass sulphur estimation technique discussed here used bacteria and fungal mycelium only. Fungal spores, yeasts and microfauna known to be present in soils are not included.

Equipments and Materials

Moisture box, separatory funnel, glass vacuum desiccator and vacuum pump, shaker, Spectrophotometer or colorimeter with blue filter, hot plate, hot water bath, glass beads, beaker (50 and 100 ml capacity), 250 ml conical flask, watch glass, volumetric flask (250, 100 and 50 ml capacity), pipette.

Reagents

- Distilled chloroform
- Concentrated sulphuric acid
- Calcium chloride solution, 0.01 **M**: Dissolve 1.47 g of calcium chloride ($CaCl_2 . 2H_2O$) in about 800 ml of distilled water and make up volume to 1 litre.
- Plus reagents as in (*See section 24.1.Page no 347*) .

Procedure

- For fumigation follow the same steps as were followed during measurement of soil microbial biomass carbon (Section 19.2.4).
- Extract both fumigated and unfumigated soil samples with 0.01 **M** $CaCl_2$ (1:5 soil: extractant ratio) by shaking for 1 hour.
- Filter through a 0.45 µm Millipore filter and analysed for total S as mentioned in (*See section 24.2.1 Page no 350*)

Calculation

$$\text{Microbial biomass sulphur (MBS), in } \mu g/g \text{ soil} = \frac{S_{EF} - S_{EUF}}{K_{ES}}$$

Where, S_{EF} = Total S released from the fumigated soil extract, µg/g soil

S_{EUF} = Total S released from the unfumigated soil extract,

K_{ES} = Efficiency of extraction of microbial S from soil; usually the value 0.35 is considered as ideal (considering 25% of the microbial population is bacteria and 75% fungi)

Questions and Hints

Q.1 Which method is most suitable for estimating available S of soil containing appreciable amount of organic matter?

Hint: Heat soluble sulphate method. The method involves gentle hydrolysis, as a result of special heat treatment (sequential wet and dry heating) and extraction with 1.0% NaCl, inorganic sulphate and a portion of organically bound sulphate S are quantified.

Q.2 What is the role of buffer salt solution in total S determination?

Hint: The precipitation of $BaSO_4$ and the turbidity is a sensitive process; hence it is necessary to standardize condition which should be strictly adhered to. Barium sulphate tends to precipitate if solution is not properly acidified. Again, if the solution is too acidic, the formation of precipitate of $BaSO_4$ will be low. Buffer salt solution prevents the wide fluctuation of pH of the medium.

Q.3 Why field moist sample are used for determination of available soil S?

Hint: Air drying of soil causes conversion of lower oxidation state S compounds to higher oxidation state S (SO_4^{2-}), thus increases the amount of extractable sulphate in soil.

Q.4 What is the function of gum-acacia or NaCl plus glycerol solution in S determination?

Hint: Gum-acacia or NaCl plus glycerol solution prevents the quick settling down of $BaSO_4$ precipitate formed on reaction of soluble SO_4^{2-} ions with $BaCl_2$ and turbidity of solution is maintained for its determination.

Q.5 What are the forms the term available soil S includes?

Hint: Sulphate present in soil solution plus a portion of adsorbed sulphate on the colloidal surface are primarily considered as available pool of S.

25

Micronutrients in Soil

25.1 Determination of Available Micronutrient cations (Zn, Cu, Fe and Mn) in Soil

The available micronutrient cations (Zn, Cu, Fe and Mn) can be estimated in a single extraction with diethylene triamine pentaacetic acid (DTPA) as proposed by Lindsay and Norvell (1978). DTPA has the excellent property to combine with free metal ions in the solution forming soluble complexes. The stability constants of metal-DTPA complexes make DTPA as a most suitable extractant for simultaneous complex formation with Zn, Cu, Fe and Mn. In calcareous soil DTPA extractant may result overestimation due to excess dissolution of $CaCO_3$, causing release of occluded micronutrient cations which are not plant available. To avoid this excessive dissolution of $CaCO_3$, DTPA is buffered at pH 7.3 using triethanol amine (TEA). At this pH major part (3/4th) of TEA is present as $HTEA^+$ (protonation) and exchanges for Ca and Mg ion from the soil exchange site. This increases concentration of Ca and Mg in the solution phase and suppresses dissolution of $CaCO_3$. The DTPA has the capacity to form complex with each of the micronutrient cations 10 times of its atomic weight and it ranges from 550 to 650 ppm depending on the micronutrient cation.

The amount of cation in the extract is determined on an atomic absorption spectrophotometer (AAS). The experimental condition such as shaking time, DTPA concentration, pH and temperature during shaking influence the amount of Zn, Cu, Fe and Mn extracted. Increase in shaking time and temperature markedly influence the extractability of these cations.

Equipments and Materials

AAS with hollow cathode lamp for respective elements or ICP-AES, analytical balance, shaker, Whatman No. 42 filter paper, 20 ml one mark pipette, graduated pipette, 50 and 100 ml volumetric flasks, 100 ml conical flask.

Reagents

- Extracting solution, 0.005**M** DTPA + 0.01 **M** $CaCl_2.2H_2O$ + 0.1**M** TEA (pH 7.3): Dissolve 1.967 g of DTPA and 1.470 g of $CaCl_2.2H_2O$ in about 100 ml of double distilled water. Add 13.3 ml of triethanol amine (TEA), mix well and make volume to about 900 ml with double distilled water. Adjust pH at 7.3 using dilute HCl and finally make up volume to 1 litre.
- Standard stock solutions

For Zn: Weigh 1.0 g of AR grade pure Zn metal and dissolve in about 10 ml of dilute HCl (1:1) and make up volume to 1 litre with double distilled water. It gives 1000 ppm standard Zn solution. Dilute 5 ml of 1000 ppm Zn solution to 100 ml to get 50 ppm standard Zn solution. Pipette 0, 0.5, 1.0, 1.5, 2.0, 2.5 and 5.0 ml of 50 ppm Zn stock solution to a series of 50 ml volumetric flask and dilute to the mark with DTPA extracting solution. The working standard solutions are now corresponding to 0, 0.5, 1.0, 1.5, 2.0, 2.5 and 5.0 ppm Zn solution.

For Cu: Weigh 1.0 g of AR grade pure Cu metal and dissolve in 50 ml of dilute HNO_3 (1:1) and make up volume to 1 litre with double distilled water. It gives 1000 ppm Cu solution. Dilute 5 ml of 1000 ppm Cu solution to 100 ml to get 50 ppm standard Cu solution. Pipette 0, 0.25, 0.5, 1.0, 1.5, 2.0 and 2.5 ml of 50 ppm Cu stock solution to a series of 50 ml volumetric flask and dilute to the mark with DTPA extracting solution. The working standard solutions are now corresponding to 0, 0.25, 0.5, 1.0, 1.5, 2.0 and 2.5 ppm Cu solution.

For Fe: Weigh 1.0 g of AR grade pure Fe metal and dissolve in 50 ml of dilute HNO_3 (1:1) and make up volume to 1 litre with double distilled water. It gives 1000 ppm Fe solution. Dilute 50 ml of 1000 ppm Fe solution to 500 ml to get 100 ppm standard Fe solution. Pipette 0, 0.5, 1.0, 1.5, 2.5 and 5.0 ml of 100 ppm Fe stock solution to a series of 50 ml volumetric flask and dilute to the mark with DTPA extracting solution. The working standard solutions are now corresponding to 0, 1.0, 2.0, 3.0, 5.0 and 10.0 ppm Fe solution.

For Mn: Weigh 1.583 g of AR grade MnO_2 or 1.0 g of AR grade pure Mn metal and dissolve in 50 ml of dilute HNO_3(1:1) and make up volume to 1 litre with double distilled water. It gives 1000 ppm Mn solution. Dilute 5 ml of 1000 ppm Mn solution to 100 ml to get 50 ppm standard Mn solution. Pipette 0, 0.5, 1.0, 1.5, 2.0, 2.5 and 5.0 ml of 50 ppm Mn stock solution to a series of 50 ml volumetric flask and dilute to the mark with DTPA extracting solution. The working standard solutions are now corresponding to 0, 0.5, 1.0, 1.5, 2.0, 2.5 and 5.0 ppm Mn solution.

Procedure

a) Preparation of Standard Curve

- Record the absorbance reading of the working standard solutions of the element concerned after setting zero with the blank.
- Draw the standard curve plotting absorbance readings along Y axis and respective concentrations (ppm) along X axis.

b) Extraction from soil and Estimation

- Weigh accurately 10 g air dry processed soil sample in a 100 ml conical flask and add 20 ml of DTPA extracting solution.
- Stopper the flask and shake on a mechanical shaker for exactly 2 hours.
- Filter the suspension through Whatman No. 42 filter paper.
- Measure the concentration of the element in the extract after necessary aliquot dilution on AAS (see section 16.5.2.2).

Calculation

Say,

Weigh of soil sample = W g

Volume of DTPA extractant added = V_C ml

Aliquot dilution (if required) = D

Absorbance reading of the final solution = A

Concentration of the element in the final solution against the absorbance, A = B ppm

Available micronutrient in the soil, ppm $= B \times D \times \frac{V_C}{W}$

25.2 Fractionation of Micronutrient Cations in Soil (Shuman, 1985)

Micronutrient cations (Zn, Cu, Fe and Mn) occur in soil in different chemical pools namely water soluble, exchangeable, sorbed fractions including manganese oxide, crystalline and amorphous iron oxide, carbonate, sulphide, organically bound and residual fraction present in sand, silt and clay particles, etc. The different fractions can be sequentially extracted from the soil by different specific extractants as given in the following flow chart.

Equipments and Materials

AAS with hollow cathode lamp for respective element or ICP-AES, centrifuge, mechanical shaker, hot water bath, 250 ml centrifuge bottle

Reagents

- Magnesium nitrate, 1**M** (pH 7.0): Dissolve 256 g of magnesium nitrate $[Mg(NO_3)_2.\ 6H_2O]$ in 800 ml of water and adjust pH at 7.0 with $Mg(OH)_2$ or HNO_3. Make up volume to 1 litre
- Sodium hypochlorite, 0.7**M** (pH 8.5): Dissolve 115.1 g of sodium hypochlorite ($NaOCl.\ 5H_2O$) in about 800 ml of double distilled water. Adjust pH at 8.5 just before use with NaOH or HCl and dilute to 1 litre.
- Hydroxyl amine hydrochloride, 0.1 **M** (pH 2): Dissolve 6.95 g of hydroxyl amine hydrochloride ($NH_2OH.HCl$) and adjust pH with NH_4OH and HCl. Make up volume to 1 litre.
- Acidified ammonium oxalate, 0.2 **M** (pH 3.0): Dissove 28.4 g of ammonium oxalate $[(NH_4)_2C_2O_4.H_2O]$ and 25.2 g of oxalic acid ($H_2C_2O_4.\ 2H_2O$) in about 800 ml of double distilled water. Adjust to pH 3.0 using NH_4OH or oxalic acid and make up volume to 1 litre.
- Ascorbic acid, 0.1**M**: Dilute 17.6 g of ascorbic acid ($C_6H_8O_6$) to 1 litre with distilled water.
- Sodium pyrophosphate, 0.11**M:** Disolve 49.06 g of Sodium pyrophosphate ($Na_4P_2O_7.10H_2O$) in distilled water and make up the volume to 1 litre.

Procedure

Micronutrient fractions present in soil separates is determined through following steps:

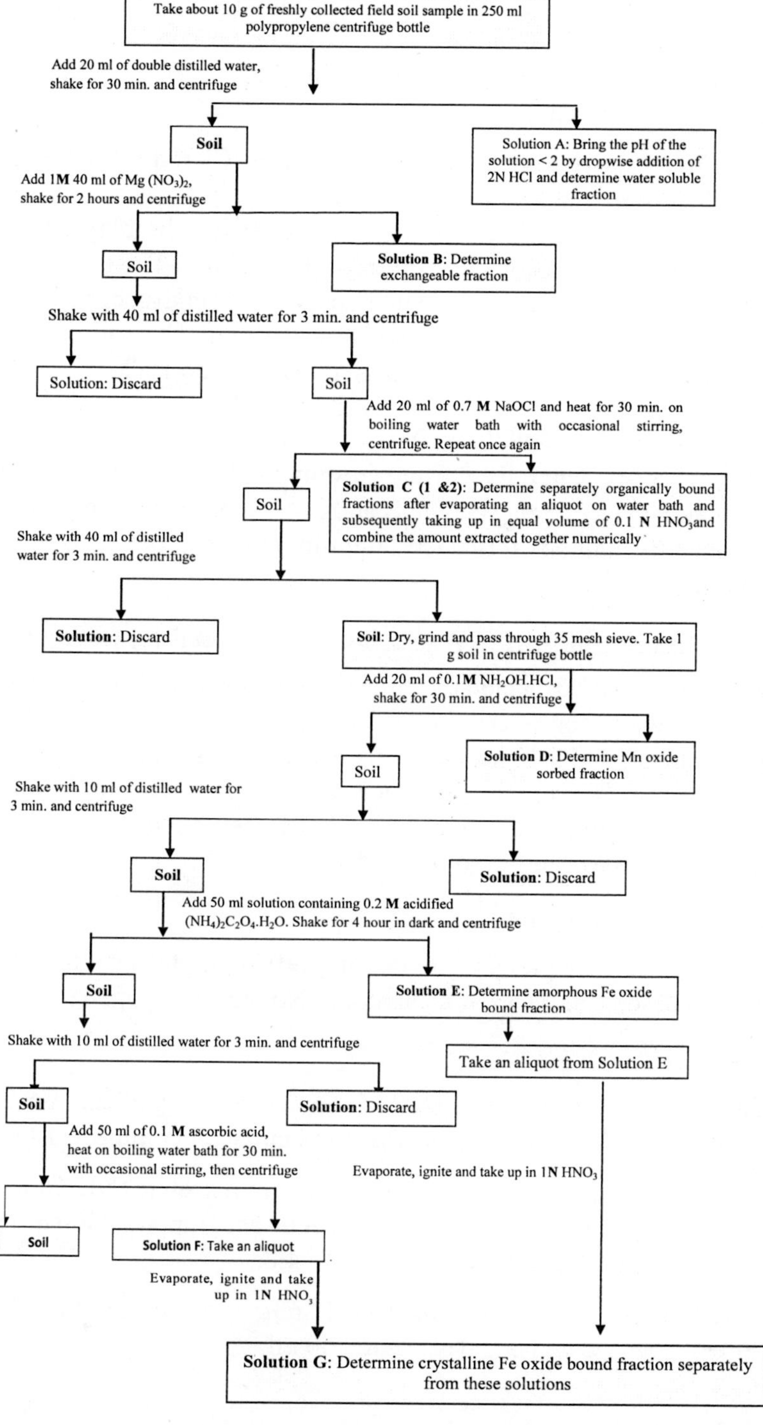
Take about 10 g of freshly collected field soil sample in 250 ml polypropylene centrifuge bottle
Add 20 ml of double distilled water, shake for 30 min. and centrifuge
Soil
Solution A: Bring the pH of the solution < 2 by dropwise addition of 2N HCl and determine water soluble fraction
Add 1M 40 ml of Mg $(NO_3)_2$, shake for 2 hours and centrifuge
Soil
Solution B: Determine exchangeable fraction
Shake with 40 ml of distilled water for 3 min. and centrifuge
Solution: Discard
Soil
Add 20 ml of 0.7 M NaOCl and heat for 30 min. on boiling water bath with occasional stirring, centrifuge. Repeat once again
Soil
Solution C (1 &2): Determine separately organically bound fractions after evaporating an aliquot on water bath and subsequently taking up in equal volume of 0.1 N $HNO_3$and combine the amount extracted together numerically
Shake with 40 ml of distilled water for 3 min. and centrifuge
Solution: Discard
Soil: Dry, grind and pass through 35 mesh sieve. Take 1 g soil in centrifuge bottle
Add 20 ml of 0.1M $NH_2OH.HCl$, shake for 30 min. and centrifuge
Soil
Solution D: Determine Mn oxide sorbed fraction
Shake with 10 ml of distilled water for 3 min. and centrifuge
Soil
Solution: Discard
Add 50 ml solution containing 0.2 M acidified $(NH_4)_2C_2O_4.H_2O$. Shake for 4 hour in dark and centrifuge
Soil
Solution E: Determine amorphous Fe oxide bound fraction
Shake with 10 ml of distilled water for 3 min. and centrifuge
Take an aliquot from Solution E
Soil
Solution: Discard
Add 50 ml of 0.1 M ascorbic acid, heat on boiling water bath for 30 min. with occasional stirring, then centrifuge
Evaporate, ignite and take up in 1N HNO_3
Soil
Solution F: Take an aliquot
Evaporate, ignite and take up in 1N HNO_3
Solution G: Determine crystalline Fe oxide bound fraction separately from these solutions

- After estimating organically bound micronutrient fraction take 8 g soil and add 10 ml of distilled water, shake end-over-end for 16 hours, wash silt and clay through 270 mesh screen with a jet of distlled water to separate out the sand fraction.
- Wash from above step is flocculated with NaCl and heated on a steam bath. The supernatant is decanted, and the solids are placed in 250 ml centrifuge bottles, centrifuged, and decanted. The clay and silt are separates by centrifuging and decanting five times. The silt is dried, and the clay transferred to 50 ml centrifuge tubes for high speed centrifuging. The salt is washed from clay before it is dried. The soil separates are then digested in Tefzel tubes using HF, HNO_3 and HCl.

25.3 Determination of Boron in Soil

Boron (B) in soil is generally present in mineral form, water soluble form, adsorbed on the surface of clay menerals and Fe/Al oxy and hydroxy compounds and in organically complex form. The later three forms in fact control the plant available B in soil.

25.3.1 Determination of Hot Water Soluble (Available) Boron in Soil

The most commonly used method for extracting easily soluble B from soil involves refluxing with water for a period of 5 minutes using a soil/water ratio of 1:2 (Berger and Troug, 1939). The following method is essentially introduced by Berger and Troug, (1939) with modification several scientists. In B analysis use of borosilicate glassware should strictly be prohibited. Plastic wares and Corning Pyrex glassware should be used.

Equipments and Materials

Spectrophotometer, hot plate, water cooling reflux condenser, 250 ml conical flask, 25 ml volumetric flasks and Whatman No. 42 filter paper.

Reagents

- Calcium chloride solution, 0.02 **M:** Dissolve 2.22 g of calcium chloride ($CaCl_2$) in 900 ml of distilled water and make up volume to 1 litre.
- Buffer solution: Dissolve 250g of ammonium acetate (NH_4OOCCH_3) in about 500 ml distilled water and add acetic acid (approx. 300 ml) to bring the pH of the final solution of 1 litre to 4.8.
- EDTA solution, 0.025 **M**: Dissolve 4.563g of disodium salt of EDTA in 200 ml distilled water and dilute to 500 ml.

- Azomethine H solution: Dissolve 0.9g of azomethine H reagent and add 2 g of L-ascorbic acid in about 200 ml of lukewarm distilled water; cool it and dilute to 250 ml.
- Standard B solution: Dissolve 5.717g of boric acid (H_3BO_3) in distilled water and dilute to 1 litre. This gives 100 ìg B/ml or 100 ppm B. Dilute 10 times (10 ml of 100 ppm solution in 100 ml) to get 10 ìg B / ml or 10 ppm B.
- Activated charcoal: Leach the activated charcoal with extracting solution (0.02 **M** $CaCl_2$) for 4 to 5 times and at last with distilled water. Dry at 60°C in an oven.

Procedure

a) Preparation of Standard Curve

- Pipette 0, 1, 2, 3, 4 and 5 ml of 10 µg B/ml or 10 ppm B stock solution to 25 ml volumetric flasks and add water to bring the volume to about 10 ml.
- Add 1 ml EDTA solution and 2 ml buffer solution and mix well.
- Then add 5 ml of azomethine H solution and adjust the volume to 25 ml with extracting solution (0.02 **M** $CaCl_2$) and mix thoroughly. The final concentrations in the volumetric flasks are 0, 0.4, 0.8, 1.2, 1.6 and 2.0 ìg B / ml.
- Keep them for 2 hours for colour development and read the absorbance in spectrophotometer at 420 nm wavelength.
- Draw the standard curve by plotting concentration in X axis and absorbance in Y axis.

b) Extraction of B from Soil and Estimantion

- Weigh 20g of air dry soil in a 250 ml conical flask and add 40 ml of calcium chloride solution and 0.5g of activated charcoal.
- Attach the water cooling reflux condenser to the flask and increase heat until initiation of boiling, reflux the suspension for 5 minutes.
- Immediately filter the suspension through Whatman No.42 filter paper.
- Pipette 5 ml of extract in 25 ml volumetric flask and follow the steps from 2 to 4 as described in the standard curve preparation.

Calculation

Say,

Weight of soil = W g

Volume of the extractant = V_A ml

Volume of aliquot taken for development = V_B ml

Volume of the final coloured solution = V_C ml

Absorbance reading of the test solution = X

Concentration of B against X absorbance obtained from standard curve = A ppm.

$$\text{Available B in soil, ppm} = A \times \frac{V_C}{V_B} \times \frac{V_A}{W}$$

25.3.2 Determination of Mannitol-$CaCl_2$ Extractable Boron in Soil

Mannitol inclusion in the extractant results the formation high affinity, soluble complex with B $(OH)_3$ and B $(OH)_4^-$, thus maintaining solution activity of these two species at low level. Boron extraction from soil with 0.01 **M** mannitol-0.01 **M** $CaCl_2$ solution removes soluble plus adsorbed B (both weakly adsorbed and specifically adsorbed) on the clay surface and associated with organic matter.

$$R_2C(OH)-C(OH)R_2 + [B(OH)_4]^- \rightleftharpoons [R_2C(O-)-C(O-)R_2\,B(OH)_2]^- + 2H_2O$$

(Mannitol)

Equipments and Materials

Polyethylene centrifuge tube (50 ml capacity), centrifuge, shaker, 25 ml volumetric flask, Whatman No.42 filter paper.

Reagents

- Mannitol–Calcium chloride solution, 0.01**M**: Dissolve 1.82 g of mannitol and 1.10 g of anhydrous $CaCl_2$ in distilled water and make up volume to 1 litre with distilled water.
- Buffer solution: Same as (*See section 25.3.1 page no 364*)
- EDTA solution, 0.025 **M**: Same as (*See section 25.3.1 page no 364*)

- Azomethine H solution: Same as (*See section 25.3.1 page no 364*)
- Standard B solution: Same as (*See section 25.3.1 page no 364*)
- Activated charcoal: Leach the activated charcoal with extracting solution for 4 to 5 times and at last with distilled water. Dry at 60ºC in an oven.

Procedure

a) Preparation of Standard Curve

- Same as section 25.3.1 except make up volume with extracting solution.

b) Extraction of B from Soil and Estimantion

- Weigh 5g of air dry soil in a 50 ml centrifuge tube and add 25 ml of mannitol-calcium chloride solution and 0.5 g of activated charcoal.
- Shake the suspension for 16 hours on a mechanical shaker.
- Centrifuge for 15 minutes at 1000 g (Appendix IV) and filter the suspension through Whatman No. 42 filter paper.
- Follow the same precess as described in (*See section 25.3.1 page no 364*)

Calculation

Same as (*See section 25.3.1 page no 364*)

***Specifically Adsorbed B = Mannitol-CaCl$_2$ extractable B – Hot water soluble B**

25.4 Determination of Available Molybdenum in Soil

Among the various extractants (acid ammonium oxalate, hot water, ammonium bicarbonate-DTPA and 1**M** ammonium acetate) used for the estimation of available molybdenum (Mo), the acid ammonium oxalate or Grigg's reagent (Grigg, 1953) is considered to be the best.

Principle

Oxalate ions replace the adsorbed molybdate ions from the soil exchange sites. Molybdate ions reacting with oxalic acid form strong Mo-oxalic acid complexes [$MoO_3.C_2H_2O_4$ and $(MoO_3)_2. C_2H_2O_4$]. The exchange reaction is irreversible, thus single extraction is sufficient. As the extractant is sufficiently buffered, it prevents any change in the pH of the soil extract.

After removing oxalate and organic matter in the extract by evaporation, the Mo content is determined by colorimetric method. Toluene-3, 4-dithiol, commonly called dithiol forms a slightly soluble dark green coloured complex $(CH_3C_6H_3S_2)_3$ with Mo^{6+} in mineral acid medium which can be extracted by organic solvent.

The resulting green coloured complex formed is used for the colorimetric determination of Mo at 680 nm wavelength. The green coloured complex is stable for at least for 24 hours.

Equipments and Materials

Analytical balance, skaker, hot plate, muffle furnace, spectrophotometer or colorimeter, centrifuge, water bath, 500 ml conical flasks, separating funnel (250 ml) with rack, porcelain dish (250 ml), volumetric flasks (50 ml and 100 ml), Whatman No. 42 filter paper.

Reagents

- Extracting solution of acid ammonium oxalate (0.2 **M** ammonium oxalate + 0.1 **M** oxalic acid, pH 3.3): Dissolve 24.9 g AR grade ammonium oxalate $[(NH_4)_2C_2O_4]$ and 12.6 g oxalic acid $(H_2C_2O_4.2H_2O)$ in double distilled water and dilute to 900 ml. Adjust pH at 3.3 with oxalic acid and NH_4OH and make up volume to 1 litre.
- Potassium iodide (KI), 50% w/v
- Ascorbic acid $(C_6H_8O_6)$, 50% w/v
- Tartaric acid $(C_4H_6O_6)$, 10% w/v
- Thiourea $[(NH_2)_2CS]$, 10% w/v
- Sulphuric acid, 4**N**: Dilute 11 ml of concentrated sulphuric acid to 100 ml.
- Hydrochloric acid, 0.1**N**: Dilute 4 ml concentrated hydrochloric acid to 500 ml.
- Hydrogen peroxide (H_2O_2), 30%
- Isoamyl acetate
- Dithiol solution (Toluene-3, 4-dithiol): Add 100 ml of 10% NaOH to 1 g of AR grade melted (51°C) dithiol $(C_7H_8S_2)$. Warm the content to 51°C on hot plate for 15 minutes with frequent stirring. Then add 1.8 ml of thioglycolic acid and mix thoroughly and store in refrigerator.
- Ferrous ammonium sulphate solution, 0.16 **N**: Dissolve 63 g of ferrous ammonium sulphate and make up the volume to 1 litre with double distilled water.
- Standard Mo solution: Dissolve 0.150 g of AR grade molybdenum trioxide (MoO_3) in 100 ml of 0.1 **N** NaOH solution, make it slightly acidic with HCl and dilute to 1 litre with double distilled water. This gives 100 ìg Mo/ml (100 ppm) stock solution. Pipette 10 ml of stock solution and dilute to 1 litre to get a working standard of 1 µg Mo/ml (1 ppm) solution.

Procedure

a) Preparation of Standard Curve

- Pipette 0, 0.5, 1.0, 2.0, 4.0, 5.0 and 8.0 ml of working standard solution (1 ìg Mo/ml or 1 ppm) of Mo in a series of 50 ml volumetric flasks and make up volume with double distilled water. These solutions correspond to Mo concentration of 0, 0.01, 0.02, 0.04, 0.08, 0.10 and 0.16 ìg Mo/ml (ppm).
- Transfer 50 ml of standard solutions in separating funnels of 250 ml capacity.
- Add 0.25 ml of ferrous ammonium sulphate solution and 20 ml of distilled water. Add excess of KI (about 2 ml) solution and neutralize the liberated iodine by adding ascorbic acid solution dropwise (to get clear solution).
- Add 1 ml of 10% tartaric acid solution, shake and add 2 ml of 10% thiourea solution and thoroughly mix the content.
- Add 5 drops of dithiol solution shake for 20 seconds and allow standing for 30 minutes.
- Finally, add 10 ml of isoamyl acetate. Shake vigorously for 2 minutes and allow standing for 1 hour for complete separation.
- Discard the aqueous phase and transfer the organic phase in a centrifuge tube and centrifuge for 15 minutes at 2000 rpm.
- Measure the absorbance of the green coloured complex at 680 nm wavelength.
- Draw the standard curve by plotting concentration in X axis and corresponding absorbance in Y axis.

b) Extraction of Mo from Soil

- Take 25 g soil in a 500 ml conical flask and add 250 ml of extracting solution.
- Shake it for 8 hours and keep overnight.
- Filter the suspension through Whatman No. 42 filter paper.
- Take 200 ml of the filtrate in a porcelain dish and evaporate to dryness on water bath.
- Ignite the residue at 500°C in a furnace for 5 hours to remove oxalate and organic matter.
- Digest it with 5 ml of diacid mixture (HNO_3: $HClO_4$ 4:1). Keep overnight and add 10 ml of 4 **N** H_2SO_4 and H_2O_2 every time taking to dryness.

- Take up the residue in 10 ml of 0.1 **N** HCl and filter.
- Wash the filter paper with another 10 ml of 0.1 **N** HCl and double distilled water till the filtrate becomes 100 ml and analyse for Mo content.

c) Estimation of Mo in the Soil Extract

- Take 50 ml of the filtrate into a 250 ml separating funnel and add 2 ml of KI solution to reduce iron present in the soil extract (no need to add ferrous ammonium sulphate solution further).
- Neutralize liberated iodine with dropwise addition of 2 ml ascorbic acid till colour disappears.
- Follow the steps 4 to 8 as described in standard curve preparation.

Calculation

Say,

Weight of soil taken = W g

Volume of extracting solution added = V_A ml

Volume of filtrate taken for removal of oxalate and organic matter = V_F ml

Volume of residue brought into solution with 0.01 **N** HCl = V_H ml

Volume of filtrate taken in separating funnel = V_S ml

Absorbance reading of the test solution = X

Concentration of Mo as obtained from standard curve against X absorbance = A ppm

$$\text{Concentration of Mo in soil, ppm} = A \times \frac{V_H}{V_S} \times \frac{V_A}{V_F} \times \frac{1}{W}$$

Critical level of available micronutrients

Elements	mg / kg soil (ppm)
DTPA Zn	0.6
DTPA Fe	4.5
DPTA Cu	0.2
DTPA Mn	2.0
Hot water soluble B	0.5
Grigg Mo	0.2

Questions and Hints

Q.1 What do you mean by the term micronutrients?

Hint: Among the essential nutrient elements, the concentration (on dry weight basis) of which in plant is less than 100 ppm i.e., 0.1% is known as micronutrient. There are 9 micronutrients namely, Fe, Mn, Zn, Cu, B, Mo, Cl, Co and Ni.

Q.2 Why the DTPA is buffered at pH 7.3?

Hint: Considerable amount of micronutrients may remain occluded in carbonates particularly in calcareous soil which is not at all plant available form. To prevent the release of occluded micronutrients through excessive dissolution of $CaCO_3$ and thereby the overestimation of available micronutrient cations in soil, the extractant (DTPA) is buffered in a slightly alkaline pH to increase the concentration of Ca in soil solution.

Q.3 Why TEA is used as buffer?

Hint: TEA burns clearly during atomization and has a pKa value of 7.8. Thus, at pH 7.3 major ($3/4^{th}$) amount of TEA is present as $HTEA^+$ which exchanges for Ca and Mg on the exchange surfaces. Thus, the concentration of Ca and Mg in soil solution phase increases which suppresses the dissolution of $CaCO_3$.

Q.4 What are the functions of tartaric acid, KI and ascorbic acid in available Mo estimation?

Hint: Interference of tungsten is prevented by adding tartaric acid, while the interference of ferric ion can be avoided by freshly prepared KI and ascorbic acid solution.

Q.5 In which form plant takes up B and Mo?

Hint: B is used as H_3BO_3, $H_2BO_3^-$, HBO_3^{2-}, and BO_3^{3-} forms while Mo in the form of MoO_4^{2-}.

26

Some Soil Polluting Heavy Metals

26.1 Plant Available Fraction of Nickel, Cadmium, Lead and Chromium

The DTPA method was originally developed to extract Mn, Cu, Fe and Zn from slightly acidic to alkaline soils (Lindsay and Norvell, 1978), but in some cases it can be used for other metals also, such as Nickel (Ni), Cadmium (Cd), Lead (Pb) and Chromium (Cr). Among them Ni is considered as an essential micronutrient. The DTPA is used to extract soil metals by forming chelates with free metals in solution, which lowers their ionic activities so that additional quantities of metal are released from the soil colloids until equilibrium is attained. Triehanolamine (TEA) is used to buffer the extracting solution at pH 7.3 in an attempt to optimize the extraction of various metals. Calcium chloride is included in the extracting solution to minimize dissolution of $CaCO_3$.

Equipments and Materials

AAS, hollow cathode lamp for respective metals, shaker, Whatman No. 42 filter paper, 100 ml conical flasks

Reagents

- DTPA extracting solution (pH 7.3), 0.005**M**: Dissolve 1.96 g of DTPA, 14.92 g of triethanolamine (TEA) and 1.47 g of $CaCl_2.2H_2O$ in 950 ml of double distilled water in 1 litre beaker. Adjust to pH 7.3 using 6 **M** HCl. Transfer the solution quantitatively to 1 litre volumetric flask and make up the volume to the mark.
- Standard solutions: To prepare 1000 ppm standard solutions

For Ni: Dissolve 1.000 g of Ni metal in 50 ml of 1:1 dilute HNO_3 and make up volume to 1 litre with double distilled water.

For Cd: Dissolve 1.000 g of Cd metal in 50 ml of 1:1 dilute HCl and make up volume to 1 litre with double distilled water.

For Pb: Dissolve 1.000 g of Pb metal in 20 ml of 6N HNO_3 or dissolve 1.5982 g of $Pb(NO_3)_2$ in double distilled water and add 20 ml of concentrated HNO_3 and make up volume to 1 litre with double distilled water

For Cr: Dissolve 1.000 g of Cr metal in 50 ml of concentrated HCl or 2.8282 g of $K_2Cr_2O_7$ in double distilled water and make up volume to 1 litre with double distilled water.

Prepare working standard solution from this 1000 ppm stock solution as per specification for each metal (Appendix XI).

Procedure

- Weigh 10 g of air dry processed soil sample in 100 ml conical flask and add 20 ml of DTPA extracting solution.
- Shake for 2 hours on a reciprocating shaker.
- Filter the suspension through Whatman No. 42 filter paper.
- Analyze the filtrate for Ni, Cd, Pb and Cr on AAS (see section 16.5.2.2) as per the analytical conditions given in Appendix XI.

Calculation

Say,

Weight of the soil = W g

Volume of the extracting solution = V_E ml

Concentration of the metal in the filtrate obtained from standard curve = A ppm

$$\text{Concentration of the metal in soil, ppm} = A \times \frac{V_E}{W}$$

26.2 Determination of Arsenic in Soil

Arsenic (As) may occur in soil in different forms or oxidation states (-3, 0, +3 and +5 valence). In strongly reducing condition, it may be present as elemental Arsenic (As^0) and As (III) form, while in aerobic environment it exists as stable As (V) oxidation state.

Principle

Arsenic content in solution can be estimated by atomic absorption spectrophotometer (AAS) provided with hydride generator unit. The principle involves the generation of hydride of arsenic, arsine (AsH_3) which is transported to the quartz cell aligned with hollow cathode lamp of As. The quartz cell is heated externally by air-acetylene flame to get the measured absorption signal.

As (III) oxidation state of arsenic is readily converted to volatile hydride (AsH_3) by sodium borohydride in acid medium, but the reduction of As (V) to As (III) by sodium borohydride is a relatively slow process. Thus, As (V) is first reduced to As (III) by potassium iodide, and then the reduced As (III) is converted to its hydride by sodium borohydride.

Equipments and Materials

Atomic absorption spectrophotometer, hollow arsenic cathode lamp, quartz cell, reaction cell for producing As hydride, argon or nitrogen gas cylinder, acetylene gas cylinder, 250 ml conical flask, shaker, Kjeldahl flask, Whatman No. 42 filter paper.

Reagents

- Sodium borohydride: Dissolve 8 g of sodium borohydride ($NaBH_4$) in 200 ml of 0.1 **N** NaOH solution.
- Potassium iodide, 10%: Dissolve 10 g of potassium iodide (KI) and dilute to 100 ml with distilled water.
- Saturated ammonium oxalate solution
- Sodium bicarbonate solution, 0.5 **M**: Dissolve 41.5 g sodium bicarbonate ($NaHCO_3$) and dilute to 1 litre.
- Concentrated H_2SO_4
- Concentrated HNO_3
- Concentrated HCl
- Stock As (III) solution, 10.0 mg L^{-1} As (III): Dissolve 1.320 g Arsenic trioxide, As_2O_3, in double distilled water containing 4.000 g NaOH and dilute to I litre to get a 1 g L^{-1} As (III) stock solution. Further dilute this by 100 times with double distilled water to obtain 10 mg L^{-1} As (III) intermediate stock solution.

Procedure

A) Extraction of As from Soil

1) Extraction of Plant Available As

- Weigh 5 g air dry soil sample in a 250 ml conical flask and add 50 ml 0.5 **M** $NaHCO_3$.
- Shake the suspension for one hour and filter through Whatman No. 42 filter paper.
- Use the clear filtrate for As estimation.

2) Digestion for Total As

- Place 1g air dry pulverized (< 100 mesh) soil sample into a Kjeldahl flask, add few glass beads and 30 ml of concentrated HNO_3.
- Digest the soil (initial half an hour slow heating) till the volume reduced to 2 to 4 ml.
- Remove the Kjeldahl flask from the digester, cool and add 5 ml concentrated H_2SO_4 and heat to boiling.
- Remove from the digester after ceasation of fuming, cool and add 25 ml of saturated ammonium oxalate solution and heat to boiling.
- Continue heating until fuming ceases.

B) Preparation of Standard curve

- Prepare desired concentration of working standard solutions [10.0, 25.0, 50.0, and 75.0 $\mu g\ L^{-1}$ As (III)] daily by diluting 10.0 $mg\ L^{-1}$ As (III) stock solution with double distilled water.
- Add 5 ml of concentrated HCl and 5 ml of 10% KI solution to 50 ml of working standard solutions and wait for half an hour.
- Feed the solution to the AAS coupled with hydride generator together with sodium borohydride, concentrated HCl and purger inert gas (argon or nitrogen) to transport the arsine (AsH_3) produced to the quartz cell aligned with the hollow cathode lamp of arsenic.
- Record absorbance readings and draw the standard curve by plotting concentration in X axis and corresponding absorbance in Y axis.

C) Estimation of As in the extract

- Follow the same steps from 2 to 4 as described in standard curve preparation.

Calculation

Say,

Weight of soil taken = W g

Volume of the extractant or digest = V_L ml

As concentration in the final solution as obtained from standard curve, µg/ ml = A

$$\text{As concentration in soil, } \mu g/g = A \times D \times \frac{V_L}{W}$$

[D is the aliquot dilution (if required)]

Questions and Hints

Q.1 What is heavy metal?

Hint: The metal having specific gravity more than 5.0 or having atomic number higher than 20 is considered as heavy metal. As a corollary, any metal heavier than Ca is a heavy metal.

Q.2 According to World Health Organisation (WHO) what is the maximum allowable As concentration in drinking water?

Hint: 0.01 mg/L.

Q.3 Why $CaCl_2$ is incorporated with DTPA in the extracting solution?

Hint: $CaCl_2$ is incorporated to increase the ionic activity of Ca and thereby to reduce the dissolution of $CaCO_3$ to prevent the release of occluded heavy metal cations.

Q.4 What is the function of hydride generator in As estimation?

Hint: The function of hydride generator is to convert As (III) to its volatile hydride (AsH_3) reacting with sodium borohydride in acidic medium. The carrier gas (N_2 or Ar) then transports the AsH_3 to the heated quartz cell aligned with hollow cathode lamp for absorption of characteristic resonance line of As.

Q.5 What role does KI perform during As determination?

Hint: Production of arsine or hydride of arsenic (AsH_3) is necessary for As estimation. As (III) is easily converted to AsH_3 by sodium borohydride, but conversion of As (V) to As (III) is very slow process. KI is used to reduce As (V) to As (III) which then can be readily converted to AsH_3 by sodium borohydride.

Q.6 Name some arsenic bearing minerals in soil.

Hint: Arsenopyrite (FeAsS), real gar (AsS), orpiment (As_2S_3).

Section -IV

PLANT ANALYSIS

27

Plant Analysis

Different diagnostic tools are designed to avert nutrient deficiency or sufficiency and if properly used both loss of yield and quality of crop can be minimized. Among several diagnostic tools, nutrient analysis of standing crop seems to be an efficient method in arriving at the need-based fertilizer scheduling for various crops.

Nutrient uptake by plants indicates the actual accessibility of nutrients in soil to the plants. Also, concentration of certain heavy metals in plant indicates their solubility in soil which leads to poor crop quality. Although, different plant species and even different varieties of same species may differ in their nutrient requirements, the concentration of nutrient element in plant can act as guide to support the soil test result in assessing nutrient supplying capacity of a soil and as a tool for correcting any deficiency if carried out early enough to prevent yield loss. However, analysis of harvested or mature crop is like a post mortem as this information only can help in nutrient management planning in subsequent years.

Like soil testing plant analysis also suffers from some limitations. Among them, the most important are the risk of sampling error as well as the chances of incorrect interpretation of test report in recommending fertilizers. Selection of right plant part and right stage of crop growth are very crucial for successful plant analysis.

27.1 Plant Sampling and Sample Preparation

Success of plant analysis for determining nutrient composition towards recommending fertilizer for the standing crop depends on sampling of correct plant part at its right growth stage. To work out total nutrient uptake by plant usually nutrient removal by the above ground parts is considered. The specific plant parts along with the growth stage for nutrient diagnosis, monitoring and efficient nutrient management for optimum yield and better quality are given in Table 27.1.

Table 27.1: Plant sampling guideline for different crops

Crop	Index plant tissue	Growth stage
Field crops		
Rice	3rd leaf from apex	At tillering
Wheat	Flag leaf	Before head emergence
Maize	Ear leaf	Before tasseling
Barley	Flag leaf	At head emergence
Pulses	Recently matured leaf	At bloom initiation
Groundnut	Recently matured leaflets	At maximum tillering
Sunflower	Youngest mature leaf	At flower initiation
Mustard	Recently matured leaf	At bloom initiation
Soybean	3rd leaf from apex	2 months after planting
Cotton	Petiole, 4th leaf from apex	At flower initiation
Jute	Recently matured leaf	60 days age
Sugarcane	3rd leaf from apex	3-5 months after planting
Sugarbeet	Petiole of the youngest mature leaf	At 50-80 days age
Plantation crops		
Tea	3rd leaf from tip of young shoot	
Coffee	3rd or 4th pair of leaf from apex of lateral shoots	At bloom
Coconut	Pinnal leaf from each side of 4th leaf	
Vegetables		
Potato	Most recent fully developed leaf	Half grown
Brinjal	Leaf blades with midribs minus petiole from most recent fully developed leaf	
Tomato	Leaves adjacent to inflorescence	Mid bloom
Onion	Top non-white portion	1/3 to ½ grown
Garlic	Most recent fully developed leaf non-white portion	Pre-bulb
Beans	Uppermost, recent fully developed trifoliate leaf	
Cabbage	Wrapper leaf	2-3 months age
Cauliflower	Most recent fully matured leaf	At heading
Peas	Leaflets from most recent fully developed leaf	At first bloom
Carrots	Most recent fully developed leaf	At mid growth
Radish	Most recent fully developed leaf	30 days old
Turnip	Most recent fully developed leaf	
Cucumber	5th leaf from tip omit unfurled	Flower bud start to small fruit
Fruit crops		
Apple, Pear	Leaves from middle terminal shoot growth	8-12 weeks of after full bloom 2-4 weeks after formation of terminal buds in bearing tree
Almond	3rd leaf from top	At the beginning of bloom
Peach	Mid-shoot leaves. Fruiting or non-fruiting	Mid-summer leaves, fruiting spurs

Plum	Leaves from middle of current season's extension growth	January-February
Cherry	Fully expanded leaves, mid-shoot current growth	July-August
Strawberry	Youngest fully expanded matured leaf without petiole	At peak or harvest season
Grape	5th petiole from basePetiole opposite to bloom	At bud differentiation stage for yield forecast For quality June
Citrus	3-5 month old leaves from new flush, 1st leaf of the shoot	
Banana	Petiole of 3rd open leaf from apex	4 months after planting
Mango	Leaf with petiole	4-7 months old leaves from middle of shoot
Guava	3rd pair of recently matured leaves	At bloom stage (Aug. /Dec.)
Papaya	6th petiole from apex	6 months after planting
Pineapple	Middle 1/3rd portion of white basal portion of 4th leaf from apex	At 4-6 months age
Pomegranate	8th leaf from apex	At bud differentiation (in April for Feb. crop & Aug. for June crop)
Ber	6th leaf from apex from secondary or tertiary shoot	2 months after pruning
Watermelon, muskmelon	5th leaf from tip (except unfurled leaf)	Flower start to small fruit
Sapota	10th leaf from apex	September

Sources: Kensworthy (1964), Reuter and Robinson (1986), Bhargava and Dhandar (1987), Jones *et al.* (1991), Bhargava and Kumar (1992), Bhargava and Raghupathi (1993).

For collection of aboveground plant parts, select sufficient number (15-50) of representative plant parts at random from each treatment. Collect plant parts with the help of sharp stainless steel knife or blade. For collection of root sample, uproot whole plant with the help of spade keeping all the roots as much as possible. Dip the root portion in water for some sufficient time so that soil adhering to the roots can be removed by gentle washing.

Wash plant sample first with 0.2 percent liquid detergent to remove the waxy/oily coating and dust particle adhering to the leaf surface. Then wash with 0.1 percent HCl (dilute 8 ml concentrated HCl to 1 litre) to remove metallic contamination, if any. Finally wash the plant sample with plenty of distilled water to remove the previous two solvents.

Wipe out extra moisture with quality tissue paper and air dry the sample in a dust-free atmosphere for 2-3 days. Then, place the sample in brown paper bag and dry in a hot air oven at 70°C for 48 hours (till constant weight is achieved).

Grind oven dry sample in a stainless steel mill having 0.5 mm sieve and ensure no metallic contamination. Every time after grinding of one sample clean the cup, blade and sieve of the mill. After grinding keep sample in oven for few hours at 70^{0}C. Store the sample in air-tight glass or plastic container or polythene bag.

27.2 Analysis of Plant Sample

27.2.1 Determination of Total Nitrogen

Total nitrogen in plant sample is determined by Kjeldahl method or by colorimetric method. Estimation of total N in plant sample involves two steps: (i) digestion of plant sample to convert all nitrogenous compounds to NH_4-N form, and (ii) determination of NH_4-N in the digest by distillation-titration/in colorimeter.

A. Kjeldahl Method

Principle

Same as the principle of determination of total N in soil (section 21.1)

Equipments and Materials

Micro-Kjeldahl flask of 150 ml capacity, Kjeldahl distillation flask of 800 ml capacity, digestion chamber with heaters and fume exhaust or block digestion cum distillation set, balance, measuring cylinder, 1 liter volumetric flasks, glass beads, burette.

Reagents

- Sulphuric acid-salicylic acid mixture: Same as section 21.1.
- Digestion mixture: Same as section 21.1.
- Sodium thiosulphate ($Na_2S_2O_3$. $5H_2O$)
- Mixed indicator: dissolve 0.5 g of bromo cresol green and 0.1 g of methyl red in 100 ml of 95 percent ethanol. Adjust the solution to the bluish purple midcolour at pH 4.5 with dilute NaOH or HCl.
- Boric acid- indicator solution, 4%: Dissolve 40 g of boric acid in about 800 ml hot distilled water. Cool, then add 5 ml of mixed indicator and adjust the ph of the solution at 4.5 with dilute NaOH or HCl (bluish colour of the solution weakens towards pink). Make up the volume to 1 liter.
- Standard sulphuric acid, 0.1 **N**: Same as section 21.1.
- Sodium hydroxide (NaOH), 40%: Same as section 21.1.

Procedure

(i) Digestion on Heater/Gas Burner

- Transfer 0.25 g (grain) or 0.5 g (straw or root) dried, processed plant sample to a 150 ml micro-Kjeldahl flask.
- Add 20 ml of sulphuric acid-salicylic acid mixture. Swirl the flask to bring the dry plant sample in contact with acid. Allow it to stand for overnight.
- Then add 5 g sodium thiosulphate through a long-stemmed funnel and gently heat for 5 minutes.
- Cool the content and add about 10 g digestion mixture and few glass beads.
- Place the micro-Kjeldahl on the digestion rack in low flame for about 30 minutes, until frothing stops. Gradually raise the temperature until the acid reaches a boil and maintain digestion temperature between 360 and 410°C.
- Gently rotate the flask intermittently to bring the residues adhering to the neck of flask into the solution.
- After completion of digestion (1±0.25 hour) remove the flask from the digestion rack and keep aside to cool. Completeness of digestion is best judged by the clear grayish blue or greenish colour of the solution.
- After sufficient cooling add 20 ml of distilled water.
- Run a blank (without plant sample) under similar condition.

Distillation

- Take approximately 25 ml of 4 percent boric acid-indicator solution in a 250 ml conical flask and place in such a way that the end of delivery tube dips into the boric acid- indicator solution.
- Transfer the plant digest to an 800 ml Kjeldahl flask by repeated washing with distilled water.
- Add 80 ml of 40% NaOH to the Kjeldahl flask and immediately connect to the distillation set.
- Continue distillation for about 30 minutes to collect about 100 ml of distillate.
- Run distillation of the blank in a similar way.

Titration

- Titrate the distillate - boric acid solution with standard H_2SO_4 until blue colour just disappears and with one drop in excess turns the solution to pink.

(ii) Digestion on Digestion Block

- Switch on the digestion system and set the temperature at temperature controller at 100°C.
- Once the initial temperature of 100°C is reached reset the temperature controller at temperature 385°C.
- Weigh 0.5 g finely ground, dried plant samples and place in the digestion tubes.
- Add one tablet of catalyst/digestion mixture and 6 ml of concentrated sulphuric acid.
- Place the stainless steel insert rack with sample on the insert rack stand. Fix the exhaust manifold system on the top of the insert rack. Open the water line for suction system.
- Load the insert rack with samples and exhaust manifold system into the digestion block.
- Adjust tap water flow rate so that it creates sufficient vacuum for safe disposal of acid fumes.
- Initially dense acid fumes evolved are trapped efficiently by the exhaust system. After sometime when fumes reduced, reduce water flow rate to the suction system to reduce suction and to avoid evaporation of acid.
- Continue digestion till completion for about 60-75 minutes (no black or brown colour).
- Switch off the system, remove the insert rack with the digested samples and exhaust manifold system and place them on the insert rack stand. Leave the suction system with minimum water flow for some time.
- Remove the exhaust manifold from the tubes after sufficient cooling.
- Dilute the sample with about 50 ml of water to avoid solidification.
- Place the diluted sample in distillation system for ammonia recovery.
- Immerse the respective hoses coming out from the system in alkali and boric acid reservoir tanks.

- Select programme number by operating programme key and wait for 'READY' indication.
- Open the water inlet to water condenser and set the alkali pump to deliver 25 ml 40% NaOH solution (reagents can also be added manually).
- Run distillation for 9 minutes. During distillation the water is filled automatically through refilling pump in the steam generator.
- Collect ammonia in 10 ml of 4% boric acid solution.

Titration

- Titrate the distillate-boric acid solution with standard H_2SO_4 until blue colour just disappears and with one drop in excess turns the solution to pink either manually or using automatic titration system.

Calculation

Say,

Weight of plant sample taken = W g

Volume (ml) of H_2SO_4 required for sample titration = V_S ml

Volume (ml) of H_2SO_4 required for blank titration = V_B ml

Strength (N) of H_2SO_4 = S

$$\text{Total N in plant sample (\%)} = (V_S - V_B) \times S \times 0.014 \times \frac{100}{W}$$

B. Colorimetric Method

In this method ammonium nitrogen concentration in the plant digest can be determined either manually or by using Auto-Analyser (Baethgen and Alley, 1989).

Reagents

- Digestion mixture: Mix 30 parts of potassium sulphate (K_2SO_4) with 1 part mercuric oxide (HgO).
- Buffer solution, 0.1 m Na_2HPO_4, 5% Na-K tartarate, 5.4% NaOH: Dissolve 26.8 g $Na_2HPO_4.7H_2O$ in about 600 ml distilled water. Add 50 g of Na-K tartarate and 108 g 50% (w/w) solution of NaOH. Dilute to 1 litre with distilled water.
- Na-salicylate (15%) - Na-nitroprusside (0.03%) solution: Dissolve 150 g Na-salicylate and 0.3 g Na-nitroprusside in 1 litre of distilled water. Store in dark coloured bottle.

- Na-hypochlorite solution: Dilute 6 ml of 5.25% Na-hypochlorite solution in 100 ml with distilled water for daily use.
- Nitrogen standard solution: Dissolve 4.715 g dry $(NH_4)_2SO_4$ in a litre of distilled water to prepare 1000 ppm N solution. From this prepare 5, 10, 15, 30, 40 and 50 ppm N standard solutions.

Digestion

Digestion of plant sample is carried out in the same manner as stated earlier. However, the composition of digestion mixture is somewhat different and 1 g of digestion mixture is used for each sample.

Determination of N

- Take 1 ml plant digest in 25 ml volumetric flasks.
- Add 5.5 ml of buffer solution and stir thoroughly and then add 4 ml of Na-salicylate - Na-nitroprusside solution and mixed again.
- Add 1 ml of Na-hypochlorite solution and again mix thoroughly (add reagents in proper sequence to avoid the formation of solid residue).
- Allow the solutions to stand for 45 minutes at 25°C (or 15 minutes at 37°C) for colour development.
- Read the absorbance in a spectrophotometer at 650 nm.
- Prepare standard curve following above procedure taking 1 ml of each standard solutions in 25 ml volumetric flasks.

Calculation

Say,

Weight of the plant sample taken = W g

Volume of acid digest filtrate/dissolved ash = V_D ml

Absorbance of the sample = A

Concentration of N in the coloured solution corresponding to A absorbance = B ppm.

$$\text{N concentration of plant sample (\%)} = B \times \frac{V_D}{W} \times \frac{100}{10^6}$$

$$= B \times \frac{V_D}{W} \times \frac{1}{10^4}$$

27.2.2 Determination of P, K, Secondary and Micronutrients

Principle

The nutrient elements other than nitrogen from the plant tissues are released by two widely adopted methods: (i) Wet Oxidation, and (ii) Dry Ashing.

A. Wet Oxidation

In wet oxidation method plant samples can be digested in tri-acid mixture (HNO_3-H_2SO_4-$HClO_4$) or di-acid mixture (HNO_3-$HClO_4$). Tri-acid mixture digestion is recommended when only P and K in plant sample are to be determined. Presence of H_2SO_4 restricts its use in the estimation of S, Ca (under estimation), and micronutrients (due to impurity in H_2SO_4). On the other hand, di-acid mixture digestion is recommended for P, K, Ca, Mg, S, Fe, Cu, Mn, and Zn. Wet oxidation is not used for the estimation of B and Mo in plant samples.

Tri-acid Mixture Digestion

Equipments and Materials

Digestion chamber with fume exhaust, hot plate, 100 ml conical flask, 100 ml volumetric flask, Whatman No. 1 filter paper.

Reagents

- Tri-acid mixture: Mix AR grade concentrated HNO_3, H_2SO_4 and $HClO_4$ in the ratio of 9:4:1 on volume basis and cool it.

Procedure

- Transfer 0.5 to 1.0 g of dried, processed plant sample to a 100 ml conical flask.
- Add 5 ml of concentrated HNO_3 (if sample contains high fats and oils) and heat at 100^{0}C temperature on hot plate in a digestion chamber having fume exhaust system for 30 minutes.
- Raise the temperature to about 200°C and continue boiling to near dryness (but not completely dry).
- Cool and then add 10 ml of tri-acid mixture.
- First heat at low temperature (180-200°C) and then at higher temperature until production of brown fumes of NO_2 ceases.
- Continue evaporation until the volume is reduced to about 3-4 ml (but not completely dry) and only clear colourless content is left.
- Cool, add 20 ml of distilled water and filter into 100 ml volumetric flask through Whatman No. 1 filter paper.

- Make up the volume to the mark with the washings of conical flask.
- Use filtrate for analysis of P and K content in plant sample.

Di-acid mixture Digestion

Equipments and Materials

Same as tri-acid mixture digestion method. acid

Reagents

- Di-mixture: Mix AR grade concentrated HNO_3, and $HClO_4$ in the ratio of 9:4 on volume basis and cool it.

Procedure

- Same as tri-acid mixture digestion except add 10 ml di-acid mixture in place of 10 ml tri-acid mixture.
- Use the filtrate for analysis of P, K, Ca, Mg, S, Fe, Cu, Mn, and Zn.

B. Dry Ashing

This method can be used for extraction of K, Ca, Mg, Fe, Cu, Mn, Zn, B and Mo from plant tissues and preferentially for B and Mo. It is easy, rapid and precise method usually without involvement any reagent. However, the shortcoming of this method is that it cannot be directly used for estimation of P, S, Cl besides N, because of their volatilization loss during ignition. Volatilization loss of P can be prevented by adding alcoholic solution of $Mg(NO_3)_2$ or $Mg(OAc)_2$, while the loss of S and Cl can be prevented by adding Na_2CO_3.

Equipments and Materials

Silica or platinum crucible or dish, muffle furnace, Whatman No. 1 filter paper, 50 ml volumetric flask.

Reagents

- HCl solution, 20%: Dilute 20 ml of concentrated hydrochloric acid to 100 ml with distilled water (use double distilled water for micronutrient analysis).

Procedure

- Transfer 1 g dried, processed plant sample in a silica crucible.
- Place the crucible in the middle of a cool muffle furnace (at least 2 cm away from the walls).

- Raise the temperature slowly to 550°C and continue ashing for 6 hours. Cool the ash and add 5 ml of 20 percent HCl to dissolve the

27.2.2.2 Determination of Plant Phosphorus

Vanadium, molybdenum and phosphoric acid react together to give a yellow colour in nitric acid medium. The exact nature of this yellow chromogen is not known, but the colour attributed to the substitution of oxyvanadium and oxymolydenum radicals for O of PO_4 to give a heteropoly compound that is chromogenic. The advantages of this method are extreme simplicity, stability of colour, lower sensitivity (permitting application on macro scale), and freedom from interferences with wide range of ionic species in concentration up to 1000 ppm, adaptability to HNO_3, HCl, H_2SO_4 or $HClO_4$.

$$(NH_4)_6Mo_7O_{24}.4H_2O + H_3PO_4 + HNO_3 \rightarrow (NH_4)_3PO_4.12MoO_3 + NH_4NO_3 + H_2O$$

Molybdophosphoric acid
(Yellow colour)

$$(NH_4)_6Mo_7O_{24}.4H_2O + NH_4VO_3 + H_3PO_4 + HNO_3 \rightarrow (NH_4)_3PO_4NH_4VO_3.16MoO_3 + NH_4NO_3 + H_2O$$

Vanadomolybdophosphoric acid
(Yellow colour)

The colour develops in 30 minutes and remains stable for 2-4 weeks.

Equipments and Materials

Colorimeter or spectrophotometer, 50 ml volumetric flasks, single mark pipettes of 5, 10 ml.

Reagents

- Vanadate-molybdate reagent: Dissolve 25 g ammonium molybdate, tetrahydrate [$(NH_4)_6Mo_7O_{24}.4H_2O$] in 400 ml distilled water. Dissolve 1.25 g ammonium metavanadate (NH_4VO_3) in 300 ml boiling distilled water. Add the vanadate solution to the molybdate solution and cool to room temperature. Add 250 ml of concentrated HNO_3 and make volume to 1 litre.

- Standard P solution: Dissolve 0.2195 g dried potassium dihydrogen phosphate (KH_2PO_4) in distilled water and make up the volume to 1 litre. The solution contains 50 ppm P.

Procedure

A. Preparation of standard curve

- Transfer 0, 1, 2, 3, 4 and 5 ml of standard P solution (50 ppm) to 50 ml volumetric flasks.

- Add 10 ml of vanadate-molybdate reagent to each volumetric flask and make up the volume with distilled water. These solutions correspond to 0, 1, 2, 3, 4 and 5 ppm of P, respectively.
- Read the absorbance of the solution after 30 minutes at 420 nm with spectrophotometer or colorimeter using blue filter.
- Prepare standard curve plotting absorbance against concentration of P (Fig. 27.1)

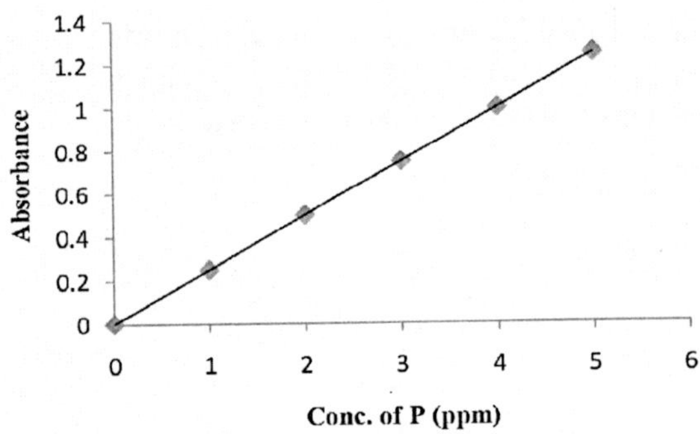

Fig. 27.1: Standard curve for P

B. Estimation of P in acid digest

- Pipette 10 ml or any suitable aliquot of the digest filtrate or dissolved ash into 50 ml volumetric flask, so that the final 50 ml solution contains 0.05 to 1.0 mg of P and the acid equivalent will be above 0.2 N, but not over 1.6 N.
- Add 10 ml of vanadate-molybdate reagent and make up the volume with distilled water.
- Read the absorbance of the plant digest.

Calculation

Say,

Weight of the plant sample taken = W g

Volume of acid digest filtrate/dissolved ash = V_D ml

Volume of aliquot taken for colour development = V_A ml

Final volume of the coloured solution = V_F ml

Absorbance of the sample = A

Concentration of P in the final solution corresponding to A absorbance = B ppm.

$$\text{P concentration of plant sample (\%)} = B\times\frac{V_F}{V_A}\times\frac{V_D}{W}\times\frac{100}{10^6}\times D$$

$$= B\times\frac{V_F}{V_A}\times\frac{V_D}{W}\times\frac{D}{10^4}$$

[D = dilution factor, if further dilution of acid digest/dissolved filtrate is required]

27.2.2.3. Determination of Plant Potassium

Potassium in the acid digest/dissolved ash of the plant sample can be determined using flame photometer exactly in the same manner as described for water soluble K in soil (section 23.1.1). Alternately, potassium in the acid digest can also be determined using AAS with higher precision.

$$\text{K concentration of plant sample (\%)} = B\times\frac{V_D}{W}\times\frac{100}{10^6}\times D$$

Where, W = Weight of the plant sample taken for digestion, g

V_D = Volume of acid digest filtrate/dissolved ash, ml

B = Concentration of K in the digest filtrate/dissolved ash, in ppm, obtained from standard curve.

[D = dilution factor, if further dilution of acid digest/dissolved filtrate is required]

27.2.2.4. Determination of Plant Calcium and Magnesium

Calcium and magnesium content in plant sample are determined from di-acid mixture digestion or dissolved ash using AAS following specification as led down in Appendix XI.

- Standard Ca solution: Weigh 2.247 g of AR grade calcium carbonate ($CaCO_3$) in 100 ml beaker and add 5 ml of distilled water. Now, add concentrated HCl drop by drop (approx. 10 ml is required) to completely dissolve $CaCO_3$. Transfer the content to 1 litre volumetric flask and make up the volume to the mark with distilled water. This will give 1000 ppm Ca solution. Dilute 10 ml of 1000 ppm Ca solution to 100 ml to get 100 ppm Ca solution. Prepare working range standard Ca solution of 1-10 ppm from 100 ppm standard stock Ca solution.
- Standard Mg solution: Dissolve 10.141 g of MgSO4.7H2O on 200 ml of distilled water and dilute to 1 litre to get 1000 ppm Mg solution. Dilute 10

ml of 1000 ppm Mg solution to 1 litre to get 10 ppm Mg Mg solution. From 10 ppm standard stock Mg solution prepare working range standard Mg solution of 0.1 to 2 ppm.

$$\text{Ca (or Mg) concentration of plant sample (\%)} = B \times \frac{V_D}{W} \times \frac{100}{10^6} \times D$$

$$= B \times \frac{V_D}{W} \times \frac{D}{10^4}$$

Where, W = Weight of the plant sample taken for digestion, g

V_D = Volume of di-acid digest filtrate/dissolved ash, ml

B = Concentration of Ca/Mg in the digest filtrate/dissolved ash, in ppm, obtained from standard curve.

[D = dilution factor, if further dilution of acid digest/dissolved filtrate is required]

27.2.2.4. Determination of Plant Sulphur

Dry ashing as well as wet oxidation with tri-acid mixture are not suitable for plant S determination. Plant sulphur from di-acid digest is determined following turbidimetric method as described for the determination of total soil sulphur content (*See section 24.2.1 page no 350*).

$$\text{S concentration of plant sample (\%)} = B \times \frac{V_D}{W} \times \frac{100}{10^6} \times D$$

$$= B \times \frac{V_D}{W} \times \frac{D}{10^4}$$

Where, W = Weight of the plant sample taken for digestion, g

V_D = Volume of di-acid digest filtrate/dissolved ash, ml

B = Concentration of S in the digest filtrate, in ppm, obtained from standard curve.

[D = dilution factor, if further dilution of acid digest is required]

27.2.2.6. Determination of Plant Iron, Copper, Zinc and Manganese

These four cationic micronutrients are determined in the di-acid digest (for high silica containing plant tissues, *e.g.* rice, wheat, barley, maize, sugarcane, etc) or dissolved ash (for low silica containing plant tissues, *e.g.* legumes) of plant sample using AAS in the same manner as described for determination of these available micronutrients in soil sample (*See section 25.1 page no. 359*) following specification mentioned in Appendix XI.

27.2.2.7. Determination of Plant Boron

Boron content in plant sample is determined from dry-ashing of plant sample.

Equipments and Materials

Spectrophotometer, 25 ml volumetric flasks

Reagents

- Buffer solution: Dissolve 250 g of ammonium acetate (NH_4OOCCH_3) and 15 g of EDTA (disodium salt) in 400 ml of double distilled water. Slowly add 125 ml glacial acetic acid and mix thoroughly.
- Azomethine-H reagent: dissolve 0.45 g of azomethine-H in 100 ml of 1 percent L-ascorbic acid solution. Store in polypropylene bottle in a refrigerator. Prepare fresh solution every week.
- Standard B solution: (*See section 25.3.1 page no 364*).

Procedure

A. Preparation of standard curve

- Pipette 0, 1, 2, 3, 4 and 5 ml of 10 µg B/ml or 10 ppm B stock solution to 25 ml volumetric flaks and add water to bring the volume to 10 ml.
- Add 2 ml of buffer solution and mix thoroughly.
- Add 2 ml of azomethine-H reagent, mix and after 30 minutes read the absorbance at 420 nm on a spectrophotometer.
- Prepare standard curve by plotting B concentrations on X-axis and absorbance readings on Y-axis.

B. Estimation of B in dissolved ash

- Pipette 1 ml of clear aliquot of dissolved ash and add 2 ml of buffer solution and mix thoroughly.
- Add 2 ml of azomethine-H reagent, mix and after 30 minutes read the absorbance at 420 nm on a spectrophotometer.

$$\text{B concentration of plant sample (\%)} = A \times \frac{V_D}{W} \times \frac{100}{10^6} \times D$$

$$= A \times \frac{V_D}{W} \times \frac{D}{10^4}$$

Where, W = Weight of the plant sample taken for digestion, g

V_D = Volume of di-acid digest filtrate/dissolved ash, ml

A = Concentration of B in the digest filtrate obtained from standard curve, in ppm.

[D = dilution factor, if further dilution of dissolved ash is required]

27.2.2.8. Determination of Plant Molybdenum

Molybdenum content in plant sample is determined from dry-ashing followed by fusion with Na_2CO_3. Fusion of ash with Na_2CO_3 helps to bring Mo into dilute ethanol solution as soluble sodium molybdate, while Fe, Ti, Ca, Mg and many other ions as precipitate.

A. Preparation of sample

Equipments and Materials

Platinum dish, muffle furnace, 250 ml beaker, graduated cylinder of 100 ml and 25 ml, Whatman No.40 filter paper.

Reagents

- HCl solution, 20%: Dilute 20 ml of concentrated hydrochloric acid to 100 ml with double distilled water.
- Sodium carbonate (Mo free), Na_2CO_3
- Ethanol, 2% (v/v): Dilute 2 ml absolute ethanol (C_2H_5OH) in 100 ml with double distilled water.

Procedure

- Perform dry-ashing in a platinum dish as described in the section 27.2.2.
- Fuse the ash with 2 g anhydrous Na_2CO_3 at 1000°C in the muffle furnace.
- Cool the dish and transfer the cake by inverting the dish over a 250 ml beaker. If the cake does not detach, place the dish in the beaker and disintegrate the cake in 100 ml of water containing 2 percent ethanol with the help of a rubber policeman on a glass rod and warming on a steam hot plate for several hours.
- Measure the volume of suspension in a graduated cylinder and filter through Whatman No. 40 filter paper into a 250 ml beaker.
- The volume of clear undiluted filtrate is measured and all (or an aliquot) is used for Mo determination.

[*Note*: If silica content in plant ash is high, then silica should be removed from the clear filtrate by following steps:]

- Acidify the clear filtrate with concentrated HCl (2 ml per gm of Na_2CO_3) and evaporate to dryness.
- Take the residue in exactly 20 ml of 3 N HCl.
- Warm the suspension on a steam hot plate, transfer to a 25 ml centrifuge tube and centrifuge to remove silica and undissolved salt.
- Use this clear supernatant solution for Mo determination].

B. Estimation of Mo in the filtrate

Determine Mo content in the filtrate as described for the preparation of standard curve and estimation of Mo in the soil extract in the section 25.4.

Calculation

$$\text{Mo concentration of plant sample (\%)} = A \times \frac{V_D}{V_S} \times \frac{1}{W \times 10^4}$$

Where, W = Weight of the plant sample taken for digestion, g

V_D = Volume of the suspension before filtration, ml

V_S = Volume of the aliquot taken in the separating funnel, ml

A = Concentration of Mo in the aliquot obtained from standard curve, in ppm.

27.3 Interpretation of Plant analysis Results

Plant analysis results indicate the nutrient composition of the specific plant parts at the time of sampling. The nutrient concentration values can be used as a supplementary to the soil testing report. As a diagnostic tool, it is of tremendous value as it can be the basis for nutrient corrective measure and to avoid the loss of yield and quality of the standing crop which even does not show any characteristic visual deficiency symptom. Plant analysis values can be interpreted in various ways, among them critical concentration concept is the most commonly used one. In DRIS approach a set of nutrient ratios rather than a single nutrient concentration are taken into consideration.

27.3.1 Critical Concentration Approach

Critical concentration level or range is the nutrient concentration indicates whether the level of nutrient is in the deficiency, sufficiency (optimum) or excessive and toxic range. Both yield and quality of the crop in deficient nutrient concentration are unsatisfactory and will respond positively to the addition of nutrient. On the other hand, at nutrient concentration above the high level will cause toxicity (except luxury consumption of K) and thereby decline the yield

Table 27.2: Deficiency, sufficiency and high limits of nutrient in some crops

Nutrient	Low*	Optimum	High	Nutrient	Low	Optimum	High
Rice				**Wheat**			
N (%)	2.40-2.50	2.60-3.20	> 3.20	N (%)	1.25-1.74	1.75-3.0	> 3.0
P (%)	0.07-0.08	0.09-0.18	> 0.18	P (%)	0.11-0.20	0.21-0.50	> 0.51
K (%)	0.80-0.90	1.00-2.20	> 2.20	K (%)	1.00-1.50	1.51-3.0	> 3.0
Ca (%)	< 1.20	1.20-1.40	> 1.40	Ca (%)	0.10	0.21-1.0	> 1.0
Mg (%)	< 0.20	0.20-0.30	> 0.30	Mg (%)	0.1-0.15	0.16-1.0	> 1.0
Fe (ppm)	60-69	70-150	> 150	Fe (ppm)	<10	10-300	> 300
Mn(ppm)	100-149	150-800	> 800	Mn(ppm)	10-15	16-200	> 200
Zn (ppm)	16-17	18-50	> 50	Zn (ppm)	11-20	21-70	> 70
Cu (ppm)	6-7	8-25	> 25	Cu (ppm)	3-5	5-50	> 51
B (ppm)	4-5	6-7	> 7	B (ppm)	-	-	-
Maize				**Barley**			
N (%)	< 3.5	3.50-5.0	> 5.0	N (%)	1.25-1.74	1.75-3.0	> 3.0
P (%)	< 0.3	0.30-0.50	> 0.50	P (%)	0.15-0.19	0.20-0.50	> 0.50
K (%)	< 2.5	2.50-4.00	> 4.00	K (%)	1.25-1.49	1.50-3.0	> 3.0
Ca (%)	< 0.3	0.30-0.70	> 0.70	Ca (%)	< 0.3	0.30-1.20	> 1.20
Mg (%)	< 0.15	0.15-0.45	> 0.45	Mg (%)	< 0.15	0.15-0.5	> 0.5
S (%)	< 0.15	0.15-0.50	> 0.50	S (%)	< 0.15	0.15-0.40	> 0.40
Fe (ppm)	< 50	50-250	> 250	Fe (ppm)	-	-	-
Mn(ppm)	< 20	20-300	> 300	Mn(ppm)	5-24	25-100	> 100
Zn (ppm)	< 20	20-60	> 60	Zn (ppm)	< 15	15-70	> 70
Cu (ppm)	< 5	5-20	> 20	Cu (ppm)	< 5	5-25	> 25
B (ppm)	< 5	5-25	> 25	Mo (ppm)	< 0.11	0.11-0.18	> 0.18
Mustard				**Soybean**			
N (%)	2.7-3.19	3.20-5.50	> 5.5	N (%)	3.10-4.00	4.01-5.50	> 5.51
P (%)	0.2-0.29	0.30-0.75	> 0.75	P (%)	0.16-0.25	0.26-0.50	> 0.51
K (%)	1.5-1.99	2.00-4.00	> 4.00	K (%)	1.26-1.70	1.71-2.50	> 2.51

Ca (%)	0.7-0.99	1.00-2.50	> 2.50
Mg (%)	0.2-0.24	0.25-0.75	> 0.75
S (%)	0.25-0.29	0.30-0.75	> 0.75
Fe (ppm)	50-69	70-300	> 300
Mn(ppm)	20-24	25-200	> 200
Zn (ppm)	20-34	35-200	> 200
Cu (ppm)	3-5	5-15	> 15
B (ppm)	25-29	30-100	> 100
Sugarcane			
N (%)	1.6-1.9	2.00-2.60	> 2.6
P (%)	0.15-0.17	0.18-0.30	> 0.30
K (%)	0.9-1.0	1.10-1.80	> 1.80
Ca (%)	0.1-0.19	0.20-0.50	> 0.50
Mg (%)	0.08-0.09	0.1-0.35	> 0.35
S (%)	-	-	-
Fe (ppm)	20-39	40-250	> 250
Mn(ppm)	20-24	25-400	> 400
Zn (ppm)	15-19	20-100	> 100
Cu (ppm)	3-4	5-15	> 15
B (ppm)	2-3	4-30	> 30
Mo (ppm)	< 0.05	0.05-4.0	> 4.0
Coffee			
N (%)	< 2.50	2.50-3.50	> 3.5
P (%)	< 0.12	0.12-0.15	> 0.15
K (%)	< 1.5	1.5-2.5	> 2.5
Ca (%)	< 1.0	1.0-2.5	> 2.5
Mg (%)	< 0.3	0.3-0.4	> 0.4
S (%)	< 0.15	0.15-0.25	> 0.25
Fe (ppm)	< 50	50-200	> 200
Mn(ppm)	< 50	50-150	> 150
Zn (ppm)	< 12	12-30	> 30

Ca (%)	0.21-0.35	0.36-2.00	> 2.01
Mg (%)	0.11-0.25	0.26-1.00	> 1.01
S (%)	0.16-0.20	0.21-0.40	> 0.40
Fe (ppm)	31-50	51-350	> 350
Mn(ppm)	15-20	21-100	> 101
Zn (ppm)	10-20	21-50	> 51
Cu (ppm)	5-9	10-30	> 31
B (ppm)	10-20	21-55	> 56
Tea			
N (%)	< 3.80	3.80-4.80	> 4.80
P (%)	< 0.19	0.19-0.25	> 0.25
K (%)	< 1.80	1.80-2.0	> 2.0
Ca (%)	< 0.40	0.40-0.60	> 0.60
Mg (%)	< 0.15	0.15-0.30	> 0.30
S (%)	< 0.10	0.10-0.30	> 0.30
Fe (ppm)	< 500	500-1000	> 1000
Mn(ppm)	< 30	30-50	> 50
Zn (ppm)	10-20	21-50	> 51
Cu (ppm)	-	-	-
B (ppm)	-	-	-
Cl (ppm)	< 30	30-50	> 50
Potato			
N (%)	3.0-3.29	3.30-4.50	> 4.50
P (%)	0.20- 0.22	0.23-0.5	> 0.5
K (%)	0.28-3.0	3.10-4.5	> 4.5
Ca (%)	0.50-0.69	0.70-1.20	> 1.20
Mg (%)	0.30-0.34	0.35-1.00	> 1.00
S (%)	-	-	-
Fe (ppm)	30-39	40-100	> 100
Mn(ppm)	30-39	40-250	> 250
Zn (ppm)	18-19	20-50	> 50

Cu (ppm)	-	-	-
B (ppm)	< 40	40-75	> 75
Mo (ppm)	< 0.1	0.1-0.5	> 0.5
Brinjal			
N (%)	3.5-3.99	4.0-6.0	> 6.0
P (%)	0.25- 0.29	0.3-1.2	> 1.2
K (%)	3.0-3.49	3.50-5.0	> 5.0
Ca (%)	0.80-0.99	1.0-2.5	> 2.5
Mg (%)	0.25-0.29	0.30-1.0	> 1.0
S (%)	-	-	-
Fe (ppm)	40-49	50-300	> 300
Mn(ppm)	35-39	40-250	> 250
Zn (ppm)	18-19	20-250	> 250
Cu (ppm)	5-7	8-60	> 60
B (ppm)	20-24	25-75	>75
Onion			
N (%)	4.5-4.99	5.0-6.0	> 6.0
P (%)	0.25- 0.34	0.35-0.5	> 0.5
K (%)	3.5-3.99	4.0-5.5	> 5.5
Ca (%)	0.80-0.99	1.0-2.0	> 2.0
Mg (%)	0.22-0.24	0.25-0.4	> 0.4
S (%)	0.3-0.49	0.5-1.0	> 1.0
Fe (ppm)	50-59	60-300	> 300
Mn(ppm)	30-49	50-250	> 250
Zn (ppm)	20-24	25-100	> 100
Cu (ppm)	< 15	15-35	> 35
B (ppm)	18-22	22-60	> 60
Cabbage			
N (%)	2.8-3.59	3.6-5.0	> 5.0
P (%)	0.27- 0.32	0.33-0.75	> 0.75

Cu (ppm)	-	-	-
B (ppm)	20-24	25-75	>75
Tomato			
N (%)	2.5-3.99	4.0-6.0	> 6.0
P (%)	0.20- 0.24	0.25-0.75	> 0.75
K (%)	1.05-2.89	2.9-5.0	> 5.0
Ca (%)	0.80-0.99	1.0-3.0	> 3.0
Mg (%)	0.25-0.39	0.4-0.6	> 0.6
S (%)	0.25-0.39	0.4-1.2	> 1.2
Fe (ppm)	30-39	40-200	> 200
Mn(ppm)	30-39	40-250	> 250
Zn (ppm)	18-19	20-50	> 50
Cu (ppm)	3-4	5-20	> 20
B (ppm)	20-24	25-60	> 60
Beans			
N (%)	4.24-4.99	5.0-6.0	> 6.0
P (%)	0.25- 0.34	0.35-0.75	> 0.75
K (%)	2.0-2.24	2.25-4.0	> 4.0
Ca (%)	1.0-1.49	1.5-2.5	> 2.5
Mg (%)	0.25-0.29	0.3-1.0	> 1.0
S (%)	-	-	-
Fe (ppm)	40-49	50-300	> 300
Mn(ppm)	15-49	50-300	> 300
Zn (ppm)	18-19	20-200	> 200
Cu (ppm)	4-6	7-30	> 30
B (ppm)	15-19	20-75	> 75
Cauliflower			
N (%)	2.8-3.29	3.3-4.5	> 4.5
P (%)	0.28- 0.32	0.33-0.8	> 0.8

K (%)	2.5-2.99	3.0-5.0	> 5.0
Ca (%)	0.80-1.09	1.1-3.0	> 3.0
Mg (%)	0.25-0.39	0.4-0.75	> 0.75
S (%)	0.25-0.29	0.3-0.75	> 0.75
Fe (ppm)	25-29	30-200	> 200
Mn(ppm)	20-24	25-200	> 200
Zn (ppm)	15-19	20-200	> 200
Cu (ppm)	3-4	5-15	> 15
B (ppm)	20-24	25-75	> 75
Mo (ppm)	0.2-0.3	0.4-0.7	> 0.7
Radish			
N (%)	2.8-2.99	3.0-6.0	> 6.0
P (%)	0.25- 0.29	0.3-0.7	> 0.7
K (%)	3.5-3.99	4.0-7.5	> 7.5
Ca (%)	2.0-2.99	3.0-4.5	> 4.5
Mg (%)	0.3-0.49	0.5-1.2	> 1.2
Fe (ppm)	40-49	50-200	> 200
Mn(ppm)	35-49	50-250	> 250
Zn (ppm)	15-18	19-250	> 250
Cu (ppm)	3-4	5-25	> 25
B (ppm)	20-24	25-125	> 125
Apple			
N (%)	1.07-1.89	1.9-2.6	> 2.7
P (%)	0.10- 0.13	0.14-0.4	> 0.4
K (%)	1.0-1.49	1.5-2.0	> 2.0
Ca (%)	< 1.2	1.2-1.6	> 1.6
Mg (%)	< 0.25	> 2.5	-
Fe (ppm)	-	-	-
Mn(ppm)	< 20	> 20	-
Zn (ppm)	< 18	> 18	-

K (%)	2.0-2.59	2.6-4.2	> 4.2
Ca (%)	1.50-1.99	2.0-3.5	> 3.5
Mg (%)	0.22-0.26	0.27-0.5	> 0.5
S (%)	-	-	-
Fe (ppm)	25-29	30-200	> 200
Mn(ppm)	20-24	25-250	> 250
Zn (ppm)	17-19	20-250	> 250
Cu (ppm)	2-3	4-15	> 15
B (ppm)	25-29	30-100	> 100
Mo (ppm)	0.3-0.4	0.5-0.8	> 0.8
Carrot			
N (%)	1.8-2.09	2.1-3.5	> 3.5
P (%)	0.17- 0.19	0.20-0.5	> 0.5
K (%)	2.5-2.79	2.8-4.0	> 4.0
Ca (%)	0.8-1.39	1.4-3.0	> 3.0
Mg (%)	0.25-0.29	0.3-0.5	> 0.5
Fe (ppm)	40-49	50-300	> 300
Mn(ppm)	40-59	60-200	> 200
Zn (ppm)	20-24	25-250	> 250
Cu (ppm)	3-4	5-15	> 15
B (ppm)	25-29	30-100	> 100
Mo (ppm)	0.4-0.49	0.5-1.5	> 1.5
Cherry			
N (%)	2.1-2.59	2.6-3.0	> 3.0
P (%)	0.11- 0.15	0.16-0.22	> 0.22
K (%)	1.2-1.59	1.6-2.1	> 2.1
Ca (%)	1.0-1.49	1.5-2.6	> 2.6
Mg (%)	0.20-0.29	0.3-0.75	> 0.75
Fe (ppm)	60-99	100-200	> 200
Mn(ppm)	25-39	40-60	> 60
Zn (ppm)	15-19	20-50	> 50

Cu (ppm)	< 4	> 4	-
B (ppm)	< 30	30-60	> 60
Peach			
N (%)	2.4-2.9	3.0-3.5	> 3.5
P (%)	0.09- 0.13	0.14-0.25	> 0.25
K (%)	1.0-1.99	2.0-3.0	> 3.0
Ca (%)	1.0-1.79	1.8-2.7	> 2.7
Mg (%)	0.2-0.29	0.3-0.8	> 0.81
Fe (ppm)	60-99	100-250	> 250
Mn(ppm)	20-39	40-160	> 160
Zn (ppm)	15-19	20-50	> 50
Cu (ppm)	3-4	5-16	> 16
B (ppm)	15-19	20-60	> 60
Mango			
N (%)	0.7-0.99	1.0-1.5	> 1.5
P (%)	0.05- 0.07	0.08-0.25	> 0.25
K (%)	0.25-0.39	0.40-0.9	> 0.9
Ca (%)	1.0-1.99	2.0-5.0	> 5.0
Mg (%)	0.15-0.19	0.2-0.5	> 0.5
Fe (ppm)	25-49	50-250	> 250
Mn(ppm)	25-49	50-250	> 250
Zn (ppm)	15-19	20-200	> 200
Cu (ppm)	5-6	7-50	> 50
Grape (var.Thomson seedless)			
N (%)	0.87-1.31	1.32-2.21	2.22-2.6
P (%)	0.19- 0.38	0.38-0.75	0.76-0.95
K (%)	0.6-1.13	1.14-2.2	2.21-2.73
2.73			
Ca (%)	0.53-0.73	0.74-1.14	1.15-1.35
Cu (ppm)	4-7	8-28	> 28
B (ppm)	15-19	20-55	> 55
Plum			
N (%)	1.7-2.39	2.4-3.0	> 3.0
P (%)	0.09- 0.13	0.14-0.25	> 0.25
K (%)	1.0-1.59	1.6-3.0	> 3.0
Ca (%)	1.0-1.49	1.5-3.0	> 3.0
Mg (%)	0.2-0.29	0.3-0.8	> 0.8
Fe (ppm)	60-99	100-250	> 250
Mn(ppm)	20-39	40-160	> 160
Zn (ppm)	15-19	20-50	> 50
Cu (ppm)	4-5	6-16	> 16
B (ppm)	20-24	25-60	> 60
Banana			
N (%)	2.0-2.49	2.5-3.0	> 3.0
P (%)	0.14- 0.17	0.18-0.4	> 0.4
K (%)	2.0-2.29	2.3-4.0	> 4.0
Ca (%)	0.4-0.69	0.7-1.4	> 1.4
Mg (%)	0.2-0.24	0.25-0.4	> 0.4
S (%)	0.2-0.25	0.26-0.5	> 0.5
Fe (ppm)	80-99	100-300	> 300
Mn(ppm)	150-199	200-2000	> 2000
Zn (ppm)	10-12	13-50	> 50
Cu (ppm)	4-5	6-30	> 30
Pineapple			
N (%)	< 1.5	15-1.7	> 1.7
P (%)	-	0.1	> 0.1
K (%)	< 2.2	2.2-3.0	> 3.0
Ca (%)	< 0.8	0.8-1.2	> 1.2

S (%)	0.07-0.13	0.14-0.27	0.28-0.34	Mg (%)	-	0.3	> 0.3
Mn(ppm)	26-75	76-174	175-223	Fe (ppm)	< 100	100-200	> 200
Zn (ppm)	13-52	53-132	133-171	Mn(ppm)	< 50	50-200	> 200
				Cu (ppm)	-	< 10	> 10
				B (ppm)	-	< 30	> 30
Papaya				**Water melon**			
N (%)	0.8-1.0	1.01-2.5	> 2.5	N (%)	3.5-3.99	4.0-5.5	> 5.5
P (%)	0.18- 0.21	0.22-0.4	> 0.4	P (%)	0.25- 0.29	0.3-0.8	> 0.8
K (%)	2.8-3.2	3.3-5.5	> 5.5	K (%)	3.5-3.99	4.0-5.0	> 5.0
Ca (%)	< 1.0	1.0-3.0	> 3.0	Ca (%)	1.0-1.69	1.7-3.0	> 3.0
Mg (%)	< 0.4	0.4-1.2	> 1.2	Mg (%)	0.3-0.49	0.5-0.8	> 0.8
Fe (ppm)	20-24	25-100	> 100	Fe (ppm)	40-49	50-300	> 300
Mn(ppm)	10-19	20-150	> 150	Mn(ppm)	< 50	50-250	> 250
Zn (ppm)	10-14	14-40	> 40	Zn (ppm)	18-19	20-150	> 150
Cu (ppm)	< 4	4-10	> 10	Cu (ppm)	4-5	6-20	> 20
B (ppm)	< 20	20-30	> 30	B (ppm)	20-24	25-60	> 60

*Concentrations less than low category denote deficiency

and quality of the crop. Although critical nutrient level acts a general guideline for nutrient management (Table 27.2) it varies among the crops and needs area specific refinement it varies with a change of large number of factors including varieties of the crop, yield potentials, status of other nutrients, etc.

27.3.2. Diagnosis and Recommendation Integrated System (DRIS)

Beaufils (1973) developed the DRIS concept for better diagnosis of nutrient deficiency wherein more than one nutrient concentration ratios are taken into consideration at a time. DRIS is a system that identifies all yield limiting factors and, thus increases the chances of getting higher crop yield through improved fertilizer recommendation. Index values indicating how far the nutrient concentrations in the plant are from their levels, are used in the calibration to classify yield factors in order of limiting importance. To develop DRIS norm for a given crop, the following requirement must be met:

1. All factors suspected of having influence in crop yield must be defined.
2. The relationship between these factors and yield must be described.
3. Calibrated norms must be established.
4. Recommendation based on judicious use of norms and suited to the particular sets of conditions must be subsequently refined.

To establish the DRIS norms the following steps are taken:

- At first a random survey of large number of sites representing the whole production area of the crop concern within the district or state.
- Information is collected related to crop production (fertilizer dose, herbicide application, cultural practices) and climatic condition.
- Plant and soil sample collected from each site are analyzed for various nutrients.
- The entire population of observations is now divided into two groups - high yielders and low yielders based on plant vigour or yield.
- Nutrient concentrations in plant are expressed as many ways as possible like, percent N in plant, or ratios N/P, N/K or products N*P, N*K and so on.
- The mean of each ratio for each population group is calculated.
- Each ratio that significantly discriminates between high- and low yielding population is retained as diagnostic parameter.

- Finally, calculate DRIS index based on significant variance ratios using formula worked out by Bhargava and Chadha (1993).

Higher value with negative sign represents the most limiting nutrient, while higher value with positive sign indicates the least required nutrient. In this way, nutrient combination which gives maximum yield and at the same time a well-balanced nutrient status is worked out.

DRIS system has several advantages over the critical level approach in making diagnosis for fertilizer recommendation:

1. The importance of nutritional balance is taken into account in deriving norms and making diagnosis.
2. The norms for nutrient content in leaves can be applied to a particular crop irrespective of locations where it is raised.
3. The norm is useful for a wide range of stage of crop irrespective of the cultivar.
4. The yield limiting nutrients can be easily identified and arranged in order of their importance in limiting crop yield.

27.3.3. Crop Logging

This technique involves keeping of graphic record of series of physical and chemical measurements like nutrient concentration and moisture content in plant, weather parameter, etc with the progress of crop growth. Here also, the critical nutrient concentration approach is used. The advantage of this method is that at any moment necessary change in nutrient management can be made to maximize crop yield.

Questions and Hints

Q.1 Why grinding of plant sample is necessary for plant analysis?

Hint: For homogenization of sample and to provide higher surface for rapid reaction.

Q.2 During digestion of plant sample temperature is maintained between 360 and 410°C-Why?

Hint: Digestion will be incomplete below 360°C and above 410°C temperature ammonia will be lost by volatilization.

Q.3 Why pre-digestion with concentrated HNO_3 is required for high fatty/oily plant sample?

Hint: Direct contact of high fatty/oily substance with $HClO_4$ may lead to explosion and fire, hence pre-digestion is required.

Q.4 What is the role of ammonium vanadate in P estimation?

Hint: Molybdenum in ammonium molybdate gives yellow coloured heteropoly phosphomolydate complex with phosphoric acid. Inclusion of vanadate as ammonium vanadate further intensifies the yellow colour with the formation of heteropoly vanadomolybdophosphoric acid.

References

Allison, L.e., Bollen, W.E. and Moodie, C.D. (1965). Total carbon. P 1346-1366. In C.A.Black *et al.* (ed) *Methods of Soil Analysis*. Part 2. Agron. Monogr. 9. ASA, Madison, W.I.

Arkley, T. H. (1961). Sulphur compounds of soil systems. Ph. D. diss. Univ. California, Berkley.

Atterberg, A. (1911). Die Plastizitat der Tone. *Int. Mitt. Bodenk.*, **1**: 10-43.

Atterberg, A. (1912). Die Konsistenz und die Bindigkeit der Boden. *Int. Mitt. Bodenk.*, **2**: 148-189.

Beckett, P. H. T. (1964). The immediate Q/I relations of labile potassium in the soil. *Journal of Soil Sci.* **15**: 9-23.

Beer, A. (1852). *Ann. Physik. Chem.* (J. C. Poggendorff). **86**: 78.

Berger, K. C. and Troug, E. (1939). Boron determination in soils and plants. *Ind. Eng. Chem. Anal. Ed.* **11**: 540-545.

Bouyoucos, G.J. (1927). The hydrometer as a new method for the mechanical analysis of soils. *Soil Sci.*, **23**: 343 -353.

Bray, R.H. and Kurtz, L.T. (1945). Determination of total, organic, and available forms of phosphorus in soils. *Soil Sci.* **59**: 39-45.

Brookes, P.C., Powlson, D.S. and Jenkinson, D.S. (1982). Measurement of microbial biomass phosphorus in soil. *Soil Biology and Biochemistry*. **14**: 319-329.

Bruce, R.R. and Klute, A. (1956). The measurement of soil moisture diffusivity. *Soil Sci. Soc. Am Proc.* **20**: 458-462.

Casagrande, A. (1932). Research on the Atterberg limits of soil. *Pub. Roads*, **13**: 121 – 130.

Chang, S.C. and Jackson, M.L. (1957). Fractionation of soil phosphorus. *Soil Sci.* **84**: 133-141.

Day, P.R. (1965). Paritcle fractionation and particle-size analysis. *Methods of Soil Analysis, Agronomy Monograph*, part 1, pp. 545 -567, Academy Press, New York.

Ensminger, L. E. (1954). Some fractions affecting the adsorption of sulphate by Alabama soils. *Proc. Soil Sci. Soc. Am.* **18**: 259-262.

Evans, C. A. and Rost, C. O. (1945). Total organic sulfur and humus sulfur of Minnesota soils. *Soil Sci.* **59**: 125-137.

Fox, R.L., Olson, R.A. and Rhoades, H.F. (1964). Evaluating the sulphur status of soils by plant and soil tests. **28**: 243-246.

Grigg, J. L. (1953). Determination of available molybdenum of soils. *New Zealand Journal of Science and Technology*, **34A**: 405-414.

Hanway, J. and Heidel, H. (1952). Soil analysis methods as used in Iowa State College Soil Testing Laboratory. Iowa State College, *Agric. Bull.* **57**: 1-13.

Hesse, P. R. (1971). *A Text Book of Soil Chemical Analysis*. John Murry (Publishers) Ltd., London, UK.

Jackson, M. L. (1973). *Soil Chemical Analysis*. Prentice Hall of India Pvt. Ltd, New Delhi.

Jenkinson, D. S. (1988). The determination of Microbial Carbon and Nitrogen in Soil, p. 368-386. In J.R. Wilson (ed). Advances in Nitrgen Cycling in Agricultural Ecosystems. C.A.B. International, Wallingford.

Joergensen, R.G. (1996). The fumigation-extraction method to estimate soil microbial biomass: Calibration of the k_{EC} value, *Soil Biology and Biochemistry*, **28**: 25-31.

Kappen, G. (1934) Pochvennaya, Kislotmost, Selkhogiz, Moscow (cf. *Journal of the Indian Society of Soil Science*, (1991). **39**: 246.

Lambert, I.H. (1760). *Photometria de Mensura et Gradibus luminous*, Colorum of Umbrae, Augsbarg; reprinted (1892) in W. Ostwald Klassiker der Exakten Wissenschaften, No. **32**: 64.

Mazurak, A.P. (1950). Effect of gaseous phase on water-stable synthetic aggregates. *Soil Sci*., **69**: 135-148.

McIntosh, J. L. (1969). Bray and Morgan soil test extractants modified for testing acid soils from different parent materials. *Agron. J.* **61**: 259-265.

McQuaker, N. R., Kluckner, P. D. and Chang, G. N. (1979). Calibration of an inductively coupled plasma-atomic emission spectrophotometer for analysis of environmental materials. *Anal. Chem*. **51**(7): 888-95.

Mehlich, A. (1953). Determination of P, Ca, Mg, K, Na and NH_4. Mimeo 1953. North Carolina Soil Test Div., Raleigh, NC.

Mehlich, A. (1984). Mehlich-3 soil test exractant: A modification of Mehlich-2 extractant. *Commun. Soil Sci.Plant Anal*. **15**: 1409-1416.

Morgan, M. E. (1941). Chemical soil diagnosis by the universal soil testing system. *Connecticut Agric. Exp. Sta. Bull*. 450.

Murphy, J. and Riley, J.P. (1962). A modified single solution method for the determination of phosphate in natural water. *Anal. Chem. Acta*. **27**: 31-36.

Nelson, D. W. and Sommers, L. E. (1982). Total carbon, organic carbon and organic matter. In *Methods of Soil Analysis*. Part 2 (Ed. Page *et al.*) Agron. 1, Am. Soc. Agron., Madison, Wisconsin: 539-579.

Olsen, S.R., Cole, C.V., Watanabe, F.S. and Dean, L.A. (1954). Estimation of available phosphorus in soils by extraction with sodium bicarbonate. *USDA Circ*. 939. USDA, Washington, DC.

Petersen, G.W. and Corey, R.B. (1966). A modified Chang and Jacksom procedure for routine fractionation of inorganic soil phosphates. *Proc. Soil Sci. Soc. Am*. 30: 563-565.

Philip, J.R. (1957). Theory of infiltration: 4. Sorptivity and algebraic infiltration equation. *Soil sci.,* **84**: 257-264.

Rabenhorst, M.C. (1988). Determination of organic and carbonate carbon in calcareous soils using dry combustion. *Soil Sci. Soc. Am.J*. **52**: 965-969.

Robinson, G.W. (1922). A new method for the mechanical analysis of soils and other dispersions. *J. Agr. Sci*., **12**: 306 -321.

Russel, E.W. (1938). *Measurement of Pore-space and Crumb or Aggregate Structure of Soils*. Proc. third conf. Cotton Growers problem, Rothamsted Exp. Stn.

Saggar, S., Bettany, J.R. and Stewart, J.W.B. (1981). Measurement of microbial sulphur in soil. *Soil Biology and Biochemistry*. **13**: 493-498.

Schllenberger, C. J. (1927). A rapid approximate method for determining soil organic matter. *Soil Sci*. **24**: 65-68.

Schofield, R.K. and Taylor, A.W. (1955). The measurement of soil pH. *Soil Sci. Soc. Am*. Proc. **19:** 164-167.

Shoemaker, H. E., McLean, E.O. and Pratt, P.F. (1961). Buffer methods for determining lime requirement of soils with appreciable amount of extractable aluminium. *Proc. Soil Sci. Soc. Am*. **25:** 274-277.

Shoonover, W.R. (1952). *Examination of Soils for Alkali*. Bull. Univ. of California Extension Services, Berkeley, California (mimeographed publication).

Shuman, L. M. (1985). Fractionation method for soil microelements. *Soil Sci.* **140**: 11-22.

Sokonov, A.V. (1939). Klmizastia sots Zamledelia, 7 (cf. *Journal of the Indian Society of Soil Science*, 11: 329-332).

Soltanpour, P.N. and Schwab, A.P. (1977). A new soil test for simultaneous eztraction of macro- and micro-nutrients in alkaline soils. *Comm. Soil Sci. Plant Anal.* **8**: 195-207.

Soltanpour, P. N., Johnson, G. W., Workman, S. M., Jones, J. B. Jr. and Miller, R. O. (1998). Advances in ICP emission and ICP mass spectrophotometry. *Adv. Agron.* **64**: 27-113.

Stevenson, F.J. (1996). Nitrogen –Organic Forms. In: *Methods of Soil Analysis*, part 3-Chemical Methods (Ed. D.L.Sparks). SSSA Book Series: 5. Chapter **39**: 1185-1200.

Stokes, G.G. (1851). On the effect of the internal friction of fluids on the motion of pendulums. *Trans. Cambridge Phil. Soc.*, **9**: 8 – 106.

Subbaih, B. V. and Asija, G.L. (1956). A rapid procedures for the determination of available nitrogen in soils. *Current Science.* **25**: 259 – 260.

Tyurin, I. V. (1931). A new modification of the volumetric method of determining soil organic matter by means of chromic acid. *Pochvovedenie.* **26**: 36-47.

USSLS (1954). *Diagnosis and Improvement of Saline and Alkali Soils*. United Rtate Department of Agriculture. Agriculture Handbook No. 60 (Reprint). Oxford and IBH Publishing Co. New Delhi.

van Bavel, C.H.M. (1949). Mean weight diameter of soil aggregates as a statistical index of aggregation. Soil Sci. Sco. Am. Proc., **14**: 20-23.

Walkley, A. and Black, I. A. (1934). An examination of the Degtjareff method for determining soil organic matter, and a proposed modification of the chromic acid titration method. *Soil Sci.* **34**: 29-38.

Waugh, D.L. and Fitts, J.W. (1966). *Soil Test Interpretation Studies*: Laboratory and potted plant. Tech. Bull. N. Carol. State Agric. Exp. Stn. (ISTP Series) No. 3.

Williams, C. H. and Steinbergs, A. (1959). Soil sulphur fractions as chemical indices of available sulphur in some Australian Soils. *Aus. J. Agric. Res.* **10**: 340-352.

Wood, L. K. and DeTurk, E. E. (1940). The absorption of potassium in soils in non-replaceable forms. *Proc. Soil Sci. Soc. Am.* **5**: 152-161.

Yoder, R.E. (1936). A direct method of aggregate analysis of soils and a study of the physical nature of erosion losses. *J. Am. Soc. Agron.*, **28**: 337-351.

Yuan, T.L. (1959). Determination of exchangeable hydrogen in soils by titration method. Soil Sci. **88**: 164-167.

Appendices

Appendix I: Sieve Size

The relationship between sieve opening size (mm) and mesh per inch

$$\text{Opening diameter in mm (approx)} = \frac{16}{\text{Mesh per inch}}$$

Opening diameter (mm)	Mesh / inch (approx)
6	4
4	6
2	10
1	20
0.25	60
0.16	100
0.1	140

Appendix II: Density and Viscosity of Water at Different Temperature

Temperature (°C)	Density (g/cc)	Viscosity (millipoise)	Temperature (°C)	Density (g/cc)	Viscosity (millipoise)
4	1.0000	15.67	23	0.9976	9.36
16	0.9990	11.11	24	0.9973	9.14
17	0.9988	10.83	25	0.9971	8.94
18	0.9986	10.56	26	0.9968	8.74
19	0.9984	10.30	27	0.9966	8.54
20	0.9982	10.05	28	0.9963	8.36
21	0.9980	9.81	29	0.9960	8.18
22	0.9978	9.58	30	0.9957	8.01

Appendix III: Factor for converting Non- SI units to SI units

Non- SI unit	Multiply by*	SI unit
Length		
inch,in	2.54	centimeter, cm (10^{-2}m)
foot, ft	0.304	meter, m
mile	1.609	kilometer, km (10^{3}m)

micron,ì	1.0	micrometer,ìm (10^{-6}m)
Angstrom unit,A°	0.1	nanometer,nm (10^{-9}m)
Area		
acre, ac	0.405	hectare, ha
square foot,ft^2	9.29 x 10^{-2}	square meter,m^2
square inch,in^2	645	square millimeter,mm^2
square mile, ml^2	2.59	square kilometer,km^2
Volume		
cubic foot, ft^3	2.83 x 10^{-2}	cubic meter, m^3
cubic inch, in^3	1.65 x 10^{-5}	cubic meter, m^3
gallon (U.S.), gal	3.78	litre,l (dm^3)
acre-foot, ac-ft	12.33	Hectare-centimeter, ha-cm
acre-inch, ac-in	1.03 x 10^{-2}	Hectare-meter, ha-m
Mass		
ounce, oz	28.4	gram, g
pound, lb	0.454	kilogram, kg (10^3 g)
quintal, q	100	kilogram, kg
tonne (metric)	1000	kilogram, kg
Yield and rate		
pound per acre, lb/ac	1.121	kilogram per hectare, kg/ha
miles per hour, mph	0.447	meter per second, m/s
gallon per minute, gpm	0.227	cubic meter per hour, m^3/h
cubic feet per second, cfs	101.9	cubic meter per hour, m^3/h
Pressure		
atmosphere, atm	0.101	megapascal, MPa (10^6 Pa)
bar	0.1	megapascal, MPa
pound per square foot, lb/ft^2	47.9	pascal, Pa
pound per square inch, lb/in^2	6.9 x 10^3	pascal, Pa
torr	133.3	pascal, Pa
Temperature		
degrees Fahrenheit (°F- 32)	0.556	degrees, °C
degrees Celcius, (°C + 273)	1	Kelvin, K
Energy		
British thermal unit,Btu	1.05 x 10^3	joule, J
calorie, cal	4.19	joule, J
dyne, dyn	10^{-5}	newton, N
erg	10^{-7}	joule, J
foot-pound, ft-lb	1.36	joule, J
Concentration		
percent, %	10	gram per kilogram, g/kg
part per million, ppm	1	milligram per kilogram, mg/kg
milliequivalents per 100 grams	1	centimole per kilogram, cmol/kg
Electrical conductivity		
millimho per centimeter, mmho/ cm	0.1	Siemen per meter, S / m

*To convert from SI to non-SI units, divide by the factor

Appendix IV. Relation between RCF and RPM in Centrifugation

Relative centrifugal force (RCF) x g = 1.12 n^2r x 10^{-5}

where, n = RPM (Revolutions per minute); r = rotational radius, in cm

Appendix V: Conductivity of 0.01N KCl Solution at Different Temperature

Temperature (°C)	Conductivity (mmhos/cm)	Temperature (°C)	Conductivity (mmhos/cm)
15	1.147	25	1.413
16	1.173	26	1.441
17	1.199	27	1.470
18	1.225	28	1.498
19	1.251	29	1.526
20	1.278	30	1.554
21	1.305	31	1.583
22	1.332	32	1.611
23	1.359	33	1.639
24	1.386	34	1.668

Appendix VI: International Atomic Number and Atomic Weight of Elements Important in Soil Analysis

Name (symbol)	Atomic number	Atomic weight	Name (symbol)	Atomic number	Atomic weight
Aluminium (Al)	13	26.98	Magnesium (Mg)	12	24.32
Antimony (Sb)	51	121.79	Manganese (Mn)	25	54.94
Arsenic (As)	33	74.91	Mercury (Hg)	80	200.61
Barium (Ba)	56	137.36	Molybdenum (Mo)	42	95.95
Boron (B)	5	10.82	Nickel (Ni)	28	58.69
Bromine (Br)	35	79.916	Nitrogen (N)	7	14.008
Cadmium (Cd)	48	112.41	Oxygen (O)	8	16.00
Calcium (Ca)	20	40.08	Phosphorus (P)	15	30.975
Carbon ©	6	12.01	Platinum (Pt)	78	195.23
Cerium (Ce)	58	140.13	Potassium (K)	19	39.100
Chlorine (Cl)	17	35.457	Radium (Ra)	88	226.05
Chromium (Cr)	24	52.01	Selenium (Se)	34	78.96
Cobalt (Co)	27	58.94	Silicon (Si)	14	28.09
Copper (Cu)	29	63.54	Silver (Ag)	47	107.88
Fluorine (F)	9	19.00	Sodium (Na)	11	22.991
Helium (He)	2	4.003	Sulphur (S)	16	32.066
Hydrogen (H)	1	1.008	Tin (Sn)	50	118.70
Iodine (I)	53	126.91	Titanium (Ti)	22	47.90
Iron (Fe)	26	55.85	Uranium (U)	92	238.07
Lead (Pb)	82	207.21	Zinc (Zn)	30	65.38

Appendix VII: Some Common Commercial Concentrated Acids and Bases

Reagent	Approx. weight (%)	Density	Molarity*	Normality*	ml required to prepare 1L of 1M solution	ml required to prepare 1L of 1N solution
Acetic acid, glacial (CH_3COOH)	99.7	1.05	17.4	17.4	57.5	57.5
Hydrochloric acid (HCl)	36.0	1.18	11.65	11.65	85.8	85.8
Hydrofluoric acid (HF)	48.0	1.15	27.6	27.6	36.0	36.0
Nitric acid (HNO_3)	90.0	1.48	21.1	21.1	47.4	47.4
Nitric acid (HNO_3)	70.0	1.42	15.8	15.8	63.4	63.4
Perchloric acid ($HClO_4$)	70.0	1.67	11.6	11.6	86.2	86.2
Perchloric acid ($HClO_4$)	60.0	1.54	9.2	9.2	108.7	108.7
Phosphoric acid (H_3PO_4)	85.0	1.69	14.7	44.0	68.2	22.7
Sulphuric acid (H_2SO_4)	96.0	1.84	17.8	36.0	55.5	27.8
Ammonium hydroxide (NH_4OH)	57.6	0.90	14.8	14.8	67.6	67.6

***Way to calculate normality/ morality of any concentrated acid/base:**

Say, for phosphoric acid,

Density = Mass/Volume; so, mass = 1000 × 1.69 = 1690 g/ litre

Actual mass of phosphoric acid present per litre = 1690 × (% weight)

= 1690 × (85/100) = 1437 g/litre

Equivalent weight of phosphoric acid = 32.67

So, 32.67 g phosphoric acid per litre of solution ≡ 1**N**

1437 g of phosphoric acid/litre of solution ≡ (1437/32.67) **N** = 43.98 **N**

Thus, normality of concentrated phosphoric acid = 44

Molecular weight of phosphoric acid = 98

So, 1437 g of phosphoric acid/litre of solution ≡ (1437/98) **M** = 14.66 **M**

Thus, molarity of concentrated phosphoric acid = 14.7

Appendix VIII: Equivalent Weight of Some Common Acids, Bases and Salts

Name, formula	Mol. Wt	Basicity or acidity	Total valence of metal atoms /molecule	Equivalent weight
		Acids		
Acetic acid, CH_3COOH	60	1		60
Hydrochloric acid, HCl	36.5	1		36.5
Sulphuric acid, H_2SO_4	98	2		49
Nitric acid, HNO_3	63	1		63
Perchloric acid,$HClO_4$	84.5	1		84.5
Oxalic acid,$H_2C_2O_4.2H_2O$	126	2		63
Benzoic acid,C_6H_5COOH	122.22	1		122.22
Succinic acid,$(CH_2COOH)_2$	118.08	2		59.04
Phosphoric acid,H_3PO_4	98	3		32.6
		Bases		
Sodium hydroxide, NaOH	40	1		40
Potassium hydroxide, KOH	56	1		56
Sodium carbonate, Na_2CO_3	106	2		53
Sodium bicarbonate, $NaHCO_3$	84	1		84
Calcium hydroxide, $Ca(OH)_2$	74	2		37
		Salts		
Copper sulphate, $CuSO_4.5H_2O$	249.69		2	124.84
Ferric chloride, $FeCl_3.6H_2O$	270.33		3	90.11
Borax, $Na_2B_4O_7.10H_2O$	381.34		2	190.67
Silver nitrate, $AgNO_3$	169.9		1	169.9

Appendix IX: Equivalent Weight of Some Common Oxidants and Reductants

Substance	Radical or element involved	*O.N. of the effective element	Reduction product	New O.N.	Change of O.N per molecule	Equiv. weight (Mol. Weight / change in O.N)
Oxidants						
Potassium permanganate, $KMnO_4$ (acid)	MnO_4^-	+7	Mn^{2+}	+2	-5	158.03 / 5 = 31.61
Potassium permanganate, (neutral)	MnO_4^-	+7	MnO_2 or Mn^{4+}	+4	-3	158.03 / 3 = 52.68
Potassium permanganate (alkaline)	MnO_4^-	+7	MnO_4^{2-}	+6	-1	158.03/1= 158.03
Potassium dichromate, $K_2Cr_2O_7$	$Cr_2O_7^{2-}$	+6	Cr^{3+}	+3	- 6	294.22 / 6 = 49.04
Chlorine, Cl_2	Cl	0	Cl^-	-1	- 2	71 / 2 = 35.5
Ferric chloride, $FeCl_3.6H_2O$	Fe^{3+}	+3	Fe^{2+}	+2	-1	270.22/ 1= 270.22
Potassium iodate, KIO_3	IO_3^-	+5	I^-	-1	-6	214.01 / 6= 35.67
Hydrogen peroxide, H_2O_2	O_2	-1	O^{2-}	-2	-2	34.02 / 2= 17.01
Cerium sulphate, $CeSO_4$	Ce^{4+}	+4	Ce^{3+}	+3	-1	332.26/ 1= 332.26
Reductants			**Oxidation product**			
Ferrous sulphate, $FeSO_4$	Fe^{2+}	+2	Fe^{3+}	+3	+1	278.02/1= 278.02
Stannous chloride, $SnCl_2$	Sn^{2+}	+2	Sn^{4+}	+4	+2	225.70/2=112.85
Metals, e.g.Zn	Zn	0	Zn^{2+}	+2	+2	65.38/2= 32.69
Oxalic acid, $H_2C_2O_4$	$C_2O_4^{2-}$	+3	CO_2	+4	+2	126.04/2=63.02
Sodium thiosulphate, $Na_2S_2O_3.5H_2O$	$S_2O_3^{2-}$	+2	$S_4O_6^{2-}$	+2.5	+1	248.21/1=248.21
Potassium iodide, KI	I^-	-1	I^0	0	+1	166.01/1=166.01

***Rules for the determination of oxidation number (O.N)**

- The O.N of the free or uncombined element is zero.
- The O.N of a compound is always zero.
- The O.N of a radical or ion is that of its electronic charge with the correct sign attached.

Appendix X: Conversion Factor of Plant Nutrient Elements

Form of nutrients	To convert from column 1 to column 3, multiply by*	Form of nutrients
P	2.29	P_2O_5
K	1.20	K_2O
S	3.00	SO_4
Ca	1.40	CaO
Ca	1.85	$Ca(OH)_2$
Ca	2.50	$CaCO_3$
Mg	3.50	$MgCO_3$
$CaCO_3$	0.56	CaO

*To convert from column 3 to column 1, divide by the factor

Appendix XI: Analytical Condition for AAS Analysis

Element	Resonance line (nm)	Slit width (nm)	Working range (ppm)	Sensitivity (ppm)	Lamp current (mA)	Flame type*
Ca	422.7	0.5	1-10	0.09	10.0	AA (stio)
Cd	228.8	0.5	0.5-5	0.03	3.0	AA (stio)
Cr	357.9	0.2	2-20	0.2	6.0	AA (red)
Cu	324.7	0.5	1-20	0.1	3.0	AA(oxi)
Fe	248.3	0.2	2-20	0.1	7.0	AA(oxi)
Mg	285.2	0.5	0.1-2	0.01	3.0	AA(oxi)
	202.6	1.0	5-20	0.1	3.0	AA(oxi)
Mn	279.5	0.2	1-10	0.06	5.0	AA (stio)
Ni	232.0	0.2	2-20	0.2	4.0	AA(oxi)
Pb	283.3	0.5	4-40	0.2	5.0	AA(oxi)
Zn	213.9	0.5	0.5-5	0.03	5.0	AA(oxi)

*oxi- oxidizing flame (lean) - A very stiff flame with small inner blue cone,
stio- Stiochiometric flame - A very stiff flame with bigger blue cone and an almost luminous appearance, red- Reducing flame (rich) – Extremely luminous flame almost scooty at the top.

Appendix XII: Primary Standard Solution Preparation for ICP-AES

Element	Metal / Compound	Weight (g)	Solvent
Al	Al	1.0000	6 **M** HCl
Sb	Sb	1.0000	Aqua regia
As	As	1.0000	4 **M** HNO_3
Ba	$BaCl_2$	1.1516	Water
Be	Be	1.0000	0.5 **M** HCl
B	H_3BO_3	5.7195	Water
Cd	Cd	1.0000	4 **M** HNO_3
Ca	$CaCO_3$	2.4972	0.5 **M** HNO_3
Cr	Cr	1.0000	4 **M** HCl
Cu	Cu	1.0000	4 **M** HNO_3
Fe	Fe	1.0000	4 **M** HCl
Pb	Pb	1.0000	4 **M** HNO_3
Mg	MgO	1.6581	0.5 **M** HCl
Mn	Mn	1.0000	4 **M** HNO_3
Hg	$HgCl_2$	1.3535	Water + 1 g $(NH_4)_2S_2O_8$
Mo	Mo	1.0000	Aqua regia
Ni	Ni	1.0000	4 **M** HCl
P	NaH_2PO_4	3.7137	Water
K	KCl	1.9067	Water
Si	Na_2SiO_3	10.1190	Water
Ag	Ag	1.0000	4 **M** HNO_3
Na	NaCl	2.5421	Water
Sr	$SrCO_3$	1.6849	1 **M** HNO_3
Ti	Ti	1.0000	4 **M** HCl
V	V	1.0000	4 **M** HNO_3
Zn	Zn	1.0000	4 **M** HNO_3

Use 100-150 ml of solvent to dissolve and bring to 1 litre volume with double distilled water to give a concentration of 1000 ppm.

Appendix XIII: Background of the Formula used in Calculations

A. In Spectrophotmetric/ Flame photometric analysis

Say, for calculation of available P in soil (Section 21.1.1)

If, weight of the soil taken = W g

Volume of the Olsen's extractant added to the soil = V_D ml

Volume of the aliquot taken for colour development = V_A ml

Final volume of the coloured solution = V_C ml

Absorbance reading of the test solution = X

Absorbance reading of the blank solution = Y

Concentration of P corresponding to X absorbance (from standard curve) = A ppm

Concentration of P corresponding to Y absorbance (from standard curve) = B ppm

Thus, (A-B) ppm is the actual concentration of available P in final coloured solution.

(A-B) ppm of available P means (A-B) mg available P per litre of final solution

1000 ml of solution contains (A-B) mg of available P

V_C ml solution contains $\frac{(A-B)}{1000} \times V_C$ mg of available P

Again, $\frac{(A-B)}{1000} \times V_C$ mg of available P was originally present in V_A ml of aliquot.
Therefore, we can write,

V_A ml soil extract contains $\frac{(A-B)}{1000} \times V_C$ mg of available P

$\therefore V_D$ ml of soil extract contains $\frac{(A-B)}{1000} \times \frac{V_C}{V_A} \times V_D$ mg of available P

Again, $\frac{(A-B)}{1000} \times \frac{V_C}{V_A} \times V_D$ mg of available P was originally present in W g of soil.
So, we can write,

W g soil contains $\frac{(A-B)}{1000} \times \frac{V_C}{V_A} \times V_D$ mg of available P

1000 g or 1 kg soil contains $\frac{(A-B)}{1000} \times \frac{V_C}{V_A} \times \frac{V_D}{W} \times 1000$ mg of available P

Therefore, the concentration of available P in soil = (A-B) × $\frac{V_C}{V_A} \times \frac{V_D}{W}$ mg/ kg or

(A-B) × $\frac{V_C}{V_A} \times \frac{V_D}{W}$ ppm.

B. In Titrimetric analysis

Say, for calculation of Ca plus Mg in soil solution (Section 17.2.4)

If, weight of the soil taken = W g

Volume of water added for extraction of salt = V_W ml

Volume aliquot taken for Ca plus Mg estimation = V_1 ml

Volume of EDTA consumed for Ca plus Mg estimation= V_X ml

Volume of EDTA consumed for blank (Ca plus Mg titration) = V_A ml

Strength (N) of EDTA = S

Thus, (V_X - V_A) is the volume of S (N) EDTA actually consumed for titration of Ca+Mg present in V_1 ml aliquot.

We know,

1000 ml 1 (N) EDTA ≡ 1 Eq. wt of any cation

Or, 1000 ml 1 (N) EDTA ≡ Eq.wt of (Ca+Mg)

1 ml 1 (N) EDTA meq. wt of (Ca+Mg))

(V_X - V_A) ml S (N) EDTA ≡ (V_X - V_A) × S meq. wt of (Ca+Mg)

Again, (V_X - V_A) × S meq. wt of (Ca+Mg) was originally present in V_1 ml soil solution

V_1 ml soil solution contains (V_X - V_A) × S meq. wt of (Ca+Mg)

1000 ml soil solution contains (V_X - V_A) meq. $\times S \times \frac{1000}{V_1}$ wt of (Ca+Mg)

And, V_W ml soil solution contains (V_X - V_A) $\times S \times \frac{V_W}{V_1}$ meq. wt of (Ca+Mg)

Again, (V_X - V_A) $\times S \times \frac{V_W}{V_1}$ × S meq. wt of (Ca+Mg) was originally present in W g of soil.

Therefore, we can write,

W g of soil contains (V_X - V_A) $\times S \times \frac{V_W}{V_1}$ meq. wt of (Ca+Mg)

100 g of soil contains (V_X - V_A) $\times S \times \frac{V_W}{V_1} \times \frac{100}{W}$ meq. wt of (Ca+Mg)

Appendix XIV. Cleaning Solution

Chromic–sulphuric acid cleaning solution is very useful for final washing of glasswares. Visible materials and organic solvents should be rinsed out with water before using the cleaning solution.

Preparation

Chromic–sulphuric acid cleaning solution: Dissolve 80 g of potassium dichromate ($K_2Cr_2O_7$) or sodium dichromate ($Na_2Cr_2O_7$) in 300 ml of water (with heating). Place the aqueous solution in a Pyrex container and cautiously add one litre of laboratory grade H_2SO_4 with stirring. Considerable amount of red chromic oxide will precipitate.

Appendix XV: Cleanliness and Orderliness in the Laboratory

The student should follow the following general rules as a guide for cleanliness and orderliness in the laboratory:

- All the glass and other containers containing solution must bear a label of its content, strength, date of preparation and initial of the user.
- Avoid unnecessary storing of chemicals. Discard the solution and clean the containers after the analysis.
- Dirty glassware should be washed immediately and never allow it to dry on the wash basin.

Appendix XVI: First Aid in the Laboratory Accidents

(A) Accident caused by acid

1. Burning on the body

- First wash the burnt portion of the body with plenty of water.
- Then wash with 2% solution of sodium bicarbonate ($NaHCO_3$).
- Wash with water and apply a paste of sodium bicarbonate –petroleum jelly mixture on the burnt portion.
- Keep it for 15-20 minutes and then wash off with water.

2. Spillage on the eye

- Follow the above first two steps and apply 2-3 drops of olive oil.

3. Oral intake

- Gargle with water and drink sufficient amount of water.

(B) Accident caused by alkali

1. Burning on the body

- First wash the burnt portion of the body with plenty of water.
- Then wash with 0.5% solution of boric acid (H_3BO_3).
- Wash with water and apply a paste of boric acid –petroleum jelly mixture on the burnt portion.
- Keep it for 15-20 minutes and then wash off with water.

2. Spillage on the eye

- Follow the above first two steps and apply 2-3 drops of olive oil.

3. Oral intake

- Same as accident caused by acid

(C) Accident caused by fire

- Immediately apply ice on the burnt portion (no rubbing) or place the burnt portion in the flow of cold tap water.
- Apply burnol on the affected portion.

(D) Accident caused by bromine

- Wash the burnt portion with alcohol or petrol.
- Apply olive oil on the affected part.

(E) Accident caused by glassware breaking

- Remove the broken glass pieces from the body.
- Wash the wound with water and then apply tincture iodine.

Consult doctor as soon as possible